Landscape Performance Modeling Using Rhino and Grasshopper

This is a guidebook for landscape architects to learn the fundamental practices and use of the computational software Rhino 3D and the plugin Grasshopper for parametric modeling, landscape inventory, and performative analysis. This process visually connects intangible and abstract information with physical and spatial relationships to signify the impact ecological, climate, and cultural factors have on landscape performance and decision making.

Each chapter begins with a summary of the performance method and its application in different projects, outlining the expected goals from industry standard equations and operations. Chapters cover parametric modeling scripts to measure ecosystem services of stormwater management, erosion control, tree benefits, outdoor comfort, accessibility, and many others. Using photographs, tables, and parametric scripts to create qualitative and quantitative representations of landscape performance and ecosystem services, readers will learn to communicate the impact and significance of their outputs.

This book will be beneficial to educators, students, and professionals interested in using computational modeling as a performance assessment and graphic visualization tool.

Phillip Zawarus is an Associate Professor at the University of Nevada, Las Vegas. He utilizes technology and computational modeling methods as decision-making tools for the analysis and design response to sensitive ecological systems. His approach attempts to bridge the gap between abstract conventions and perceptual experiences within our living environment to bring comprehension of intangible ecosystem services. His exploration and use of graphic visualizations, interactive media, parametric modeling, and augmented reality create dynamic datascapes demonstrating landscape performance methods and metrics in the field of landscape architecture.

Landscape Performance Modeling Using Rhino and Grasshopper

Phillip Zawarus

Routledge
Taylor & Francis Group

NEW YORK AND LONDON

Designed cover image: Phillip Zawarus

First published 2023
by Routledge
605 Third Avenue, New York, NY 10158

and by Routledge
4 Park Square, Milton Park, Abingdon, Oxon, OX14 4RN

Routledge is an imprint of the Taylor & Francis Group, an informa business

Library of Congress Cataloging-in-Publication Data
Names: Zawarus, Phillip, author.
Title: Landscape performance modelling using
Rhino and Grasshopper / Phillip Zawarus.
Description: New York, NY: Routledge, 2023. |
Includes bibliographical references and index. |
Identifiers: LCCN 2022026120 (print) | LCCN 2022026121 (ebook) |
ISBN 9781032076331 (hbk) | ISBN 9781032076300 (pbk) |
ISBN 9781003208020 (ebk)
Subjects: LCSH: Landscape design—Computer simulation. |
Landscape design—Graphic methods.
Classification: LCC SB472.47 .Z39 2023 (print) |
LCC SB472.47 (ebook) | DDC 712.01/13—dc23/eng/20220928
LC record available at https://lccn.loc.gov/2022026120
LC ebook record available at https://lccn.loc.gov/2022026121

ISBN: 9781032076331 (hbk)
ISBN: 9781032076300 (pbk)
ISBN: 9781003208020 (ebk)

DOI: 10.4324/9781003208020

Typeset in Univers
by codeMantra

Contents

■ Contents

Figures

SECTION 1

1 Introduction

Ecological design and landscape performance attempt to moderate the accelerating interest from the global community to minimize the disruptive impact of urban sprawl on the environment leading to climate change, urban heat islands, social inequity, and fragmented wildlife habitat. Because of this, nature-based solutions are being further appreciated beyond their aesthetics and are being recognized for their ability to sustain, mitigate, and service our progressive intentions of preservation and enhancement of the environment (Beck, 2015). For the profession of landscape architecture to properly evaluate the necessary preservation and enhancement procedures, our process must shift to a divergent method of asking questions to direct solutions so outcomes are not limited or preconceived (Lahaie, 2017) with conventional thinking.

Through the emergence of landscape performance and the integration of quantitative metrics into the qualitative aspects of outdoor spaces, landscape benefits can become more attainable and tangible for comprehending the complex ecological, social, and economic relationships of different environments. This book aims to communicate these dynamic benefits using parametric modeling as a hybrid platform of digital and physical methods to evaluate outdoor spaces. These demonstrated methods can bring students, professionals, community members, and stakeholder groups together to learn the impact of their decisions as their actions generate computational reactions through a responsive modeling interface.

Landscape performance is heavily reliant on measurements, metrics, and methods to assess and evaluate the effectiveness of outdoor spaces as contributors to ecosystem services through environmental, social, and economic benefits. This analytical process utilizes quantified information, datasets, surveys, and sensory devices to calculate outputs for cross-referencing with project goals and objectives. This book will demonstrate how this process is translated to a digital parametric model to measure a variety of landscape performance topics. Although the bulk of the modeling methods focus on external datasets, simulations, and precedent analysis, the book will also demonstrate how it will serve as a tool that can adapt and be refined from field surveys and ground truthing for an integrated approach to link the digital and physical environment.

Parametric modeling has been more commonly used for form-finding of objects, facades, and other figural design elements. The benefit of utilizing this modeling technique within landscape architecture beyond those outcomes is by generating an analytical and responsive digital model that relies on the complex

DOI: 10.4324/9781003208020-2

measurables of rich environments and ecologies. This is not entirely a new idea of modeling with data as many professionals have pioneered similar methods such as the work by Nadia Amoroso in Figure 1.1. More specifically in this book, the parametric modeling process will use spatial and mathematical calculators to perform these measurements based on the input of data and information relative to performance metrics. This process affords the opportunity to visually connect intangible and abstract information with physical and spatial relationships to signify the impact ecological, climate, and cultural factors have on landscape performance and decision making.

As a guidebook, the chapters will instruct the reader on the fundamental practices and use of the computational software Rhinoceros 3D (Rhino) and the integrated Grasshopper plugin for parametric modeling. The demonstrations will layout algorithmic scripts to model landscape characteristics for performative analysis as it relates to landscape architecture, urban design, and environmental science. Detailed written and visual reference material in the form of photographs, tables, and parametric scripts will instruct the reader on the development of their landscape performance model. The parametric scripts will utilize internal and external data

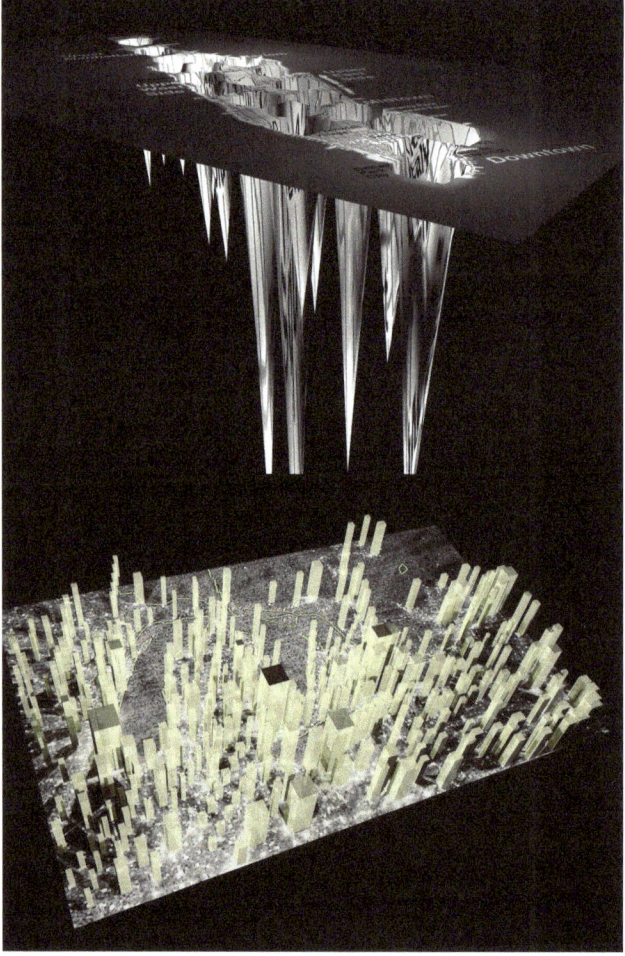

Figure 1.1
Work by Nadia Amoroso. The top image is a datascape representation of the landscape based on crime. Here there's a negative 'pulling' to create a dark, eerie landscape dotted with 'depressions' in the ground (Crimescape, NYC 2005). The bottom image is a datascape of light pollution as buildings within the Great Lakes region (Skyscrapers of Light Pollution, 2008).

inputs, mathematical operations, and analytical outputs to create qualitative and quantitative representations of landscape performance and ecosystem services.

This book can benefit educators, students, and professionals that are interested in using computational modeling as a performance assessment and graphic visualization tool. It will be using industry standard calculators, metrics, and methods from the Landscape Architecture Foundation's (LAF's) *Evaluating Landscape Performance: A Guidebook for Metrics and Methods Selection* (Figure 1.2), the *Site Engineering for Landscape Architects* book, and other sources to implement outputs as both quantitative and qualitative outcomes. Historically, these methods have often resulted in tabular and abstract outputs without the spatial incorporation of a project site; however, this book will demonstrate opportunities to effectively communicate the integration of data and space. It will also add a valuable skillset that is becoming more commonly practiced in the profession of landscape architecture during the analysis and post-construction phase.

The demonstrated modeling methods throughout the book will fit with the many aligning interests and goals of ecological and performative design for students and practitioners within landscape architecture, urban planning, and architecture. Evaluating performance in both the natural and built environment is a professional standard for accredited programs, with some of the examples directly implementing the specific methods used for Sustainable Sites and LEED certifications. This makes the value of the lessons learned more universally significant for any unique site condition. Although some of the performance modeling examples will occur in an arid environment and other parts of the United States, the primary objective of the book is to disseminate this performative modeling globally as a relevant and replicable methodology. For most landscape performance assessments, it is necessary to incorporate third-party resources and datasets to complete the evaluation process so references to external resources are also provided as they relate to climate, soil, and topographic data for the reader's needs.

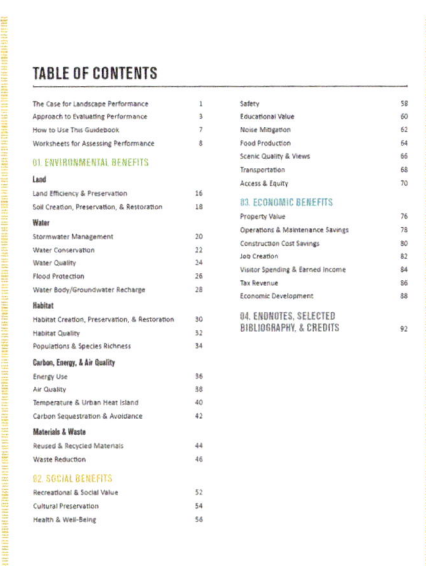

Figure 1.2
LAF's Evaluating Landscape Performance guidebook on different methods and metrics to measure environmental, social, and economic benefits.

PURPOSE

There are a wide range of landscape representation books but as the field is shifting from the scenic and picturesque conventional landscape design typologies to regenerative and ecological design, decision-making tools will take precedent for analytical representations, performance modeling, landscape metrics, and a platform for other advanced digital media. In the field of landscape architecture, technical books on representation and software are often limited to conventional and static two-dimensional drawings, whereas examples in this book push the opportunity of utilizing temporal and dynamic aspects to reinforce the constant fluctuations in the environment. Some scripts will even be primed and formatted to create animations of these dynamics, since that is also one of the beneficial capacities of the software and modeling method.

Parametric modeling is becoming less of a novel concept within the profession of landscape architecture and the allied fields; however, it is often used for generative and figural design, and this book embeds performance measurements as a key component to the intended outcome. The Sustainable Sites handbook along with the Landscape Performance Guidebook both provide methods and metrics for measuring the landscape that this book will build on but will also introduce the spatial correlation between physical, biological, and cultural attributes for ecosystem services. Also, the three-dimensional environment of parametric modeling serves as a platform to transition the information and representation to other media devices that include virtual and augmented reality interfaces, digital fabrication, multimedia graphics, and conventional representation. All these distinct features provide both a unique and integrated outcome for landscape architecture audiences.

One of the primary outcomes from this book is to determine if parametric modeling can effectively integrate landscape performance metrics into a responsive platform between the analytical and design process of a project. By working within the responsive platform of parametric modeling, design outcomes can adapt and be evaluated for specific environmental, social, and economic measurables in relation to a site's goals and objectives. Due to this dynamic workflow within a digital three-dimensional environment, parametric modeling can be explored to communicate analytical metrics into relative and visual measurements, reducing the abstraction of landscape performative benefits. Methods of representation must transform to address landscape performance complexities by using dynamic visualization techniques that can include analytical datascapes, immersive experiences, interactive media, and animated scenarios, in response to spatial and temporal conditions.

Parametric modeling creates a responsive workflow by taking a site's measurable conditions such as hydrology, soil, vegetation, slope, and climate, then outputting a framework of constraints and opportunities on site. Since the calculations are performed from a digital site model, the analytical output will respond to changing conditions. By having the quantitative aspects combined with the digital model into a collective aggregation, dynamic representations can be extracted from the digital source model in the form of datascapes and immersive experiences.

The digitally modeled project sites will investigate characteristics of topography, surfacing, soil, and drainage along with some of the intangible inventorying of elevation, slope, aspect, and climate conditions. The relationship of these elements will be analyzed to measure different landscape performance metrics related to hydrology, erosion, tree benefits, outdoor comfort, and accessibility. The results from these methods will provide opportunities and constraints to develop a responsive design solution as it relates to stormwater management, outdoor comfort, and others. Design strategies can be tested for their performance to validate the findings or propose alternative strategies. Utilizing parametric modeling in the design process is becoming common practice; however, integrating it into a responsive design outcome from landscape performance metrics has been limited within calculators and spreadsheets. Integrating those calculators and formulas used for performance metrics into a 3D design environment can potentially communicate the outcomes more effectively through dynamic visualizations of data and design.

The value of using parametric modeling to quantify environmental, social, and economic is the ability to generate responsive outcomes that adhere to the necessities of a site's conditions, commonly known as evidence-based design. The advantage of modeling environmental performance benefits with parametric modeling is in the communication of analytics within specific site conditions that relate to location, materials, and vegetation by removing the abstraction and relativity of numeric data and replacing it with comprehensive outputs.

STRUCTURE

Although there are a number of literature pieces on the topic of landscape performance from ASLA, CELA, and LAF, there are limited resources on applying these methods to computational and parametric modeling. The book's first section will begin by providing a broader overview of the topic and how it can be integrated into the modeling workflow. This will first require an introduction to data resources relevant to landscape performance in Chapter 2, while Chapter 3 focuses on the benefits of computational modeling. Familiarizing the reader to this specific workspace will be one of the first steps in preparation for the following chapters and will cover elementary operations, data structure, parameters, transformations, and supplemental software plugins. Chapter 4 will introduce effective plugins that can be utilized to seamlessly integrate external resources and datasets to the scripting procedures or display information through alternative means such as supplemental charts.

Chapters 5 and 6 in section two of the book will then demonstrate how to extract basic information from a project site's topography and land cover to inventory other critical systems needed to measure landscape performance, similar to the examples in Figure 1.3. Both those chapters will provide introductory scripting processes to measure and visualize a specific site characterization or inventory topic. It is important to have these fundamental operations, as they will be key components for more complex and dynamic parametric scripts in the next section. The more advanced scripting in the next chapters will often contain elements from previous chapters, so it is important to treat the book as a sequence of operations. In other words, it will be difficult to start at Chapters 6 and 7 without referring to scripts from Chapters 3, 4, and 5.

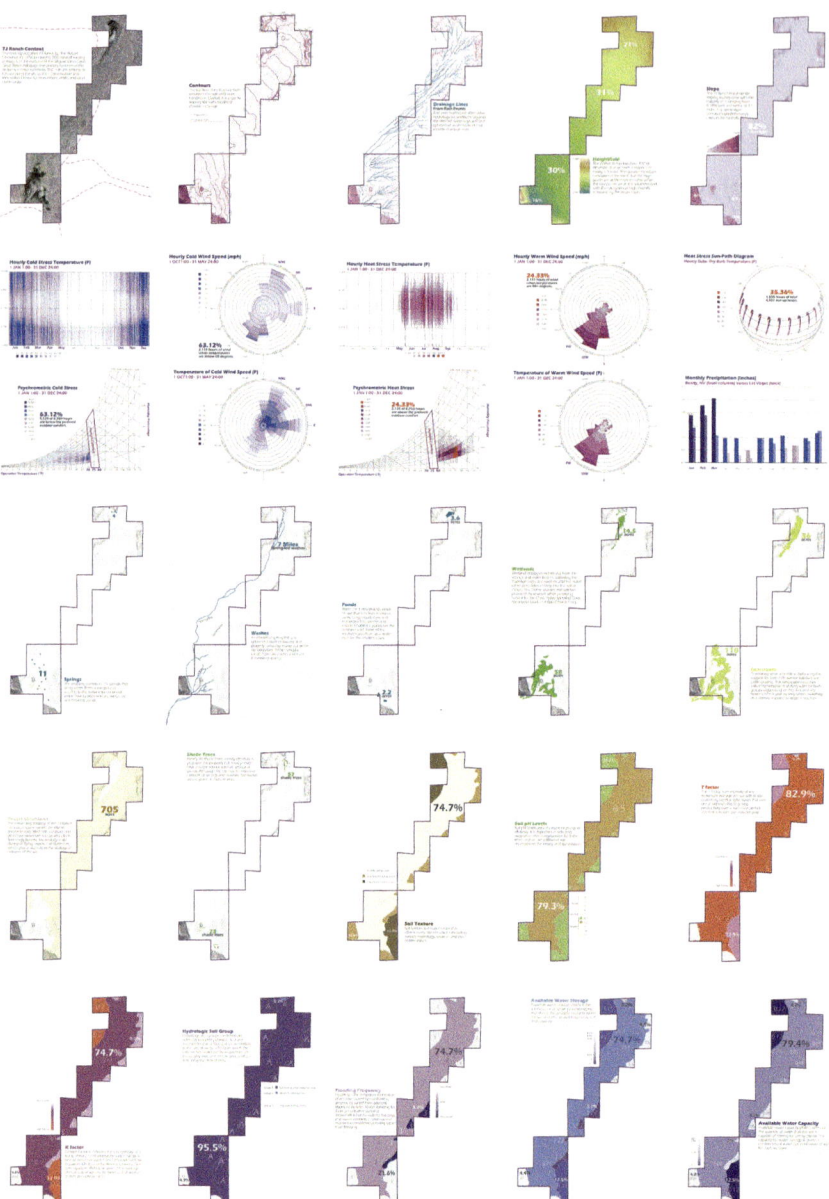

Figure 1.3
Collection of
inventoried systems
for a project site by
students and edited
by Phillip Zawarus.

The remaining chapters in the third and final section will cover various landscape performance modeling methods that utilize and compute the previously documented landscape characteristics, essentially transitioning the site inventory into site analysis. Landscapes provide a variety of ecosystem services by managing stormwater runoff, mitigating erosion potential, creating wildlife habitat, cleaning the air, and creating microclimates for outdoor comfort and accessibility. These are the types of services that will be measured with specific modeling methods of environmental, social, and economic outputs to compare with the reader's goals and objectives. These chapters will cover the diverse range of measurables and methods applicable to landscape architecture and allied professions in pursuit of

Figure 1.4
A composition
of a parametric
hydrology model
and supporting
factors created by
Sonny Geronimo
and Mey Fa Choy.
Courtesy of Nadia
Amoroso.

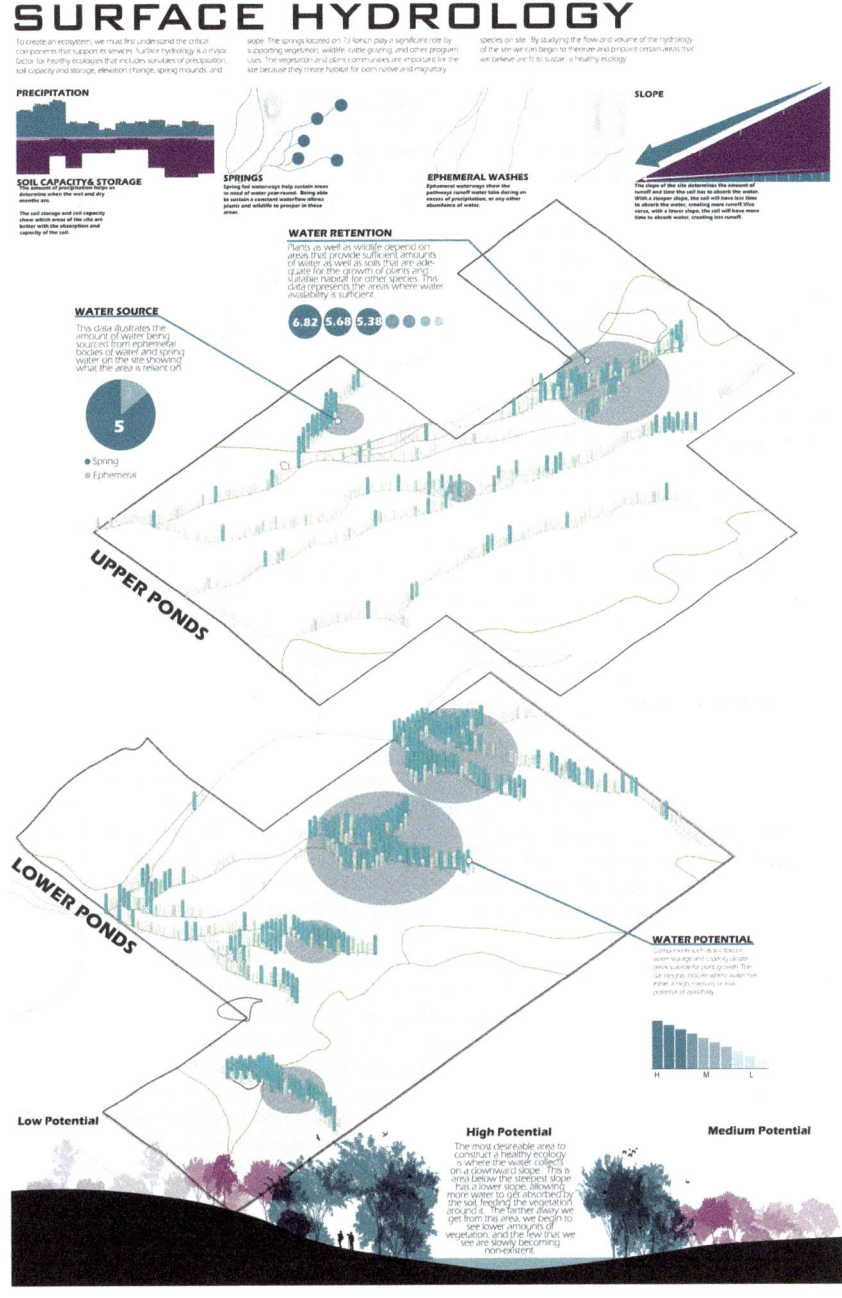

landscape performance evaluations. The resulting models can be incorporated into a more comprehensive communication of the topic with other supplemental graphics and information from other software as shown in Figure 1.4.

In conclusion, the book will offer future opportunities to explore with parametric modeling and landscape performance as technology continues to advance for responsive design opportunities, city planning, and policy amending. It will address the overarching intentions of how these modeling methods can be pushed and explored further by incorporating responsive design solutions to the evaluation process to foster scenario building for decision making and how the advancement of technology, such as augmented reality, can harness this process as an immersive user experience. As stated throughout, these modeling methods deliver outputs to inform how specific goals and outcomes may be achieved. Additionally, appropriate strategies and solutions will be recommended in response to the analytical data as an evidence-based design process, like the image shown in Figure 1.5. With a variety of different outcomes due to temporal and other dynamic factors, it is key to understand that diverse and multiple conclusions can be derived from the process to suggest various opportunities for decision making and engagement with design.

Each chapter will begin with a summary of the performance method and its application for different projects. It will also outline the expected goals and outcomes from these industry standard equations and operations. Next, the performance metric will be elaborated on with identified equations to compute the model's measurables. This will segue into the main section, where the primary intention of the chapter demonstrates parametric modeling scripts to measure different landscape characteristics and landscape performance. As a result, the next section will communicate the impact and significance of the model's outputs to assess with intended goals and objectives. Finally, the chapter will conclude on future potential for the script to pursue other valuable outcomes related to the chapter topic.

Figure 1.5
Student work by Fausto Saavedra in response to the hydrology analysis of the project site in the previous figure.
Courtesy of Landscape Architecture Foundation.

COMMUNICATION

As performative metrics rely on measurables and calculations, it is the intention of this book's accessible structure to become replicable for others to learn, adapt, and integrate into various goals and objectives on improving our living environment. The process of communicating landscape performance goes beyond the computational model by providing seemingly limitless representation options within a three-dimensional environment. Communicating landscape performance is not graphically representing it but much more. It requires an elegant orchestration of geometry, color, and text to compose a narrative of the significance landscape performance affords to improving our ecological landscape.

The majority of the examples visualized throughout the book will be of three-dimensional caliber because it is often necessary to understand a project site as space and not a cognitive abstraction – plan, section, elevation, etc. This is, however, not a definitive stance on using only this form of graphic because every project is unique and requires the most effective form of graphic convention that only the primary user and audience can decide on. The power and advantage of using a three-dimensional workspace is that it affords the opportunity to visualize the model from endless perspectives that are most suitable for the range and scope of a project. Different cognitive and perceptual visualizations can also be generated rapidly with changing parameters and conditions for a more accurate depiction of a site's unique characteristics.

The book will cover basic and advanced computational and parametric modeling methods as a starting point for different levels of experience with the software as they explore different communication techniques. It will begin with introductory representations of data as it progresses to more robust visual outputs that utilize color, size, and annotation of geometric representations.

2 Landscape Performance

The implementation and evaluation of landscape performance is an evidence-based and responsive approach to provide ecosystem services to outdoor and indoor spaces. In the global context of our living environment, designers and the allied professions are faced with the challenge of dealing with a multitude of concerns related to climate change, environmental justice, social equity, and economic viability.

One of the best ways to confront these challenges is through a collective effort of shared interests between experts, policy makers, developers, and the community to preserve, heal, and ensure that the benefits of landscapes and the natural environment provide ecosystem services. Discussions between all these stakeholders can be managed through a universal language of sustainability that focuses on environmental, social, and economic benefits. These three categories are often relatable to some degree by all involved parties and can be a way to strategize solutions for local, regional, and global issues. This book can serve as a methodology of bringing abstract numeric data to a spatial grounding of the triple bottom line that most organizations relate to through performance modeling.

With landscape performance, project sites can be measured with both qualitative and quantitative information to expand the comprehension of space beyond anecdotal understanding. Surveys, datasets, calculators, methods, and outputs can be utilized to evaluate how a space performs toward environmental, social, and economic goals that benefit the natural and built ecology of place. These types of measurables aid in determining the impact of design interventions through baseline comparisons. Baseline comparisons serve as a reference point that can be assessed from pre-development conditions, alternative design options, or by the dynamic and fluctuating conditions through different time intervals.

Measurables and comparisons will often lead to outputs, which are valuable in themselves; however, it isn't until they are evaluated against project goals and objectives that an outcome can be determined from the performative landscape. It is from performative outcomes that design can be evaluated for its benefits to outdoor and indoor spaces.

Evaluating landscape performance is a collective and collaborative process that aggregates metrics and methods from not only accepted landscape architecture methods but also from allied professions that are working toward the same project goals. Even though this book will draw many of the metrics and methods from the Landscape Architecture Foundation's (LAF's) performance guidebook, other

DOI: 10.4324/9781003208020-3

approaches to measuring the landscape will also be demonstrated that includes work from the US Department of Agriculture (USDA), Environmental Protection Agency (EPA), and others.

One of the reasons that make the implementation of landscape performance so effective with allied professions and decision-makers is because it aligns succinctly with the widely accepted sustainability terminology of the triple bottom line's environmental, social, and economic benefits. Having a dialogue that shares in common interests and metrics makes discussions with these groups more fluid and relatable as data, methods, and outcomes are compared across disciplines. With a more universal understanding of the environment, other professions can begin to see some of the underlying values of landscape architecture beyond the aesthetics of greening a space through nature-based solutions. By integrating empirical and evidence-based approaches to the overall design process, the community, clients, and other stakeholder groups can better understand the decision making and impact of the design's outcome.

The integration of landscape performance can occur at various phases of the design process and most likely at multiple junctures through responsive investigations and revisions. Whether it's the analysis, design, planning, policy, maintenance, or cost assessment of a project, they all utilize some form of measurable data.

For the analysis, it makes sense to follow Ian McHarg's overlay method, as shown in Figure 2.1, by recreating it as a digital layering and calculating of data (Corbett, 2001). Environmental, social, and economic layers all have complex relationships of different systems that can be digitally stitched together and computed with parametric modeling to generate robust datasets to cross-reference with project goals. These resulting outputs and outcomes can then guide the strategic decision-making process to validate and assess on whether responsive design interventions fulfilled their objective. In other words, if the analytical model indicates flooding at a specific location on a project site, are the proper design procedures implemented to reduce or even eliminate the pooling of runoff? These are some of the direct and interdependent relationships that can be answered between the analytical and design workflow process.

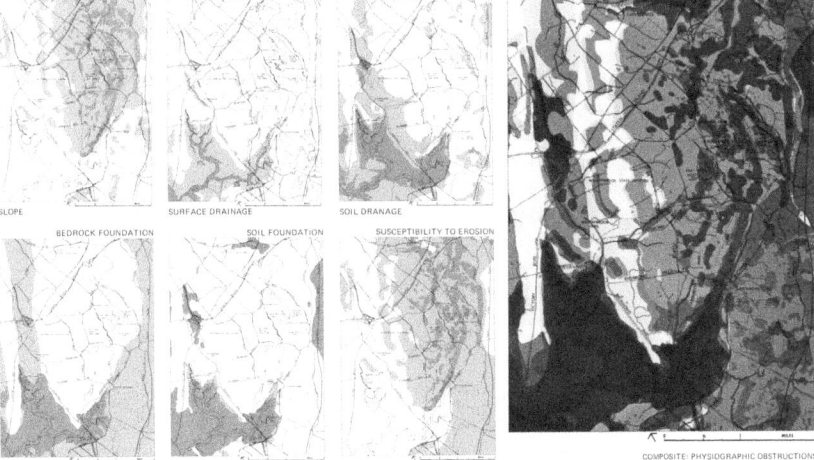

Figure 2.1
Ian McHarg's overlay method for the American Interstate Highway System's assessment report.
Source: McHarg, 1969.

The same linked relationship between a project site analysis and design can also be scaled up for large-scale planning projects. At the planning scale of landscape performance, outputs and outcomes may serve more appropriately as guidelines and recommendations. For example, it may be suggested that a specific number of trees be planted along a public right-of-way to reduce flooding, sequester carbon dioxide, and mitigate urban heat islands. It does not specifically go into which tree species or how the tree planters are designed to accommodate them since that can be solved through a more extensive design study. By measuring and evaluating this recommended number of trees or how the planter implementation may occur during the design process, a comparison can be made of tree benefits between the recommended outcome versus the municipal planting codes. This is one way in which landscape performance can begin to influence and revise policies that shape our built environment.

Lifespan and long-term cost savings are another way to employ landscape performance at another form of scale and scope of a design project. The lifespan and maintenance costs of materials in the design versus conventional materials can be compared to determine if proposal is cost effective. The cost effectiveness of a project is usually best determined when considering the longevity and endurance of materials, as some may breakdown and deteriorate at a much faster pace than others; so although they may be more affordable upfront, the replacement and maintenance may be cost prohibitive in the long term.

Although this book will focus on the analysis and planning scope of landscape performance, strategies for how the models can be utilized for other opportunities will also be discussed for future demonstrations. What makes these computational models most valuable is the responsive revisions that can be implemented through ground truthing to adapt the digital parameters and outputs to match the physical and existing findings. This dynamic will be further elaborated on in the next chapter on integrating performance metrics with computational and parametric modeling.

The intent of this book aligns well with LAF's landscape performance mission in progressing current methods and ideologies of excellent design by encompassing design goals for climate change, social equity, public health and well-being, and many more strategies. In their guidebook, it is stated:

> Embracing performance measures and evaluating the performance of built projects can increase knowledge, support innovation, and elevate the quality of designed landscapes. By validating past research and raising new questions, it also grounds and strengthens the body of more rigorous landscape performance research being conducted within a variety of disciplines and through multidisciplinary collaboration. Continuing to study the connections between landscape and the health of ecosystems, people, and economies increases our understanding and our collective capacity to create a more sustainable, just, and resilient future by using landscape solutions to their fullest potential.
>
> (Canfield, Yang, & Whitlow, 2018)

It is understood through this statement that it is a collective effort of shared interests between multiple disciplines to work toward holistic goals that utilize evidence-based knowledge and unique expertise to complex challenges in our built environment. This book not only shows how industry standards are used in measuring performative landscapes but also un-abstracts their numeric outputs into a spatial, tangible, and recognizable format for a broader audience. This approach and outcome from the computational modeling of landscape performance can provide a visual discussion based on data just as other conventional drawing methods serve at roundtable meetings.

GOALS AND OBJECTIVES

Without project goals to cross-reference with landscape performance, it becomes impossible to evaluate the impact or success of those design decisions. Goals may be initially discussed at the beginning of the project by the client and their preference for the programming of the site through activation, preservation, or economic stimulus. It could also be part of a larger collective interest to fulfill municipal or agency standards, such as runoff volume, or decarbonize our built environment initiated by the Green New Deal (Text – H.Res.109–116th Congress, 2019–2020). In more urbanized settings, a component of the design may be to reintroduce nature or mimic its ecosystem services which can only be evaluated through baseline comparisons with pre-development conditions. Although it is preferred to establish goals and objectives at the start of the design process, it should also be understood as a fluid process where they can change, adapt, or even no longer become relevant when tested during the design development process. For example, during the analysis phase of a project, new goals may be developed as part of the investigation in a more divergent process. Below are some examples on discussing or determining project goals, in which many projects may use multiple approaches.

Landscape performance objectives can demonstrate how goals are evaluated by instructing specific design interventions that need to occur in order to achieve those benchmarks. These are usually derived from rating systems and sustainability criteria like LEED and SITES, government regulations, or municipal codes. An example of a project objective is to retrofit a certain percentage or square footage of impervious surfaces to permeable surfaces to reduce runoff volume. Initially, performance objectives could be established for environmental and economic actions since they had easily defined quantifiable metrics, whereas social measurables have historically been more challenging. This has, however, begun to shift toward more attainable means as technology and tools have progressed to track and monitor social behavior.

A case study brief from the Landscape Performance Series website showcases how methods were used on visits to the New York City High Line project to track social media indicators from Flickr and Twitter (Plunz & Moskalenko, 2017). Techniques like this with social media analytic have demonstrated a new and innovative way of measuring social benefits of a project site that divert from conventionally surveying practices. This type of information is even being explored with plugins for parametric modeling software to integrate the metadata spatially more seamlessly for more comprehensive outputs.

It may also be important to consider a classification or hierarchy of goals to achieve with the project. There will often be a primary intent behind the design that can be supported or supplemented with other subsequent benchmarks. Nature rarely produces singular benefits, so there may often be unintentional benefits that occur and are discovered during the design or post-construction assessment process. People will ultimately use the space the way they want or by how it was designed. Design, whether it is performative or not, should often be resilient and adaptive to the changing cultural, environmental, and economic influencers over time.

OUTPUTS AND OUTCOMES

Like goals and objectives, outcomes and outputs work in similar ways, especially as performance metrics are integrated into parametric modeling. Outputs are like the inventory of a project, where the quantity, area size, and distance are measured but do not elaborate on their significance to the overall project goals. Outcome is the analysis of this quantified information to give meaning and evaluate the impact of a project toward project goals in the form of ecosystem services and benefits. Performance outcomes will usually demonstrate the impact through improvement, mitigation, and reduction of presented issues. The parametric models demonstrated in this book will output data related to stormwater runoff, tree benefits, and microclimates but that information cannot be interpreted as performative outcomes until cross-referenced with project goals. An example of this would be that a certain number of trees can sequester a specific amount of carbon dioxide, but how do those values compare to the national average from trees? In the end, performance models should aim for impactful outcomes, not just the outputted data.

Performance calculations and models can output primary and secondary datasets to compare with project goals. Primary data will generally be collected and measured at the source through field surveys and ground truthing, whereas secondary data is an existing dataset gathered from an external resource such as soil infiltration rates from Web Soil Survey. The performance models in this book will use a combination of both by calculating and outputting new sets of data from existing ones while also refining the parameters and metrics from ground truthing on site. The book will focus on the model calculations, but instruct and provide field methods from the LAF guidebook. External datasets should come from reliable sources such as government agencies and national organizations; however, regardless of the source, the data should be verified through additional research and visiting the site for ground truthing. Some of these reputable resources are provided later on in this chapter.

METRICS AND METHODS

Landscape performance relies on metrics and methods to evaluate the impact of various pre-development, existing, and proposed conditions on a project site. Metrics are types of measurable data that can be calculated, such as runoff volume, hours of reduced air temperature, or increase in property value, but should ultimately align with the project goals. The chapters in Section 3 of this book will demonstrate different metrics that can be calculated from modeling methods. Some of these performance metrics can be very challenging to communicate to stakeholder groups, so it is important to present the information in a comprehensive way that can be relatable to a specific audience. Some groups may want to see the economic savings of a project while others may prefer the public health factor with clean air and accessible green space.

The outputs generated from the demonstrated models in this book can ultimately be catered to different stakeholder goals or categorized into environmental, social, and economic benefits in order to become a universal tool in communicating the impact of landscape performance. Since the models will output different forms of numeric data (volume, quantity, percentage, etc.), this information can then be translated into either opportunities or benefits. For example, one of the chapters will output potential soil loss as an erodibility index where the higher the value, the more susceptible an area is to erosion. This can then be used to either inform necessary planting and preservation strategies to stabilize an area with vegetation or evaluate the economic feasibility of regrading or introducing retaining walls to alleviate the issue.

Another important factor when developing performance metrics is the variable of time – hourly, seasonal, or annual data for simulating the change of a space over time to make inform predictions and future decision making. This element works seamlessly with the performance modeling of this book as the scripts being used to measure and calculate outputs are often dictated by parameters. Parameters can essentially mimic these time intervals or any change that may occur to a project site. For example, some parameters can control different planting densities and placement within a space, which can then generate different outputs of tree benefits, stormwater management, or outdoor comfort. This type of process becomes very valuable when design decisions do not have a perceivable performance and relies on this type of predictive modeling to evaluate with project goals before construction even begins. But as mentioned earlier, these strategies should not only occur at the initial design phase but also continue post-construction to measure how benefits may change as vegetation becomes more established or how the surrounding context may influence an area with more future development.

Relying too heavily on predictive models and outcomes can be a slippery slope, as it is impossible to account for everything that factors into the changes of space over time. Any performance model, whether it be an online calculator or the ones demonstrated in this book, needs to be understood as estimations and assumptions, but not definitive. The outputs should also not be considered as an exact or precise measurement but should be used to suggest upward/downward trends or improve/diminish in performance.

Some methods to evaluate these conditions are through baseline comparisons to quantify differences and changes from performative measures. Once again, LAF's performance guidebook elaborates on three different types of comparisons that will be utilized throughout the proceeding chapters.

Before/After

This is one of the most common methods that compare a site's condition before and after design construction. Some models will expand on this even further when striving to mimic natural ecologies by also including a pre-development/natural metric for sites that are already disturbed.

Conventional/Sustainable

Performative landscapes intend to introduce sustainable environmental, social, and economic principles, so this modeled outcome can also be compared with development practices that do not align with these same objectives. This becomes a very powerful methodology in the spirit of iterations and scenario building with multiple different end products that may prioritize one goal or outcome over another for a tradeoff analysis.

Benchmarks

Municipal codes and national regulations often have a standard for performative outputs that include inches of runoff reduction and carbon dioxide sequestration. This will be very specific to the location and context of a project site, so it is necessary to research on these specific standards that should be attempted to meet, exceed, or fulfill a percentage of to evaluate.

With all the different forms of outputs and comparison, it is also important to make the information attainable versus abstract and open-ended. Data and information should be comprehended by the audience with something that is relatable through relationships, equivalencies, savings, and projections. When outputted raw information without context, it can often be perceived as incomprehensive, so it is important to relay it in terms of relative values. For example, a tree may sequester 100 pounds of CO_2 annually, which is over twice that national average. Another example, involving energy savings as an ambiguous numeric output, would be to communicate the kilowatt hours saved into how many homes does that translate to. With cost savings, it would also be helpful to elaborate on how certain design interventions may offset their reliance or dependency on a municipal utility versus a renewable energy source such as solar panels. These are the types of outcomes that can be formulated with embedded metrics and standards in the performance models or independently into a more relatable understanding.

DATA RESOURCES AND SOFTWARE

New data and resources are being discovered and created routinely, so it is rare for there to not be a good starting point. Data can be collected directly from the source using new methods or be an extension of what has already been done as it changes over time. This availability can also be provided from big open datasets from public and private sectors; however, it will still require some verification before any substantial decisions and evaluation criteria are formulated. Below are a few examples of reputable resources that provide data specifically for the measurement of landscape performance and ecosystem services.

Landscape Performance Series

The Landscape Performance Series website (https://www.landscapeperformance. org/) was launched by the LAF to develop and disseminate metrics, methods, and case studies that demonstrate effective landscape performance strategies. One of their latest contributions to this specialized field was the *Evaluating Landscape Performance: A Guidebook for Metrics and Methods Selection* in 2018 to suggest specific strategies on attaining data and deploying them onto project sites with case studies as examples. The website is constantly expanding on new projects and methods to aid in the initial process and fulfill unique needs for our built environment.

In Figure 2.2, the Case Study Briefs page from the website shows where projects can be searched by specific benefits, features, project type, location, and even budget along with other filtering options. This becomes a great resource in demonstrating how project outcomes may be referenced or emulated for other similar built projects or as part of the performance models in this book.

Stormwater Runoff Calculators

There are multiple accepted stormwater calculators that are part of site engineering books for landscape architects available from the EPA or even independently operated as an online resource. Regardless of where the calculator is provided, it is best to use either the Rational Method or Natural Resources Conservation Service (NRCS) Method for calculator stormwater runoff. When visiting an online resource that calculates runoff volume, such as EPA's or the Green Values version shown in Figure 2.3, it is important to find out which method is being used to ensure that proper procedure is being used to output runoff data.

Figure 2.2
The Landscape
Performance Series
Website homepage.
Courtesy of Landscape
Architecture
Foundation, 2021.

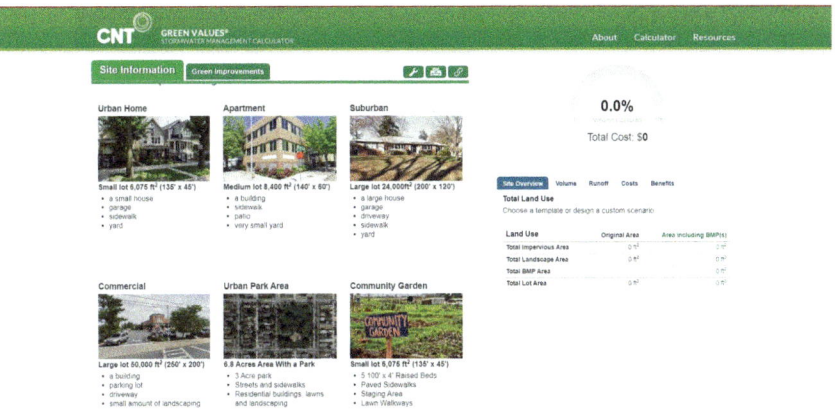

Figure 2.3
The Green Values
Stormwater
Management Online
Calculator.

The Green Values Stormwater Management Calculator uses the NRCS method, which can be found under the Resources tab of the website (Evaluate, 2006). The website has developed over time to become more user-friendly while maintaining a robust set of outputted data that relates to hydrology and cost savings. At the start of the website, a user is given the opportunity to determine a project site by type or use a custom scenario, as shown in Figure 2.3. Another valuable attribute of this website specifically is the ability to compare conventional practices with the green improvements strategy to see the tradeoffs for each. Although the chapter that models this calculator does not go into the same level of detailed outputs, it can certainly be developed to do so with additional parameters.

Geographic Information System (GIS) Software

GIS software has the ability to discover, aggregate, analyze, and format data for performance and modeling methods. GIS software can read vector, raster, and tabular data which can be converted to the software's preferred shapefile format. The Environmental Systems Research Institute (ESRI), which is responsible for the development of ArcGIS, has an extensive growing archive of data that has been mapped and even analyzed for a variety of topics that can be found on their website (https://www.esri.com/). Both their online and stand-alone mapping services allow for the layering of data in which toolboxes can be used to do numerous calculations and detailed spatial and tabular analysis.

Another valuable tool with this software is its ability to format native shapefiles to specific projections or coordinate systems that other software may require to read. This component will especially be helpful when introducing plugins to the parametric modeling aspect that can import GIS shapefiles. Although ArcGIS is a preferred industry standard software, there are several alternative open-source platforms that include QGIS (https://www.qgis.org/).

Web Soil Survey (WSS)

Whenever it is necessary to measure the performance of natural systems specifically related to the physical attributes of a project site, the Web Soil Survey website (https://websoilsurvey.sc.egov.usda.gov/) can provide very robust datasets

Figure 2.4
The USDA Web Soil Survey online mapping services for soil characteristics.
Source: NRCS, 2019.

Figure 2.5
The TNM Download
website to acquire
national, regional,
state, and local
datasets in the form
of digital elevation
models and
shapefiles.
Source: TNM Download
V2, n.d.

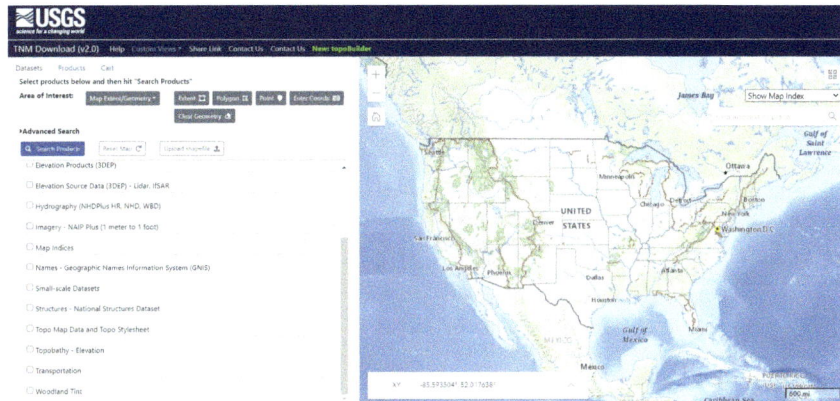

confined within soil profiles. Some of these datasets range from planning topics for development and agriculture to soil biological and chemical composites such as pH levels and erosion factors. When selecting these dataset topics, the website will output both a shapefile of the soil profile curves along with its associated metadata for each profile.

As shown in Figure 2.4, the process begins with either defining an area of interest (AOI) from its mapping service or allows the user to import an already exist-ing project boundary shapefile. Utilizing an established project boundary shapefile originally generated in GIS or provided from a client is the best practice, as that file can be utilized for other online resources besides WSS. This will maintain a cohesive and consistent overlay of data and maps for modeling and analysis. More thorough instructions with this website will be elaborated on in future chapters.

USGS TNM Download

Landscape performance models will often require site-specific and smaller-scaled datasets that can be found from local municipal and state websites; however, for larger-scale datasets, the US Geological Survey (USGS) The National Map (TNM) Download website (https://apps.nationalmap.gov/downloader/) can be used to acquire vector and raster file types. This online resource is an example of where a site boundary shapefile can be imported to search for a variety of elevation, hydrologic, transportation, and other topics within that specific location, as shown in Figure 2.5.

Once a subtopic is selected to search for, different file types will be shown that are available for that area to download.

SCALE

The scale of measuring landscape performance can be related to its physical size, time, and function. Material components of a designed element can be analyzed and measured for their lifespan, cost, and physical characteristics or an entire city can be evaluated for its urban forestry benefits or watershed management of resources. Regardless of the physical size of a project, appropriate measures and methods can be applied for an extensive and comprehensive understanding of its performance related to the project goals and objectives. The same thing can also be said about the scale and change of time of a project. When it comes to evaluating outdoor comfort or other metrics related to dynamic climate and weather conditions, these variables are constantly changing by the hour throughout the day, during seasons, and even in annual patterns.

Lastly, it is also important to consider the scale of a project as the product of its functionality. Sustainable and performative landscapes should aim for multifunctional and integrated solutions toward project goals. Elements within a project should not be a singular or siloed incident of benefits but should cross-pollinate with multiple aspects of a project's scale and scope of ecosystem services.

GROUND TRUTHING

Neither a standardized method nor a computational model should treat the process of measuring landscape performance as a static or onetime phase of a project. The frequency of collecting and reevaluating outputs may be often at the initial phase of the process, but can eventually taper off to annual or biannual assessment monitoring. As data is collected from these field monitors or from an external resource, such as census data, they need to also be routinely assessed for their validity and accuracy. When analyzing the impact of food deserts within a community, grocery stores may close and move or a farmers market may be introduced as part of a revitalization effort. These types of things are in constant flux, so therefore it is necessary to ensure the data, metrics, and methods comply with what actually exists at the time of investigation.

The monitoring of this data may not always be ideal for in-person survey or fieldwork, but luckily technology is advancing and providing tools and sensors to collect data as well. This can occur with temperature readings, video recordings, or even flow rates which may require sophisticated devices to compute data that is incapable of doing through analogous means.

This book ultimately aims to provide a new and innovative approach to perceiving the often intangible and invisible aspects of our living environment and make that information tangible, comprehensive, and relatable to a universal audience as data gets grounded to space. The modeled data will be communicated as numeric, spatial, computed, and responsive to the influencing parameters of a project site. Parameters in these models like performance measurables and metrics can be utilized to assess project's aim toward goals and outcomes. Most importantly, though, these computational performance models can guide strategies to maintain and progress a project's ecosystem services toward environmental, social, and economic benefits.

The models are dynamic to ensure adaptability to changing metrics and variables over time and programmatic use, so that new outputs and outcomes can be reevaluated on whether they are supporting or improving the performance of a project site. Non-compatible and diverse dataset types can also be managed within these models to reveal system relationships that could not be done so before. These types of innovations with modeling landscape performance aren't intended to output the same results of conventional means, but serve as a robust tool to guide the design decision-making process.

REFERENCES

Canfield, J., Yang, B., & Whitlow. H. (2018). *Evaluating Landscape Performance: A Guidebook for Metrics and Methods Selection*. Landscape Architecture Foundation. https://doi.org/10.3153/gb001

Corbett, J. E. (2001). Ian McHarg, Overlay Maps and the Evaluation of Social and Environmental Costs of Land Use Change. CSISS Classics.

Evaluate Benefits of Green Infrastructure to Prevent Urban Flooding. Green Values Stormwater Management Calculator. (2006). Retrieved from https://greenvalues.cnt.org/

McHarg, I. L., & American Museum of Natural History. (1969). *Design with Nature*. Garden City, NY: Published for the American Museum of Natural History [by] the Natural History Press.

NRCS. (2019). Web Soil Survey–Home. Retrieved from https://websoilsurvey.sc.egov.usda.gov/

Plunz, R., & Moskalenko, E. (2017). *The High Line*. Landscape Performance Series. Landscape Architecture Foundation. https://doi.org/10.31353/cs1250

Text – H.Res.109–116th Congress (2019–2020): Recognizing the duty of the Federal Government to Create a Green New Deal. (2019, February 12). https://www.congress.gov/bill/116th-congress/house-resolution/109/text

TNM Download V2. (n.d.). Retrieved from https://apps.nationalmap.gov/downloader/

3 Computational Modeling

The possibilities and options are seemingly limitless as the three-dimensional workspace provides opportunities to create realistic and detailed replications of our environment or envision completely abstract and imaginary scenes impossible to experience otherwise. No matter what the outcome is with the making of digital objects in a computational model, it can always be seen as a stepping-stone to another opportunity in the visual communication of space.

Computational models, specifically Rhinoceros 3D (Rhino) for the purpose and demonstration of this book, should be used to represent three-dimensional ideas instead of underutilizing it for conventional two-dimensional drawings. That is not to say there is no value to two-dimensional drawings but those should only serve as either a starting point or be a byproduct or extraction from a three-dimensional model.

GEOMETRIES

In its most basic and elementary application, computational models consist of points, curves, surfaces, and meshes geometries that can generate one type or be extracted from another type. It can even be interpreted as linear generation process where points inform curves that can be 'lofted' into a surface or mesh. This is certainly not always the case and doesn't always follow this order of operations, sometimes even rarely does, but it is a simplistic way to understand the basic process of progressing the complexity of a three-dimensional model. The opposite can occur as well where curves and/or points can be derived from a surface or mesh. Essentially, the more complex geometry types like surfaces, meshes, polysurfaces, and Breps can be converted into simpler geometries like curves and points.

Throughout the book, unless stated explicitly, surfaces can and may be used synonymously with polysurfaces and Breps because property wise, they are very similar but are defined more so by the operation that created them. Surfaces and meshes will often digitally appear identical; however, they contain very different attributes and characteristics. This will become even more apparent when performing parametric modeling operations with these geometry types.

These geometries can be transformed and tooled to move along vectors, copied, alter other geometry, and transition from one type to another along with many other basic and advanced operations. These actions can become very helpful in the workflow of model making as it can streamline the process in a more efficient way.

DOI: 10.4324/9781003208020-4

Each geometry will contain its own advantages and disadvantages when referencing to the parametric modeling workspace; so although it may seem counterintuitive to do so, the resulting output from the process may be more effective. It will also be learned through both computational and parametric modeling, that there is often more than one way to get the same end result with little to no advantage over the other. This book will attempt to demonstrate the most efficient ways and instruct on best practices, but at the end it is the outcome that matters the most.

PARAMETRIC MODELING

Parametric modeling is a form of visual programming or coding. It is almost identical to computational modeling in terms of geometry types and operations but instead the process is done with numeric data. It no longer becomes about modeling with 'physical' or digital objects that are visible, but about manipulating and reinterpreting intangible data to generate new sets of information that can result in geometry or other forms of output. This is why this process can be a steep learning curve due to its abstract and coding formula; however, once the process is understood to follow similar procedures of traditional modeling techniques, the focus can be placed on the advancement through big datasets.

With parametric modeling, geometries can be transformed with changing values (parameters) or have geometry characteristics create parameters to inform transformations as perceived as a closed-loop workflow process. This will often be the case in the proceeding chapters as the original referenced geometry is modified, analyzed, deconstructed, and transformed into a completely different end product. Parameters can also create restraints or restrictions on how data is computed to achieve a specific outcome or allow for a calculator to process information accurately.

INTEGRATING LANDSCAPE PERFORMANCE METRICS

The measurables and metrics that evaluate landscape performance can be understood as the parameters of computational modeling. Data that relates to different ecosystem services and landscape benefits can be calculated using the same mathematical operations and equations such as stormwater runoff, tree benefits, or walkability using components in the parametric workspace. The physical and intangible characteristics of a project site can be analyzed, evaluated, and inputted into their respective methods to output back into the model as a spatially associated outcome. This becomes a very dynamic and advanced workflow where both space and data are linked together so as parameters or measurables change, the outcomes will respond accordingly and be recomputed in a seamless workflow.

It is because of this dynamic workflow and modeling of data that makes the integration of landscape performance measurables and metrics so appropriate for integrating into parametric modeling. Data can be visualized and space can be parameterized simultaneously and be responsive.

The power of modeling landscape performance is its ability to rapidly compute large spatial and numeric datasets that would otherwise be cumbersome or impossible to do. Traditionally, the processing and computing of numeric datasets is not a new endeavor since programs like Excel and other calculators have been capable of doing these same operations far before Rhino and Grasshopper. However, the key distinction with this book's workflow and demonstration is that these datasets and calculators no longer need to be abstract or external, as they lack the spatial association to the physical modeling, analysis, and design of a project site. They can now be interdependent of one another to discover real-time outcomes and assessments as part of the process rather than supplementary.

Design is an iterative process where conditions and parameters are constantly changing as new discoveries are made through analysis and design development.

DATA REPRESENTATION

Visually communicating data is not a novel idea in the scope of historical representation of site analysis and design in landscape architecture and other design professions. Some of these range from the graphic design of Edward Tufte's original work or featuring of others in *Envisioning Information* to Keith VanDerSys's use of sensory data collection for new visions of environmental systems, as shown in Figures 3.1 and 3.2. These timeless approaches have proven to be an effective method of bringing comprehension to the abstract when important decisions need to be made instead of interpreted.

In the book *The Exposed City*, along with the images in Chapter 1, Nadia Amoroso explores different visualization techniques used by key theorists and practitioners in the profession of landscape architecture and other related fields. These techniques build upon the importance of digital representations integrating information and art.

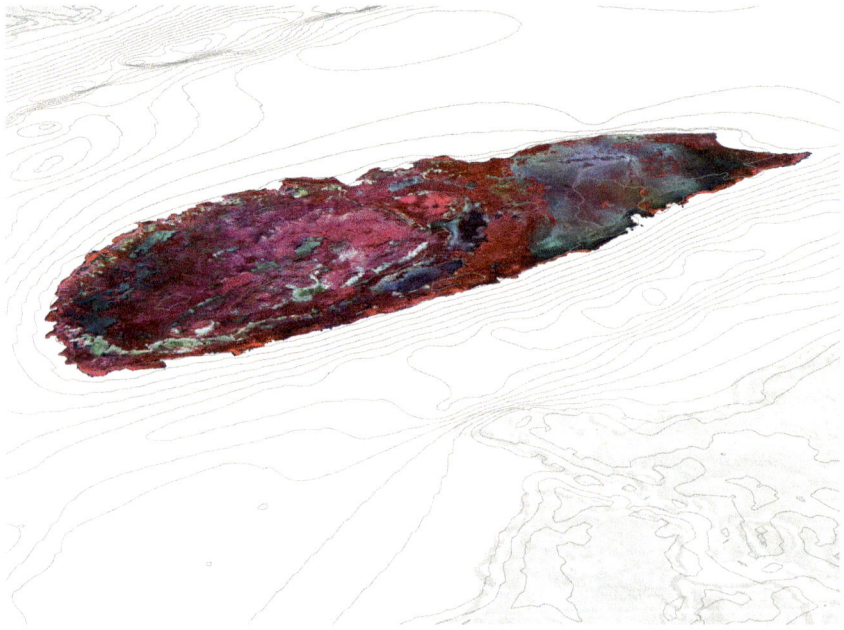

Figure 3.1
UAV 5-band color infrared multispectral imagery. Large areas of high marsh habitats were located and classified by using in situ multispectral collection to match similar spectral ranges in satellite multispectral datasets.
Courtesy of Keith VanDerSys, Karen M'Closkey, Sean Burkholder, and Leila Bahrami.

Figure 3.2
High marsh habitats play a critical role protecting coastal communities. Existing wetland datasets, however, typically exclude differences in marsh types. UAV multispectral collection was used to capture and classify high marsh and test its inclusion in sea level rise marsh modeling projections.
Courtesy of Keith VanDerSys, Karen M'Closkey, Sean Burkholder, and Leila Bahrami.

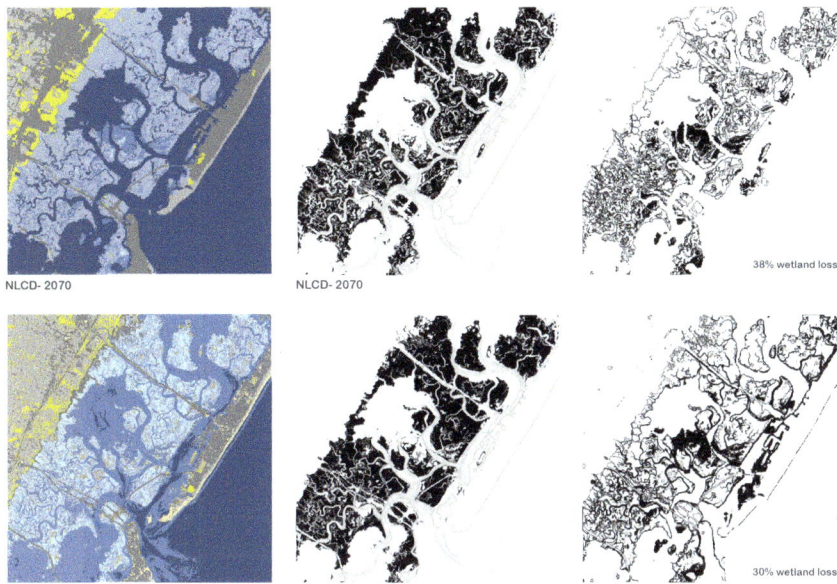

NLCD- 2070

NLCD- 2070

38% wetland loss

Custom UAV- 2070

Custom UAV- 2070

30% wetland loss

When representing data through computational and parametric modeling, it is imperative to maintain the graphic principle of hierarchy through contrasting color, size, shape, and even annotation. These three to four distinguishing factors can effectively communicate areas of importance and significance to a landscape performance metric. The amount of data that can be represented can be overwhelming, so it is important to distinguish between areas of focus and non-priority areas within landscape performance. When utilizing these contrasting characteristics, the main elements that may be contributing to a specific metric or project goal should be highlighted against the other forms of data. Distinguishing these elements will often be a direct relationship since factors that contribute more to a performance metric should be larger in value or numeric data so that it can then be used to visually represent the information.

The range of values, regardless of how big or small that may be, can be utilized as a ratio to distinguish data through its physical size. That same concept can then be applied to both color and shape as well where the significant higher values utilize a color that contrasts significantly from the rest as a gradient. This can also assist in transforming geometric shapes to make the same distinction. In some of the end products shown throughout the book, annotation and text size can also be applied to this same workflow for a fully comprehensive datascape.

Another method of creating hierarchy to communicate critical analysis and design opportunities is through formulating boundaries around clusters or hotspots of data in the model. In some instances, the performance metric may not be about individual outputs but rather congregations and compounding of multiple singular elements. This often occurs with groupings of tree species that output a variety of significant benefits to a project site that can be highlighted to ensure they are protected and monitored for the future. Subbasins or compounding variables within a watershed may warrant attention as they contribute more stormwater runoff than other areas, which will require more strategic management and mitigation practices with green infrastructure.

Much of this information can be communicated directly within the project site model itself; however, charts, graphs, and tables can also be supplemental in supporting the landscape performance outputs, outcomes, and goals. With certain plugins, data can also be represented with these more conventional methods to bring additional clarity and legibility to the overall objectives of the model. As shown in Figures 3.3–3.6, these models display a rich set of data; however, the scripting process to perform all these forms of data communication is quite extensive. In order to make chapter and script lengths more manageable, only portions of these examples will be contained within a single script; however, between all the chapters combined, the same visual communication of data can be achieved by incorporating the different parts of one script to the other.

In Figure 3.3, highly erodible land (HEL) is communicated through colorized and scaled massing volumes where the more prominent geometries indicate a higher erosion probability. Blue highlighted boundaries indicate all the areas that fall within the HEL classification, as drainage lines reinforce contributing forces to these erodibility potential. Supplemental charts break down the quantified information as well for proportions and relativity of the areas.

A terrain's aspect and orientation of faces relative to sun angles can dictate plant communities within a large project site. The graphics in Figure 3.4 show the relationship of north- and south-facing slopes and how they may impact preferred plant conditions for sun, moisture levels, and other factors.

Tree inventories can be modeled to demonstrate their extensive benefits, whereas in the case of Figure 3.5, their economic benefits of cost savings are quantified for a campus' urban forest. After considering initial purchasing costs and annual maintenance, the net profit can be calculated when compared to their annual savings on energy, air quality improvement, runoff reduction, and property value increase.

Finally, in Figure 3.6, a collection of trees can be modeled for their shade envelopes to evaluate which areas get the most consistent hours of shade throughout the day as concentrated heat maps. This form of model can be valuable in creating design responses that either strategically place additional trees for more shade or focus amenity placement within heavily shaded areas.

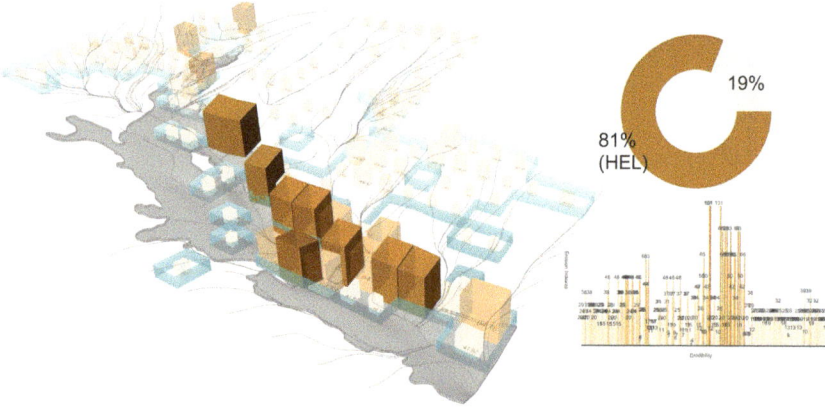

Figure 3.3
A highly erodible land (HEL) model with supplemental charts and quantified characteristics.

Figure 3.4
A terrain model
showcasing
different aspect
angles relative to
sun angles.

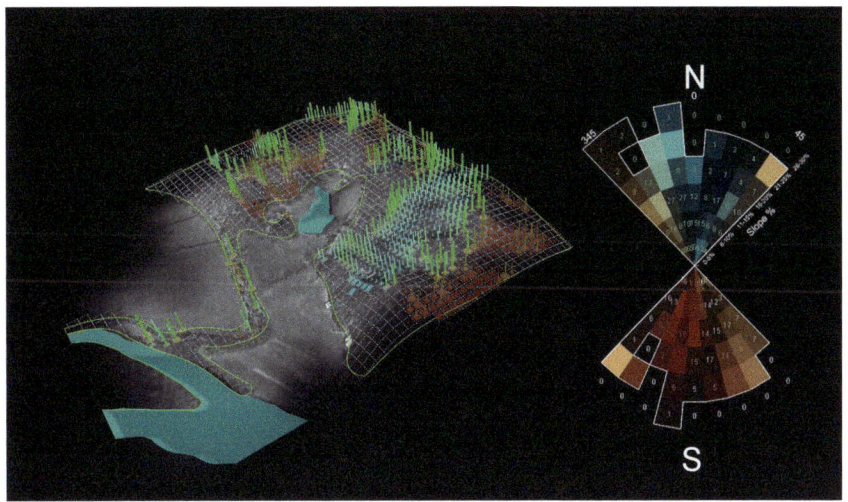

Figure 3.5
A campus tree
inventoried
modeled for its
benefit and net
annual net profit.

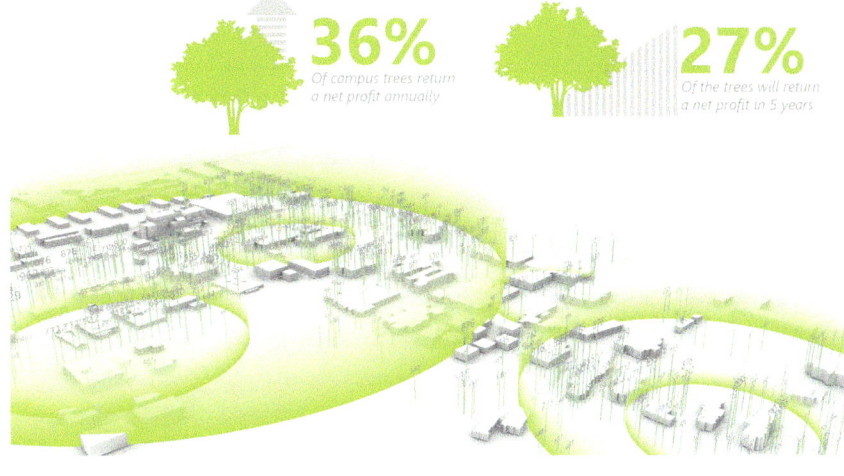

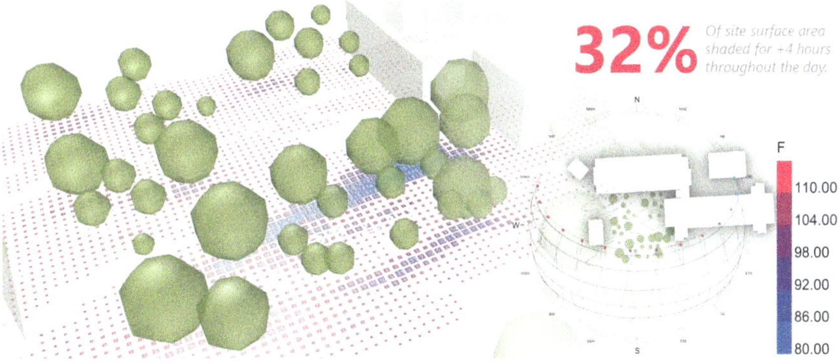

Figure 3.6
A shade envelope
model from trees
that visualize hours
of daily shade on a
project site.

Hybridizing the representation of data as geometry within the model and through supporting charts collectively, robust performance metrics can be communicated at multiple scales of information. Overall performance and outcomes can be the focal point while individual and singular datasets that contribute to these performance methods can contain the aggregated details for a comprehensive understanding of landscape performance. This array of data representation working succinctly with the model's responsive capabilities is what separates this book's methodology from previous works.

WORKSPACE

The next part of this chapter will be informative for first-time users, but not necessary for those already familiar with Grasshopper; however, it can still be valuable for discovering new techniques and workflows. It is best to first get familiar with some specific terms when modeling with Grasshopper. When learning Grasshopper from other users, there may be slightly different nomenclature or terms used, but these will be the ones used throughout the modeling process in the book.

Verbiage and Workflow

Components are what data is housed within to reference external data or perform calculations. They are the main element when modeling in Grasshopper. When modeling with Grasshopper, geometry and data can be referenced from either Rhino or another external source. Original geometry can also be created within Grasshopper. There are specific *Geometry* components under the *Params* tab in Grasshopper to *reference* their corresponding Rhino geometry by right-clicking on that component and choosing to either *Set one...* or *Set Multiple...* This will prompt the Rhino workspace to open for setting the intended geometry.

Created or referenced geometry that has been altered in Grasshopper does not physically exist in Rhino, even though it will be visually displayed in that workspace. It is not until it is *Baked* that new geometry will exist and can be selected in Rhino. If the geometry in Rhino that has been referenced is deleted, then the Grasshopper component containing that geometry will turn orange with an error message that it can no longer find that data. To avoid this issue along with the visual clipping of Rhino's geometry and the Grasshopper geometry display, it is best to turn off that geometry's layer or *Hide* it. Another option in managing geometry between the two workspaces is to *Internalise Data*. This will embed the data into the Grasshopper component, so that it is no longer dependent on the Rhino geometry. This can be performed by right-clicking on any component name or icon and selecting the *Internalise Data* option.

Each component has an *input* and *output* of data that are connected to talk with each other through *wires.* Many of the Grasshopper components will also contain multiple with initials corresponding to what is specifically being inputted and outputted; so, if it is unclear what type of function is being performed by that component, hovering the mouse over the initial will pop-up additional information. Wires can be solid or dashed, depending on the data structure of that component's function.

Solid wires mean that all the data is pooled into one list, whereas dashed wires indicate that the list of data has been separated into separate lists. This will be elaborated on further in the *Data Branching* section of this chapter. Other terms that refer to these data structures is *flattened* or *grafted*. Flattening collapses separate lists of data into one, whereas grafting separates items into separate lists. This can be done at the input or output of a component by *right-clicking* on them and choosing that option.

The collective combination of multiple components to perform specific tasks is referred to as *script*, which may be used synonymously with *definition*, so these terms may be used interchangeably throughout the book. Within the entirety of a Grasshopper script, the term *operation* may also be used in this book to refer to a small collection of computing tasks between several components. There will often be standard operations that are used repetitively throughout the chapters; so that is the intent on clarifying that term. The overall processing of these scripts will flow from left to right in the workspace and by default will display any geometry that is contained within a component, whether it is the original or altered state. Geometry that is no longer needed to be displayed can be *right-clicked* and have the *Preview* option turned off. It is best practice that as the script progresses any unnecessary component can have the *Preview* turned off.

Throughout the book, components being demonstrated will display both their icon and label for ease of finding within the Grasshopper workspace. If it is still unclear where that component may be found, then the workspace canvas can be *double-clicked* to open a search bar where the component's name can be typed out and selected from the list.

Display Mode

Like displaying geometry in Rhino, these same display options will affect the visualization of geometry from Grasshopper. The *Wireframe* option will read dark and flat and can be challenging to decipher the detail of certain models, one example being terrain models. Unless color swatches are applied to geometry, the *Rendered* display option will still read flat like the *Wireframe* display option, as shown in Figure 3.7. The workflow being used in the demonstrations throughout this book will use the *Shaded* display option to best communicate all geometry referred from Grasshopper unless otherwise instructed.

To change these display settings, either *right-click* on the Rhino viewport name and choose the *Shaded* display option or go to the *View* tap at the top and select that option from the drop-down menu.

Relays

Relay elements are a great tool to both organize and condense Grasshopper scripts when one component is being connected to multiple components. The first advantage of this tool is that when scripts become long and complex, it can bring that component output closer to where it is being inputted instead of having

Figure 3.7 Viewing Grasshopper geometry through different display modes, with the shaded option being preferred.

Rhino Viewport Display Modes

Wireframe   *Shaded*   *Rendered*

to continuously refer back to its origin. The other benefit, as already mentioned, is that it can be used to consolidate the output of a component when being used repeatedly for another series of components, as shown in Figure 3.8.

The first step in creating a *Relay* is by *double-clicking* any of wires needing relayed. The created *Relay* will initially be blank but can be *right-clicked* to rename, preferably the name of the component or data that it is connected to. The output of the *Relay* can now replace or be inputted into the other components.

Renaming Components

An additional tip when developing advanced and complex Grasshopper scripts is knowing how to rename components to help organize data. A common operation throughout the scripts in this book will use Boolean patterns to parse data, which will be elaborated on in the next chapter. The Boolean pattern can be used to not only parse the data used to create the pattern but also parse similar data lists using the *Cull Pattern* component, as shown in Figure 3.9. As shown on the left image of Figure 3.9, seeing multiple *Cull Pattern* components can get confusing since it is difficult to identify which data types are going into each *List (L)* input. Just like with the *Relay* in the previous section, these components can be renamed by *right-clicking* and typing in the field at the top of the pop-up menu the connected data type.

As already mentioned, this will be a common operation where the renaming technique can save time and help organize complex script operations. It is important to note that even though the component can be renamed even if the *Display Icon*

<u>Apply a Wire Relay</u>

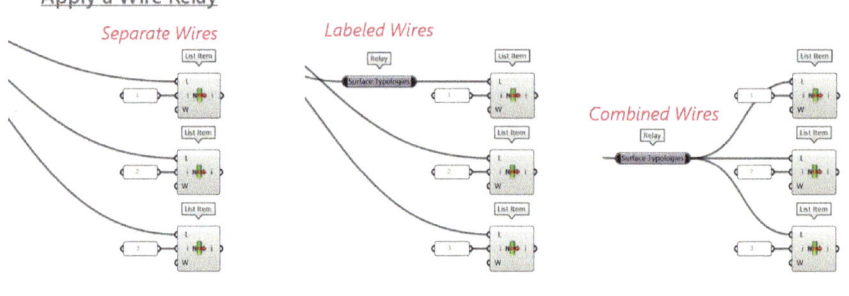

Figure 3.8
Using Relays for different wire configurations by labeling and combining inputs.

<u>Labeling Components</u>

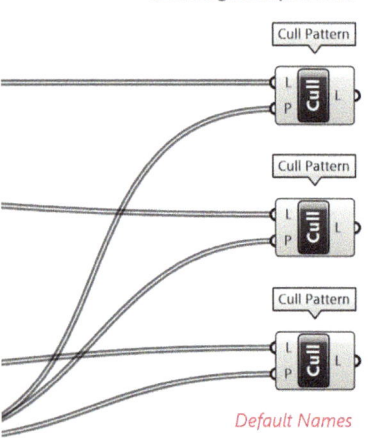

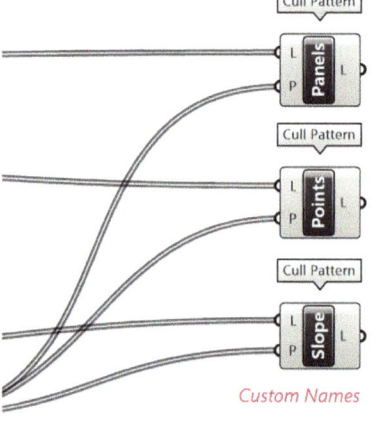

Figure 3.9
Creating custom labels for components that can best suit the user's organization needs.

option is turned on, the new name will not appear since the component icon will take priority in the workspace display.

Component Function

Another valuable organization consideration as scripts become more robust with different data parameters and transformations is processing that information for Grasshopper functions and Rhino display. In Figure 3.10, the same component for displaying/processing geometry data has three different options. The first *List Item* component is a light gray indicating that the data will be visible in Rhino, far left image, if it contains geometry such as points, curves, surfaces (Breps), or meshes. The darker grey indicates that the geometry is either turned off by *right-clicking* the name or icon and deselecting the *Preview* option or does not have any data to display. A good habit is to turn off all components that don't display vital data so that only the components showing geometry in Rhino are indicated with a light grey component and become easier to find. The component will still function, but it just won't display unnecessary data.

The faded dark grey component display indicates that it is not processing any data by *right-clicking* component and deselecting *Enabled*. This can be important for a variety of reasons. As scripts begin to process an extensive amount of data, the system may begin to slow down or lag. If additional operations need to be added prior to this component processing it, the changing of that data can be delayed or even crash a file. Another operation that may slow down the processing speed is when a number slider is being used and is changed for differing results. When a number slider is used and there is a big shift in the values being inputted, for example, going from *five* to *30*, ever value in between will also be processed which can again create issues with the processing. Essentially, anytime that inputted data is changed by these different processes, it can cause issues; so if it is known that a component takes a significant amount of time to compute, it may be advantageous to temporarily disable it until all operations are completed.

A good practice to keep in mind for these demonstrated scripts is that as it progresses from left to right in the workspace to sequential turn off the components so that there is no unnecessary overlapping and clipping of certain geometry types. This is not always the case since there may be some data that can remain turned on if its operation terminates, yet there are more steps within the script moving forward.

Combining Inputs

A simple technique if unfamiliar with scripting in Grasshopper is the ability to combine or merge data into one input of a component, as illustrated in Figure 3.11. To do so, hold *Shift* on the keyboard for the output of the component intended to be

Figure 3.10 Changing the displaying and computing of components for scripted operations.

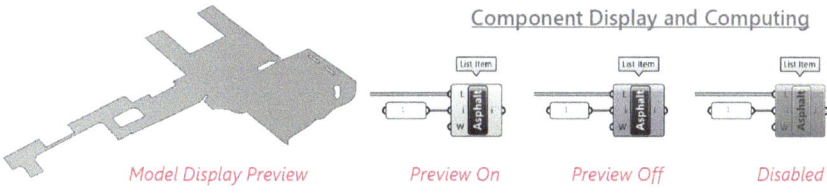

Component Display and Computing

Model Display Preview   *Preview On*   *Preview Off*   *Disabled*

merged with an already connected component. A green arrow with a + icon will appear to indicate that it will merge, not replace, a component input.

To disconnect a wire from an input, hold *Ctrl* while dragging from the input item in the same technique as merging. For this, a red - icon will appear.

Static versus Dynamic Values

When manually inputting values into components, they can either come from a *Panel* or a *Number Slider*, as shown in Figure 3.12. When using a *Panel* input, *double-clicking* workspace and beginning the command with *"*, it is best to think of this as a static value, meaning it will not be a parameter or change during the script. These are often used for known variables when doing conversions from square feet to acres or changing a decimal value to a percentage. By default, *Panels* will be colored yellow; however, that can be changed by *right-clicking* it and going to the *Colour* settings.

Number sliders can be considered dynamic values, since it is intended with this component to change or having varying options for that intended component. With the number slider selected, *clicking Ctrl (Command) G* will use the group command to give it a highlighted fill around it, which can be useful to indicate and find these dynamic input values within the complexity of a Grasshopper script. *Right-clicking* on the group will initiate a pop-up menu where there are options to remove the group or change its color. It should also be understood that if a number slider is used, it should be experimented with to achieve the most legible outcome, as demonstrated within the *Elementary Operations* section of this book.

An effective technique in creating a specifically defined number slider is by *double-clicking* within the Grasshopper workspace and entering a value with the number keypad. Some keyboard shortcuts include:

1. Entering a value between *2* and *10* for a number slider from *0–* to *10*
2. Entering a value greater than 10 but less than 100 for a number slider from 0 to 100

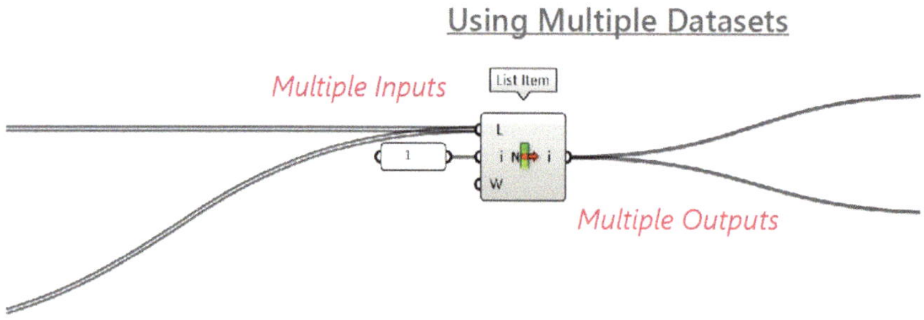

Using Multiple Datasets

Multiple Inputs

Multiple Outputs

Figure 3.11
Demonstrating the ability to have multiple inputs and outputs from a component.

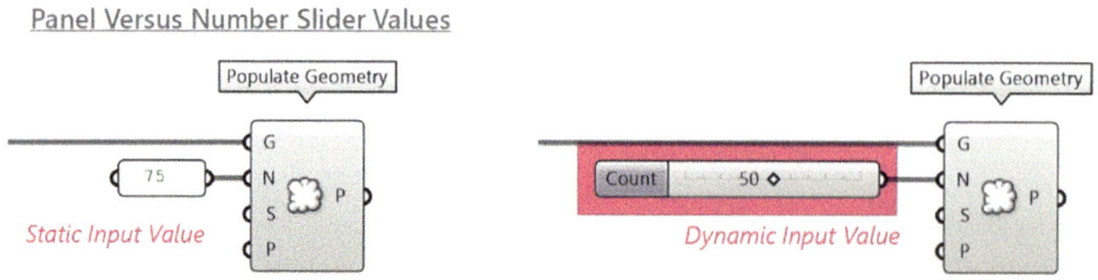

Panel Versus Number Slider Values

Static Input Value

Dynamic Input Value

Figure 3.12
Using static panel input values versus dynamic number slider inputs.

3. Entering a value with a less than key will create a specific range in the slider (5 < 50); if a middle value is used (10<25<200), then that value will be where the slider value begins.

Expressions

The use of expressions is a great way to consolidate an algebraic formula within a component instead of using additional components. These can be either simple formulas like $x/2$ or large equations. An expression can be applied at either the input or sometimes the output of certain components by *right-clicking* and choosing the *Expression* option to prompt a formula. Either the letter for the input, for example in Figure 3.13 it is U, or x can be used for variable being calculated.

Whenever an expression is used for an operation in the demonstrated scripts, it will be noted what the formula is since it cannot be displayed outside the component.

Unit Display

By default, Grasshopper will display up to ten decimal units which can be unclear and clutter visualization if data values are being displayed. To change the units for either calculations or display, go to *File > Preferences > Display* and change the decimal values to the desired unit.

In Figure 3.14, a *Panel* is used to showcase this process, but the units will not change until the wire is reconnected to the *Panel* or any other component by *clicking* and *dragging* back into said input.

Data Branching

A fundamental concept to understand with parametric modeling is how data is structured through branching or grouping of data based on the transformations and parameters. Figure 3.15 serves as a reference material for the elementary process of deconstructing a set of different surface types. The first *Surface* component contains three-sided objects and the second one contains four-sided objects. Since this is the first step of the process, these objects can be perceived as a

Figure 3.13
Creating
expressions
for additional
input and output
computations.

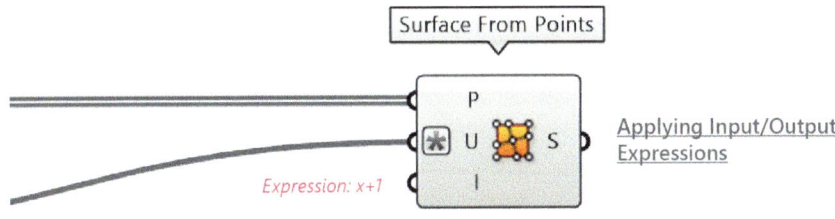

Figure 3.14
Changing
Grasshopper
decimal units for
display purpose in
the Rhino model.

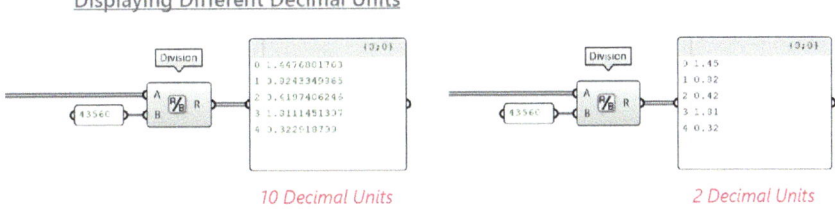

Understanding Data Structures

Figure 3.15
Demonstrating how data is structured after different forms of operations and manipulation of data and geometry.

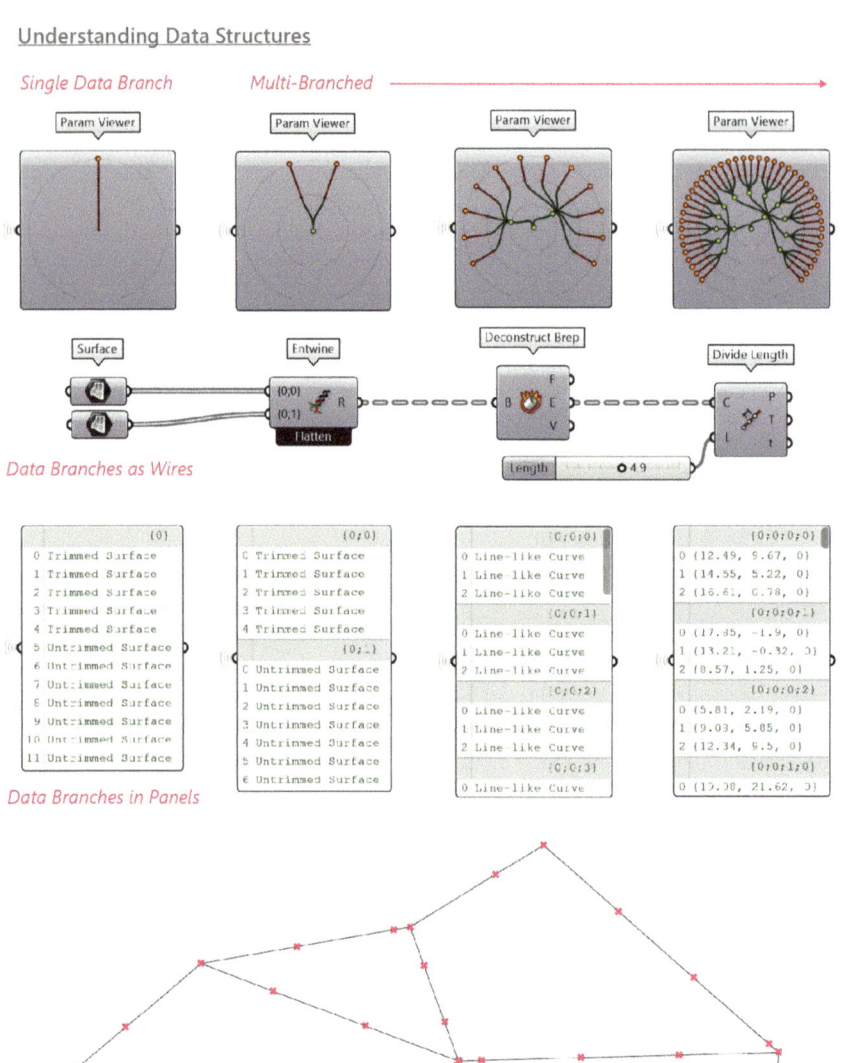

Single Data Branch   *Multi-Branched* ──────────────────→

Data Branches as Wires

Data Branches in Panels

Data Branches as Geometry

single branch of data, meaning there is no delineation between their characteristics. With an *Entwine* component, these different number sided objects are merged to create the first branching of data. Both the *Param Viewer* and *Panel* components show how these two branches contain their respective number of objects: branch one contains five three-sided surfaces and branch two contains seven four-sided surfaces.

Now, as each branch of different surface types is deconstructed to extract their edge curves with the *Deconstruct Brep* component, a new branch is added to the already existing branches as shown in the *Param Viewer* and *Panel* components. The last deconstructive process of adding varying number of points to each branched curve with the *Divide Length* component adds a new set of branches representing the unique number of points for each curve associated with the different surface types.

This abstract representation and understanding of data can be challenging to first comprehend, but as advanced scripts begin to delineate and parse data with landscape performance metrics, it will be a necessity to be attentive with.

After getting acquainted with these terms for the Grasshopper workspace, the abstract comprehension of coding with components and wires to compute data in a variety of ways can become more familiar in navigating a steep learning curve. There are certainly other nomenclatures for some of these elements as well as different procedures for the computing process, which is not the primary purpose of this book. It is ultimately about reaching the same end results and conclusion with performance metrics through parametric modeling methods. In the next section of this chapter, basic procedures to complete elementary operations that will occur frequently throughout the next sections of the book will demonstrate how datasets can be further managed and understood.

ELEMENTARY OPERATIONS

Throughout the book, when working on developing scripts or definitions, the process of computing data through several components is referred to as an *operation*. This can be a small grouping for a very simple operation or it can be a bit larger and repeated within a single script or used throughout a more complex modeling process. For the purposes of simplifying the instructional process of some of these scripts, several elementary operations are demonstrated here. These operations will not only help with future chapters, but also provide a better understanding to the significance of why certain calculations are being computed.

Calculating Geometry Area

The process of referencing Rhino geometry like curves and surfaces is quite simple, yet extracting the necessary and accurate outputs require specific procedures. When it comes to the common practice of measuring the area of a space or object in square feet, different techniques will result in different calculations, as demonstrated in the examples in Figure 3.16. This is particularly important when outer and inner curves are used in framing an enclosed space, such as a raised curb or a sidewalk in the case of the figure example.

The intention is to measure the size of the bound area between the two curves; however, in the first example of the two referenced curves, the *Area* component measures the square footage of all the space within each curve separately opposed to the difference. In Rhino, selecting the two curves and executing the *PlanarSrf* command will generate a surface between the two curves. It would seem that that would be the same case in Grasshopper where the referenced curves are converted to a surface but that would simply result in two separate surface area measurements, similar to the previous outcome. One solution to resolve this miscalculation would be to perform the *PlanarSrf* command in Rhino, then reference the generated surface as shown in the third example. This will result in the correct measurement; however, this is counterproductive to the purpose of Grasshopper and parametric modeling since there could potentially be significantly more of these cases throughout a model or the curves themselves are imported from elsewhere instead of being physical geometry in Rhino.

The fourth and final example shows how the same two initial referenced curves are connected to a *Boundary Surface* component instead to generate the same geometric and calculated result as the third example without having to contend with the additional Rhino operation.

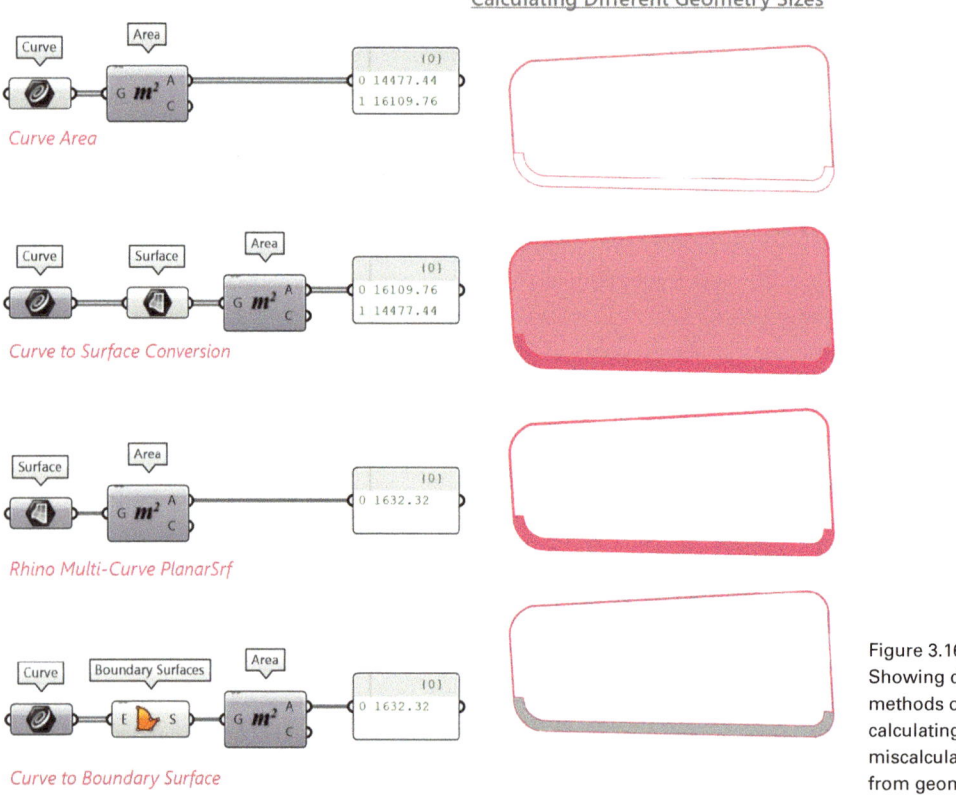

Calculating Different Geometry Sizes

Curve Area

Curve to Surface Conversion

Rhino Multi-Curve PlanarSrf

Curve to Boundary Surface

Figure 3.16
Showing different methods of calculating or miscalculating data from geometry.

Surface Terrain Detail

With parametric modeling in general, data or geometry is continuously being deconstructed and reparametrized, then reconstructed. This especially happens when working with surface terrain models as different analyses required different levels of detail within the terrain. For example, with drainage lines, a softer and less detailed terrain is preferred, resulting in accurate generalized flow patterns versus a highly detailed terrain with deficient nuances that generate precise and inaccurate drainage. When it comes to modeling slope conditions for plant communities, however, a highly detailed surface terrain model may be preferred.

Because this is all contingent on the purpose of the terrain model, the operations in Figure 3.17 show that with the use of a number slider for point densities on a surface, this decision can be quite flexible and dynamic.

When creating different point densities with the *Divide Surface* and *Surface from Points* components, the resulting terrain detail is demonstrated. For the *Surface from Points* component to function properly, the Points (*P*) being inputted need to be *flattened* and then the (*U*) input needs to be provided an *expression*

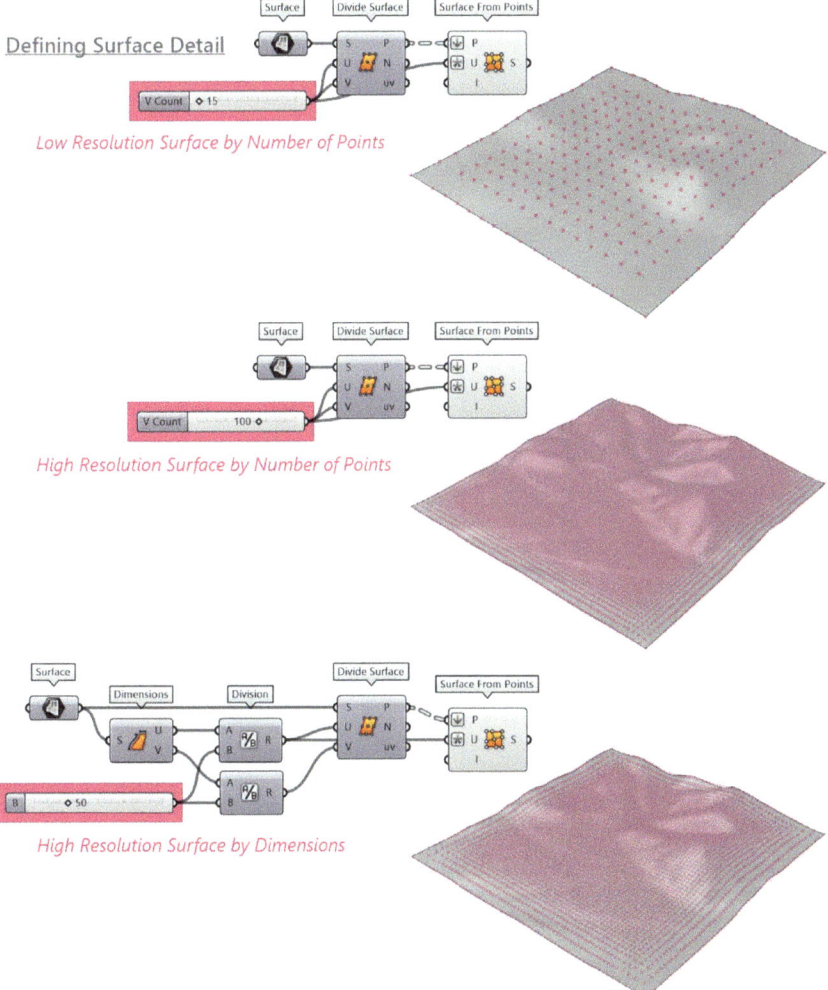

Defining Surface Detail

Low Resolution Surface by Number of Points

High Resolution Surface by Number of Points

High Resolution Surface by Dimensions

Figure 3.17 Changing the resolution or detail of surface terrain geometry by point densities or set distance spans.

of x+1. That is, because when dividing a surface into points, the (U) input for that component will actually create that number of spans between points, not a number of points. For one span to be created, two points are needed, so it will also require one additional point regardless of the number of spans. Therefore, when the *Surface from Points* component is requiring a number of points for the (U) input, the same number slider can be used since one additional value will be added from the expression.

This operation also demonstrates that if it is not preferable to model a surface terrain by a set number of points or spans, then the use of distance between points can also be utilized. For example, instead of having ten points across the surface that may have a distance of 215 feet, it may be more legible to place a point every 25 feet, as shown in the bottom image of Figure 3.17. This operation can be achieved by extracting the dimensions of the surface with a *Dimensions* component and have both the U and V outputs divided by a number slider simulating different distance values. The new divided U and V outputs will replace the previous use of a number slider for the rest of the operation.

Color Gradients

The operation of colorizing geometry based on a set of parameters is nearly used in every demonstration to ensure graphic legibility of data through one of the key elements – color. To make the explanation and process as straightforward as possible, the x value of a point's coordinate is used to define color values, since that can be easily correlated with the Gradient component function, as seen in Figure 3.18.

First, a series of random points have their x, y, and z extracted with *Deconstruct* component. The x value from this component is where the point sits on the x-axis within the model space or as it moves left and right. To colorize any set of geometry, the lowest and highest value of that parameter range (domain) needs to be determined using the *Bounds* component. This range of point parameters are then separated using the *Deconstruct Domain* component, so each can be inputted respectively into the *Gradient* component where L0 (*Lower Limit*) needs the lowest

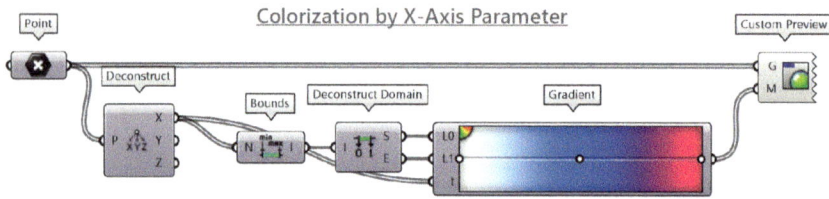

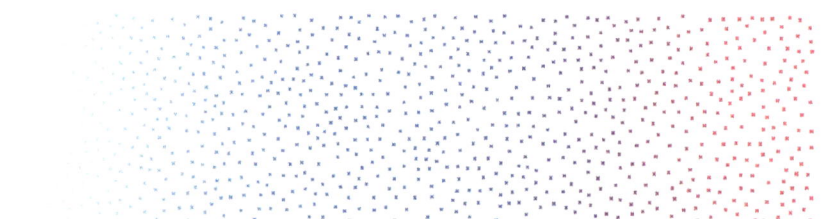

Figure 3.18
Colorizing point geometry based on their x coordinate parameters using a color gradient display.

value and *L1* (*Upper Limit*) needs the highest value. All the *x* parameters are then connected to the *t* (*Parameter*) input to determine where each *x* value fits within the entire range of values. The *Point* component is then connected to the *G* input of the *Custom Preview* component with the *Gradient* component connected to *M* input to assign that color to the point geometry, as shown in the bottom image of the figure.

Although this demonstration is specific to the *x* coordinates of points, this operation is applicable to any set of parameters such as slope, distance, curvature, size, and the many others showcased throughout the book.

Data Order and Orientation

Since parametric modeling is data driven, understanding the order of data lists is critical in refining operations and robust scripts at the starting point. The series of examples shown in Figure 3.19 demonstrate the order and arrangement of data through colorized points and vectors with a color gradient. The blue to pink spectrum that goes from left to right will simulate the same order of data where the first item will be blue and transition to pink for the last item. These images are not necessarily intended to be scripted out but rather serve as a visual reference. Some of the included operations for rotating and changing degree values will however be instrumental in the later chapters.

In the top image of the figure, a circle is divided by a series of points, 360 for each degree, to illustrate how the first point of the circle begins at the far right or essentially at the red *x*-axis and proceeds in a counterclockwise direction. This is important to know, as things may need to be oriented in accordance with cardinal directions. So the first point, in this case, represents due east (*90°*), the 90th point is due north (*0°*), the 180th point is due west (*270°*), and lastly the 270th point is due south (*180°*). This is counter to the cardinal directions and order; so that is why it is important to understand this mathematical operation to ensure accurate and correct results when measuring for the orientation of Rhino and Grasshopper geometry.

The next series of images, starting with the second from the top, demonstrate the process and logic of reorienting vector angles or degrees for future Grasshopper scripts. With the first one, vectors from the center point of a circle are created for 24 angles using the *Vector 2Pt* component. The vectors can be measured against a designated vector, in this case, a *Unit Y* value connected to the *B* input of the *Angle* component. This operation will shift the degree of measurement to due north to establish the same starting degree as conventional cardinal directions. But because there is no optimal plane connected to the *P* input of the *Angle* component, the degrees radiating out from that *0°* will be the same on both sides and conclude at *180°* or due south. Although the results from this operation will not translate correctly for cardinal directions and terrain aspects, it is useful for viewshed degrees when modeling walkability and urban settings.

Connecting a *XY Plane* to the *P* input of the *Angle* component will match vectors and degrees to cardinal directions, as shown in the third image of Figure 3.19. For this to be useful for terrain aspect where north-facing slopes should include a range of degrees from 315 to 0, and then from 0 to 45, the vectors need to first be rotated, as shown in the last image of the figure. The vectors from the *Vector*

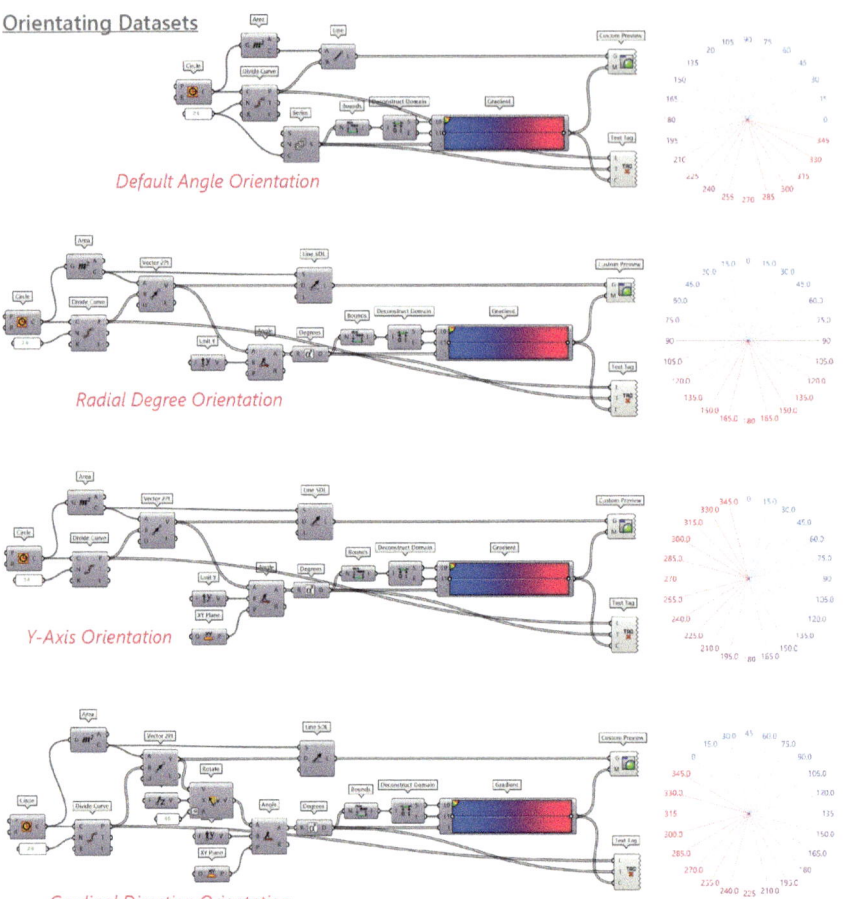

Figure 3.19 Demonstrating orientations and angles of geometry in Grasshopper with operations to adjust according to specific measurement needs.

2Pt are rotated along the z-axis with a *Unit Z* component connected to the *X* input of the *Rotate (VRot)* component. The rotational degree and direction is performed with a *Panel* input of −45 and *Degrees* assigned instead of radians.

Boolean Patterns

Boolean patterns are *True* or *False* statements that assess whether a list of data matches a specific parameter condition. *True* being that it does and *False* being that it doesn't. Some common parameter conditions involve determining whether a set of values is greater than a given value or, as in the case below, whether a point is inside, coincident, or outside a set of curves. The resulting Boolean patterns can then be used to either separate or remove different lists of data accordingly. The examples in Figure 3.20 showcase both options.

The evaluation of whether a point sits within a curve or surface is a common scenario when modeling landscape performance. These points can often refer to a tree location, a drainage line, or a surface material, and it is then necessary to determine whether that item is located within a specific spatial boundary such as a soil profile or shadow projection. For the examples below, hypothetical points are located randomly within a composition of multiple curves. This operation will determine which points are located inside the curves by outputting *True* and *False* values for each point's condition.

Both the points and curves are referenced and connected to their respective inputs in the *Point in Curves* component. There is a similar *Point in Curve* component as well that functions and outputs the same; however, it can only evaluate the points with a single curve rather than with multiple curves as is being demonstrated. Even if there is only a single curve referenced into the *Point in Curves* component, it will still function properly, so it seems unnecessary to use the other component.

The *Point in Curves* component does not directly provide Boolean patterns, but instead numeric values of *0, 1,* and *2.* The *0* value indicates that a point is outside a curve, *1* that it is coincident, and *2* means inside a curve, which can be seen in the first image of Figure 3.20. These numeric values are then inputted into a *Larger Than* component with a *0* condition, to determine which points are not outside of the curves so that it includes both coincident and inside values. In some cases, *1* may need to be the condition since coincident points are not preferred. The resulting *Larger Than* output is a series of *True* and *False* values that correspond with each point. In the example below, the two panels of numeric and Boolean pattern values show how *2*s result in a *True* value and *0*s result in a *False* value. These values can then be used with a *Dispatch* component to separate the two conditions of the

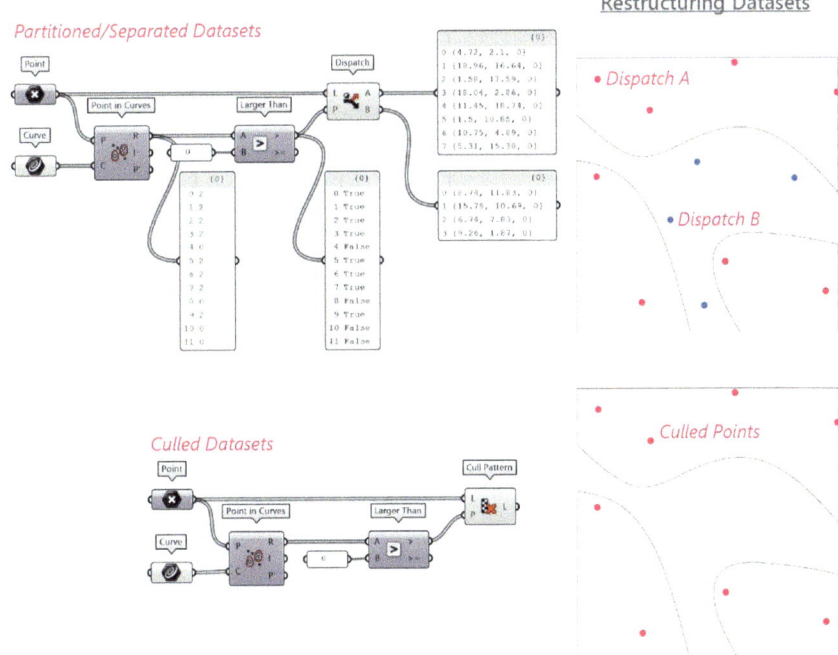

Figure 3.20
Using Boolean
Patterns to partition
and restructure
datasets by
separating or
culling (removing)
false values.

points where output *A* are points inside the curves, colorized pink, and output *B* are outside, colorized blue.

In some instances, it is necessary to keep all the data, but have it parsed out with this *Dispatch* component but as shown in the second image of the figure, if *False* values are not necessary, then a *Cull Pattern* component can be used.

In this demonstration, the referenced points are the list of items being parsed or culled; however, any list of items with the same number of items can also be applied. Examples in other chapters will demonstrate this as well.

Closest Point

Many of the processes when inventorying and analyzing a landscape for ecosystem services and landscape performance rely on distance proximity such as tree benefits, erosion control, stormwater runoff, outdoor comfort, and many others. This operation demonstrates an effective way to both measure those proximities between multiple objects and how to delineate which objects are closer to one over the other. With that basic logic, isolating different groups and their closest points can also be achieved when parsing data is necessary to attribute additional external data to these individual groups.

In the first example of Figure 3.21, a set of random points, *Populate 2D*, is used to determine which points are closest to the three referenced points. To compute this, the *Closest Point* component is used. The function of this component is to measure the distance between all the random points and the three referenced points and then determine which one is closest. As a result, both points and the index of the closest point are outputted. As a visual indicator, a *Line* component is used to create a connection between the random point and the closest point. Because the index of the closest point is outputted, that value can be used for the gradient operation to colorize the delineated groups.

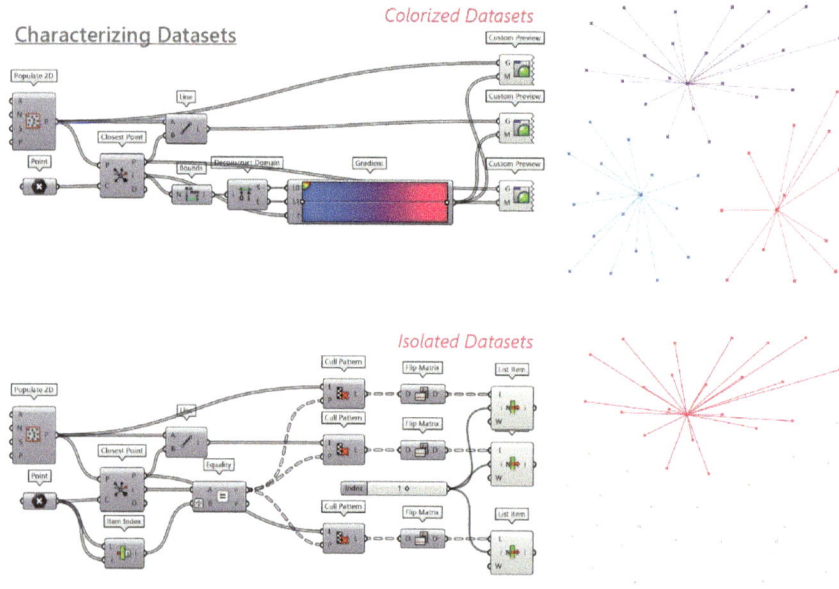

Figure 3.21
Using Boolean patterns to filter or isolate datasets according to a specific parameter(s).

Another component, *Closest Points*, performs the opposite function where the referenced points will find a defined number of closest random points. Although it may be useful in specifying a set number of points to find, if multiple referenced points are closest to the same random point, they will both connect to it, whereas this operation will delineate the random points to just one closest point. The *Closest Points* operation will be utilized for measuring tree performance since buildings can benefit from multiple shared trees instead of just the closest one.

In the second image of this figure, the process of parsing and then filtering the data can also be achieved. This becomes extremely useful for modeling stormwater runoff and erosion control, since the newly parsed data will also be required to match with external data that has the same branching structure.

Each of the referenced points is assigned an index value to match with the index values outputted from the *Closest Point* component by connecting it to both the *L* and *i* input of the *Item Index* component. This will simply create a sequence of index values starting at *0* for the number of referenced points. The *Equality* component is then used to determine which index from the *Closet Point* component matches the index from the *Item Index* component. The input for the *Item Index* component needs to be *grafted* to ensure each referenced point is matched separately with the other index values. This will first ensure each referenced index is cross-referenced with every *Closest Point* index or else only the last referenced index value will be matched with the remaining longer list of index values.

The *Equality* component will then output Boolean patterns on whether the random points match with the referenced points. Based on that pattern, the *Cull Pattern* component can then be used for both the random and referenced points along with the line connecting the two to remove all the *False* values resulting in random points parsed into their respective closest point groups. After using the *Flip Matrix* component for each one, a *List Item* component can be connected with a number slider for the *index* (*i*) input to scroll through and filter the groups.

Communicating data through different forms of representation and modeling methods has been pioneered by great designers for several decades and is continuing to build momentum with even newer ideas as the age of technology and large datasets continue to advance. From infographics to dynamic animations, datasets can shift from the abstract of tables and charts to more comprehensive forms of communication for a wider audience. The portrayal of landscape performance is no different. It is vital to take advantage of these same resources and tools to convey the significance of ecosystem services from nature and sustainable design strategies resulting in environmental, social, and economic benefits.

Landscape performance relies on measurables, methods, and metrics to evaluate how spaces are performing in comparison to baseline conditions. This cannot be achieved without the incorporation of quantifiable information to the more common qualitative aspects of space for people and ecological systems. Parametric modeling serves as a valuable mediator between the abstract construct of data with the tangible spatial aspects of our environment.

4 Plugins

Grasshopper has an extraordinary collection of components to complete the majority of any basic calculation to extensive parameterized cities. Most situations will not require anything beyond the default tools; however, as expected outcomes begin to require a litany of external resources, advanced spatial computation, or even retrieval of unordinary information, the inclusion of additional plugins becomes a necessity. Some plugins can simply turn a tedious process of repetitive operations into a single executable computation to make the workflow more efficient. Even though Grasshopper is an integrated plugin with the newer versions of Rhino, this plugin requires plugins of its own.

This chapter will cover a variety of required plugins to perform specific operations related to landscape performance that include the importing of climate data, Excel files, and geographic information system (GIS) shapefiles. Some are more elementary in providing visual aid to the Grasshopper workspace or create different types of geometry not offered with Grasshopper's components. These are why it is important to seek plugins for this book's intentions as well as for other unforeseen reasons, all while knowing just like any other form of technology, updated versions of these same plugins or entirely new ones get published for use. The majority of plugins are also open-source and shouldn't require much effort to install.

GENERAL INSTALLATION PROCESS

Most Grasshopper plugins can be directly accessed and downloaded from the *Food4Rhino* website (https://www.food4rhino.com/) after creating a free account. The installation process for the majority of them follows a common procedure outlined below; however, on their respective page installation, instructions are usually provided. Once the latest version of the plugin is downloaded as a zipped folder, follow these steps:

1. *Right-click* zipped folder and select the file's *Properties* option
2. Select the *Unblock* option at the bottom of the *General* tab (this will unblock all the files that have been zipped instead of unblocking each individually if not performed on the zipped folder)
3. *Unzip* (*Extract*) folder for access to the repository of plugin files
4. In Grasshopper, go to *File -> Special Folders -> Components Folder* to open the software's plugin directory location

DOI: 10.4324/9781003208020-5

5. Select all the previously unzipped plugin files and copy them to this directory location
6. Shut down and reboot Grasshopper and Rhino

If these steps are followed correctly, then the plugin should appear as a new tab after the *Display* tab in Grasshopper. As stated earlier, most plugins have installation instructions on their page on the website or in a file as part of the download. Refer to these first if not comfortable with these instructions or contact the plugin developer for any technical issues.

BIFOCALS (BY MARCSYP)

Bifocals is a great user interface plugin to tag components with their respective name if the *Draw Icons* display selection is turned on. This becomes a helpful tool in providing both a visual and annotated indicator for the components to help search by name or find by recognizing the icon used for the component.

The plugin also comes in handy since some components have the same name, for example the *Construct Domain²* component, but perform differently or require different sets of inputs. These ones in particular do have different icons to assist in using the correct one. Another instance in which this plugin can be useful with finding components is shown with the last image of Figure 4.1. The name of the component is *Surface from Points* but the abbreviated term on the component reads *SrfGrid* which is significantly different than the name of the component. Components can still be searched for by *double-clicking* the Grasshopper canvas and typing the abbreviated name as well.

The bifocals plugin component will need to be placed with the Grasshopper workspace to activate the labels. After installation, it can be found within the *Params* tab under the *Utility* section. This component will be used for all the script definitions demonstrated throughout the book to best communicate all the operations for ease of use and instruction.

MOSQUITO (BY STUFF1@CEEDSTUDIO.COM)

The Mosquito plugin is one of the most effective plugins and coding to depict drainage lines along a surface. It is important to note that the plugin only works on surfaces/Breps; so if a mesh terrain is being used, it needs to be converted into a surface using the *MeshtoNURB* Rhino command or the *Mesh to Surface* Grasshopper operation from the previous chapter's section "Elementary Operations".

Figure 4.1
Label tags for components displaying their icons using the Bifocal plugin.

Component Display Modes

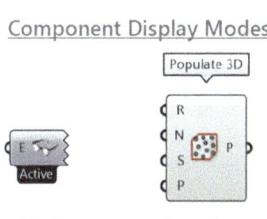
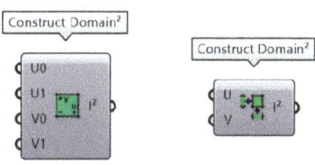
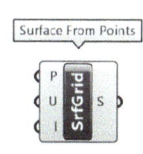

Plugin Component   *Draw Icons*   *Same Name/Different Icon*   *Default Display*

Like that operation, a grid or even randomized points can be used to determine the starting point for the drainage lines. Using a *number slider* component to divide a surface with either low or high values is useful to begin visually determining the density of drainage lines wished to be modeled. These points need to be *flattened* for the *Points* input of Mosquito's *Flow* component, as shown in Figure 4.2.

The next two settings for the *Flow* component will determine the resolution and length of the drainage (flow) lines along the terrain model. The *Res* input will dictate the length of each drainage line segment, in other words the detail, to ensure accurate representation of flow along topography by following valleys and constrained by ridgelines in the terrain. A more detailed example of these parameter differences is shown later. Essentially, the lower the number slider value for the *Res* input, the less accurate, while higher values become more precise. Because of the heightened quality of detail in the drainage lines with higher values, the number of points/data used to compute and generate the lines will grow exponentially and cause some prolonged processing time for this plugin and cause lag for any other operations reliant on this geometry.

With more detail or number of drainage line segments, the number of segments will also need to be larger to ensure that the full length of the possible drainage line is achieved or else they will end abruptly before reaching the lowest drainage point. This is controlled by the *number slider* for the *Calc* input and will ultimately be dictated by the size and area of the terrain model. So, as the *Res* value increases, so should the *Calc* value. The two examples from before show drainage line densities using different number slider values for the *Divide Surface* component. A True/False *Boolean Toggle* is an option for the *pLine* input, but rarely makes a significant difference.

The detail or resolution of the drainage lines is often challenging to notice at a larger scale, but, as shown in Figure 4.3, there is a clear distinction between a low and high value for the *Res* input of the *Flow* component. The low-resolution value

Drainage Line Densities

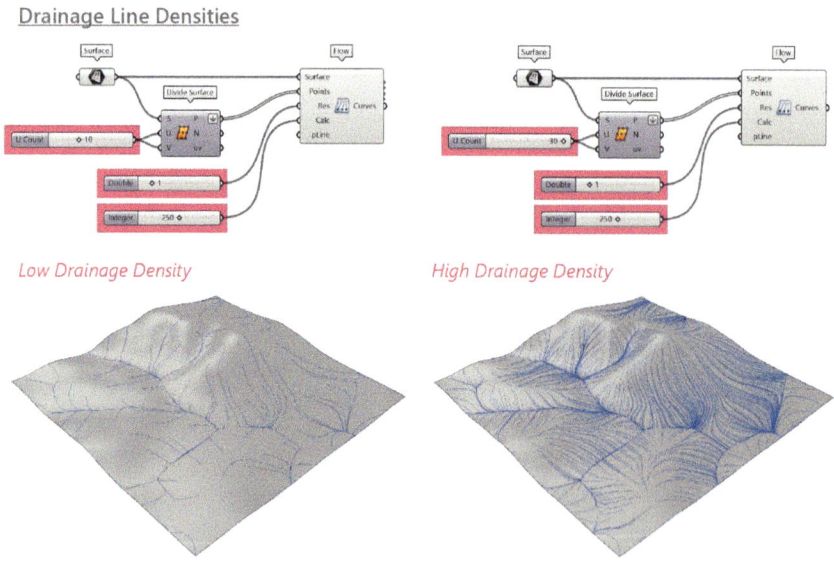

Low Drainage Density   *High Drainage Density*

Figure 4.2
Generating different densities of drainage lines with the Mosquito/Sonic plugin.

Figure 4.3 Demonstrating how different resolution settings affect drainage line details with the Mosquito plugin.

Drainage Line Detail

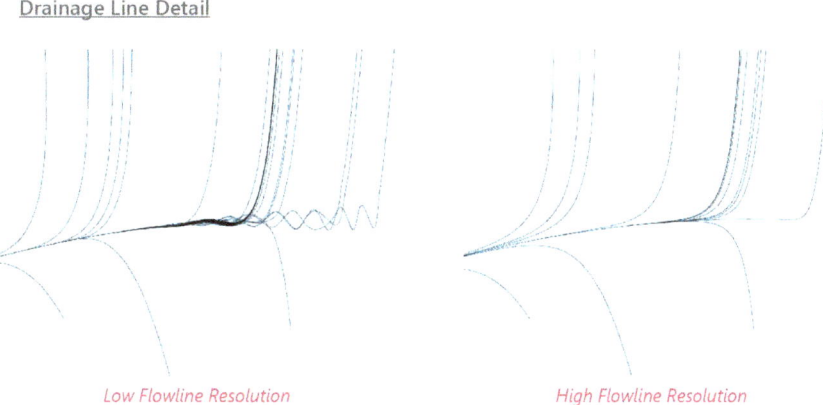

Low Flowline Resolution   *High Flowline Resolution*

represented in the left image shows that the drainage lines can become erratic and irregular to natural flow patterns, whereas the image to the right shows how the higher resolution significantly improves the accuracy of the drainage lines.

This simple yet effective plugin operation will be used for numerous demonstrations throughout the rest of the book as hydrology and the flow of water plays a major role in ecosystem services and landscape performance.

@IT (BY ELCINERTUGRUL)

With the *@it* plugin, shapefiles can be imported into the Rhino workspace based on their coordinate projection. This means that the model will not be placed at the *0, 0, 0* c-plane origin but rather its geolocation in relation to the global longitude and latitude coordinates. This can be found using a Rhino file with no other geometry and then using the Rhino *Zoom extents* command (*Shift + Ctrl + E*) to zoom to where the geometry is placed after completing the first operation in Figure 4.4. Another option if there is already geometry in the Rhino file is to bake the Grasshopper geometry to a unique layer and selecting that geometry to perform a *Zoom Selected* command.

For accurate results, it is important to open the shapefile in a GIS software and exporting it with a *WGS84* map projection. This process is elaborated on further in Chapter 5 to ensure accurate overlay of multiple shapefiles; however, for this example, it is not a necessity.

This plugin will import both the physical geometry (points, curves, and polygons) and the feature attributes for each geometry. In this example, contour lines are imported as a flat projection and then elevation values are assigned to move the contour lines to their respective height.

To begin the process, the *shp* filetype is referenced from its file location using the *File Path* component and connected to the *C:/* input of the *Imp@it* component. This component can then be connected, as shown in Figure 4.4, to the appropriate inputs of the *DataVis@it* component. The features from the *F* output of the *Imp@it* component can be filtered using a *List Item* component with an index value for the specific feature list. The panel in the image shows which index value is associated with each feature attribute.

If these features are filtered at this point in the operation, then the rest of the script is limited to only those values, whereas when multiple features will be necessary to complete the definition, it is best to use from the *Val* output of the *DataVis@it* component. With this component display on, the associated geometry can be viewed.

If the model needs to be in imperial units instead of metric, then the geometry needs to be scaled first before applying the contour elevation values. This will need to be done even if the shapefile was exported in feet from a GIS program, because metric is the default unit for the *@it* plugin regardless of the file's unit of measurement. To make this unit conversion, whichever geometry is being imported, in this

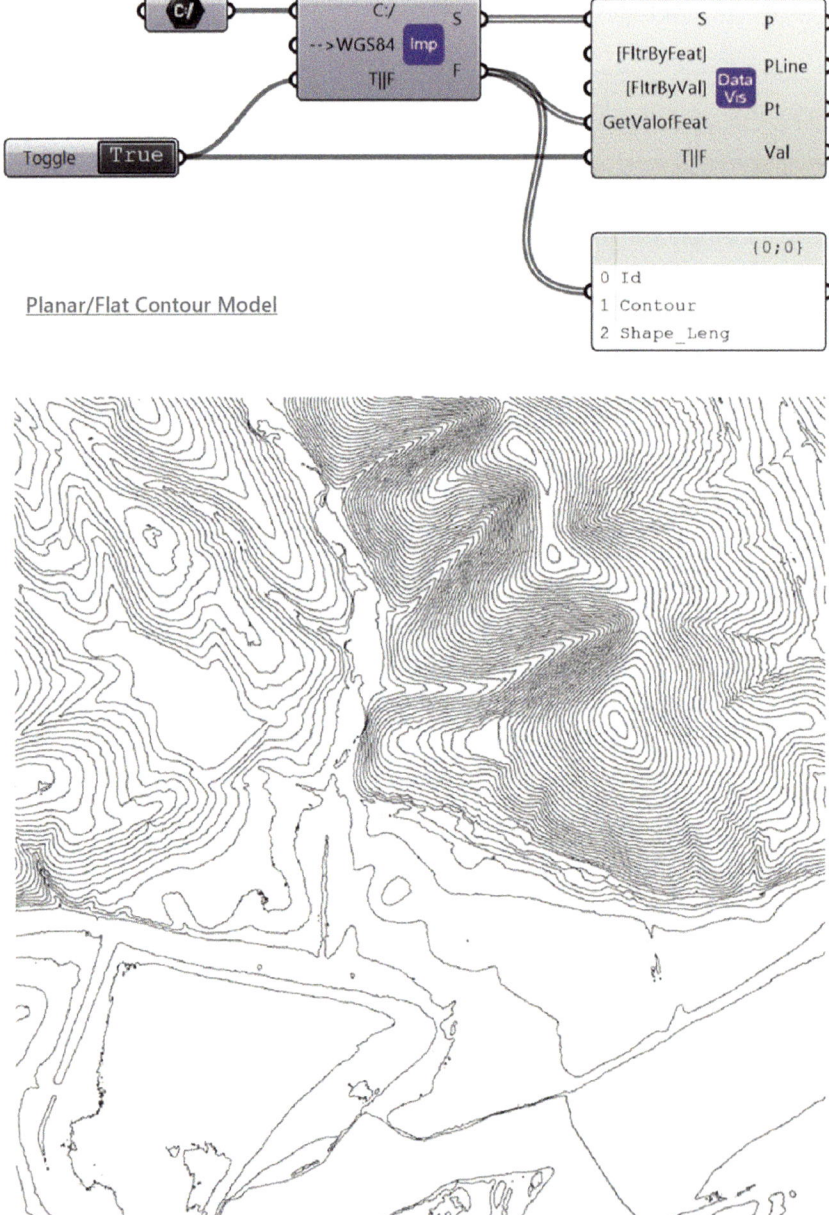

Figure 4.4
Importing contour lines from a shapefile with the *@it* plugin.

case *Plines*, needs to be *flattened* at the input of the *Bounding Box* component. The *Bounding Box* component also needs its settings changed to be a *Union Box* by *right-clicking* on it and making that selection in the pop-up window or else each individual curve or geometry type will have its own bounding box instead of a bounding box for the entire imported shapefile. This bounding box is used as a reference to scale from by extracting the *Centroid* from an *Area* or *Volume* component.

The same geometry needs to also be *flattened* at the *Geometry (G)* input of the *Scale* component to match data structures. A scale factor of *3.2808* can then be inputted to convert the model from meters to feet. The geometry still has a flat projection, so the next step is to elevate each contour line using the elevation feature from the shapefile.

As stated earlier, a *List Item* component can be used to filter data features. Having a panel connected to the *F* output of the *Imp@it* can help determine which index value needs to be used for the *List Item* component. In this example, the elevation values are housed in the *Contour* feature attribute or index item *1*. In Figure 4.5, the *List Item* component with that value is *flattened* and connected to a *Unit Z* component to elevate the contour lines to their respective height using the *Move* component.

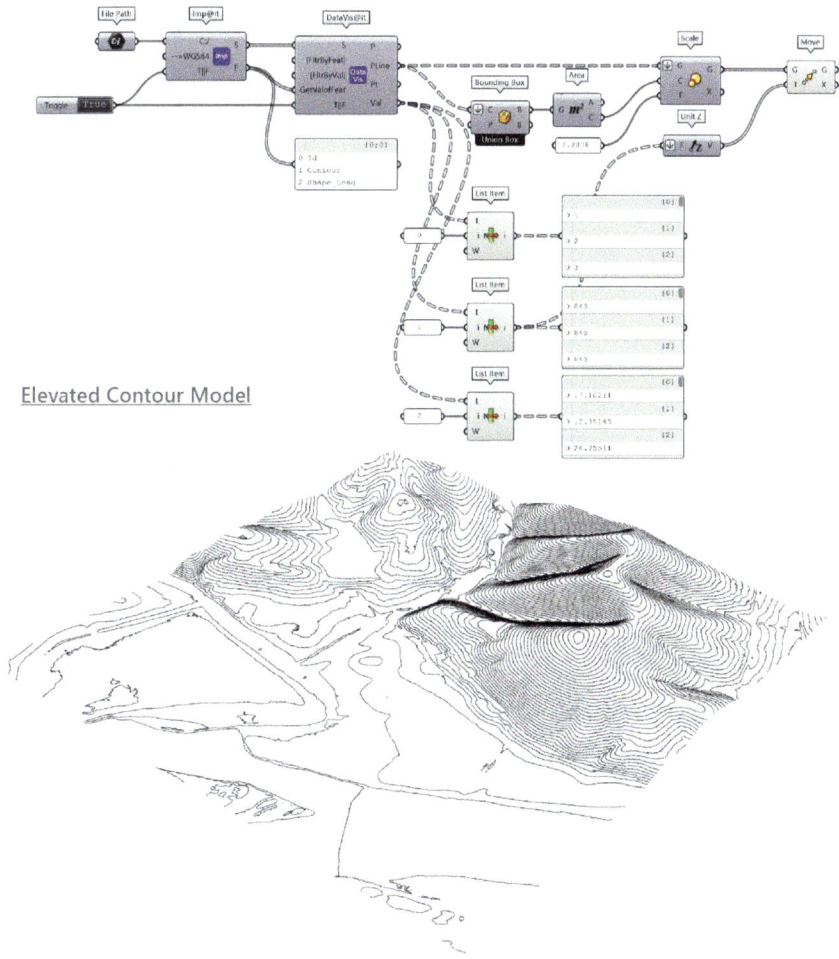

Figure 4.5
Scaling and moving contour lines to their respective elevation.

Now that a three-dimensional model has been generated from this process, it can serve as the foundation for many other future operations, scripts, and landscape performance modeling, since it can be converted into either a mesh or surface for all other processes. That will all be demonstrated, starting in Chapter 5.

Since these scripts can begin to compound quickly and begin to slow down the processing time as the script progresses, it may be important to either strategically *Internalise* the data so it's imbedded within a Grasshopper component or *Baked* into Rhino. Whichever method is applied, it is crucial to not physically move the geometry from its geolocation, since the overlaying of other future shapefiles will occur and require them to match in location, size, and orientation.

GHEXCEL (BY XIAOMING)

When it comes to modeling landscape performance, data tables, whether internal or external, are a necessity to quantify the qualitative elements of the landscape by using metrics and measurements. When data tables are robust in multiple columns of data, it is important to parse or separate the information into separate outputs for ease of calculating landscape performance.

When external data, such as Excel or csv files that commonly come from other resources, is needed for the calculating process, the plugin *GhExcel* can be used to reference a Microsoft Excel file and all the housed data. Just like the *@it* plugin, the file needs to be first located and referenced with a *File Path* component to connect with the specific *ExcelDynamicRead* component. This component versus the *ExcelStaticRead* component will open the Excel file and allow edits to be made and updated back in Grasshopper. In Figure 4.6, a panel is used to show how the data from the Excel file, on the right-hand side, is structured in branches, one for each column of data in the Excel file.

To separate the data, a simple operation is used to filter out unnecessary or unusable data. This process can also be done initially within the Excel file as well but this process, in Figure 4.7, demonstrates how all the original data can be maintained while still usable. Since the branches or columns of data contain both text and numeric data, it is important to first remove the text labels since those will

Importing Excel Tables

Excel Data Structure

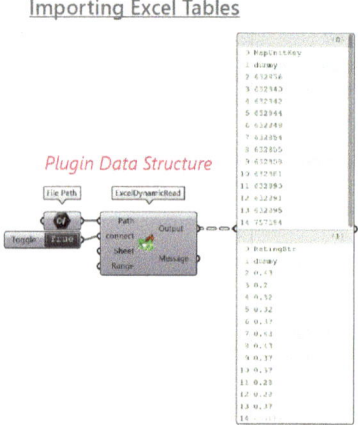

Plugin Data Structure

	A	B	C	D
1	MapUnitKey	RatingStr	RatingNum	RgbString
2	dummy	dummy	123.456	dummy
3	632836	0.43		#54ADFF
4	632840	0.2		#F0FD4F
5	632842	0.32		#77F7E2
6	632844	0.32		#77F7E2
7	632849	0.37		#32E3FF
8	632854	0.43		#54ADFF
9	632855	0.43		#54ADFF
10	632858	0.37		#32E3FF
11	632881	0.37		#32E3FF
12	632890	0.28		#B1F9B5
13	632891	0.28		#B1F9B5
14	632895	0.37		#32E3FF
15	757584			#DCDCDC

Figure 4.6
Importing Excel files with the GhExcel plugin.

Figure 4.7
Partitioning and
restructuring Excel
file to be compatible
with Grasshopper
computations.

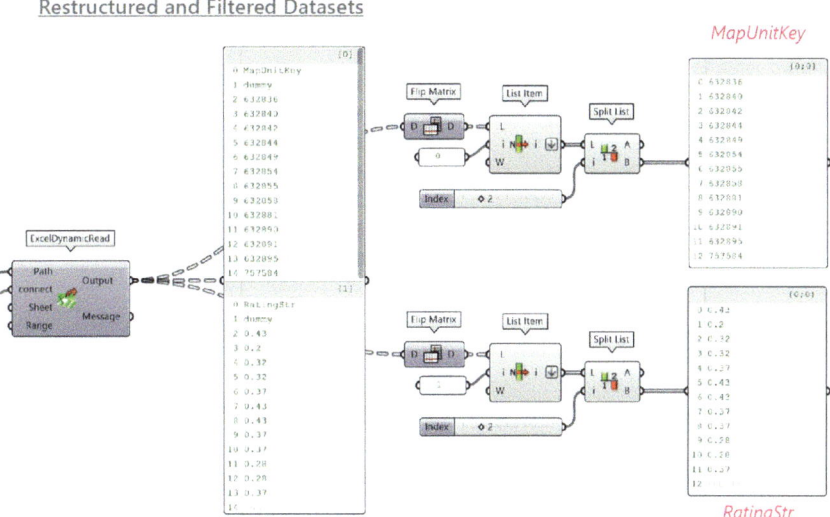

conflict with any numeric operation and lead to errors. For the functionality of the *List Item* component, the data first needs to be restructured with the *Flip Matrix* component. An *index* input of *0* is used for the first column, *MapUnitKey*, then *1* for the second, *RatingStr*, and so on. The column of both text and numeric values is then separated using the *Split List* component with an *Index (i)* input of *2* to remove the first two text labels. If the data has no text labels, then the splitting operation is not necessary. However many rows of text labels are within the Excel file is the input value to split with. The *B* output of the *Split List* component now contains only the numeric data and can be used for future calculations.

Another reason to remove the text labels from the usable data is because that can also lead to an uneven branching of data when attempting to cross-reference with physical geometry. For example, in this demonstration, each column initially has 15 list items when including the first two text labels *MapUnitKey* and *dummy* as shown for the first column. If that list of 15 items is cross-referenced with its associated 13 geometries, as is often the case, the first two text labels, instead of the first two numeric items, will be connected to the first two physical geometries leading to a critical error.

With many of the landscape performance modeling later in this book, this operation will also be used frequently, since external data is often needed to be attained as separate Excel files.

SHORTEST WALK GH (BY GIULIO@MCNEEL.COM)

Contrary to visualizing distances from a location using concentric circles representing different lengths, the *Shortest Walk* plugin provides accurate and realistic distances along existing physical routes such as roads, alleys, and trails. This component also has the ability to map the shortest route to multiple locations and colorize based on the range of distances for better comprehension of distance mapping.

The *Shortest Walk* plugin will be located in the *Curve (Crv)* tab within the *Utility* section. It will first require a curve network, either with referenced Rhino geometry or imported as shapefile as demonstrated with the *@it* plugin. The distance and routes are dictated by the detail and extent of these curves; so if the locations for mapping are not immediately next to the curve, it will not account for that missing information. If that level of detail and accuracy is necessary, then make sure the network contains that level of data. For large-scale city mapping, major roadways should be adequate.

After the curve network is connected to the *Shortest Walk* component, a simulated line segment between the start and end point(s) is needed for the plugin to understand where routes need to be created along the curve network. Figure 4.8 demonstrates the use of multiple end points to calculate the different routes for future measurement and visualization.

After these inputs are computed, routes will be generated as curves along with associated data that includes succession or the index of the curve used as part of the route directions as Boolean values, and most importantly, the length of the curve or route distance. These generated curves will overlap with the existing curve network, so it will be necessary to distinguish them through color and other means of visualization.

The first step in clarifying the mapped routes, as shown in Figure 4.9, is with the typical operation of using the associated curve lengths for a range of color values from shortest to longest distance using the *Gradient* component. The end points use the same coloring scheme to cohesively begin connecting the elements of routes, points, and annotated distances. The spatial mapping is then further enhanced by including annotated distances for each end point using the *Text Tag 3D* component.

As already noted, annotating data does require some extensive additional operations; however, this information brings additional clarity and comprehension to any parametric and landscape performance modeling process. The annotated distances first need a point for its location which is being demonstrated with the

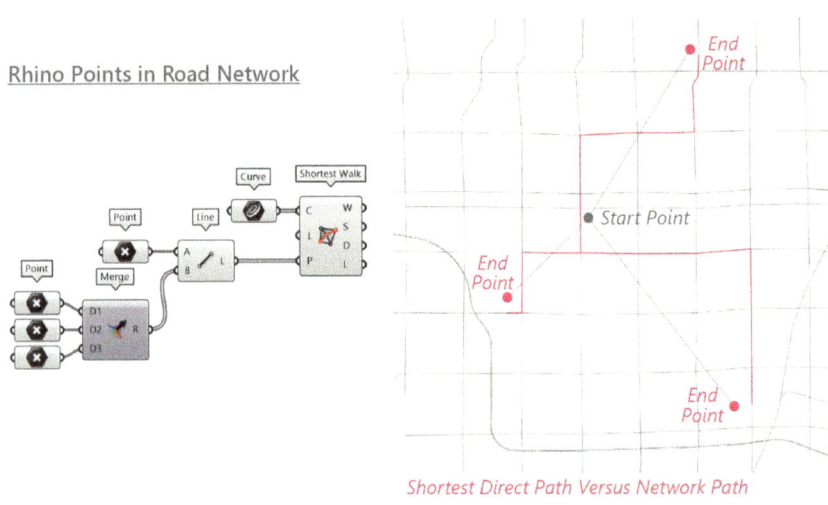

Figure 4.8
Using the Shortest Walk plugin to convert direct path to a network path along roadways.

Point on Curve component to determine a specific location along each curve that is most appropriate for the label. This will most likely require some adjustments as the additional text settings are inputted. Another option for the text location is the end points that have been combined and outputted from the *Merge* component.

The distance of each route will be in the same units as the Rhino model, in this case, feet, but it is more appropriate to convert this unit to miles due to the scale of the map example. The unit of measurement is also included as part of the text tag by using a *Panel* with the word *Miles* typed out along with a *space* before

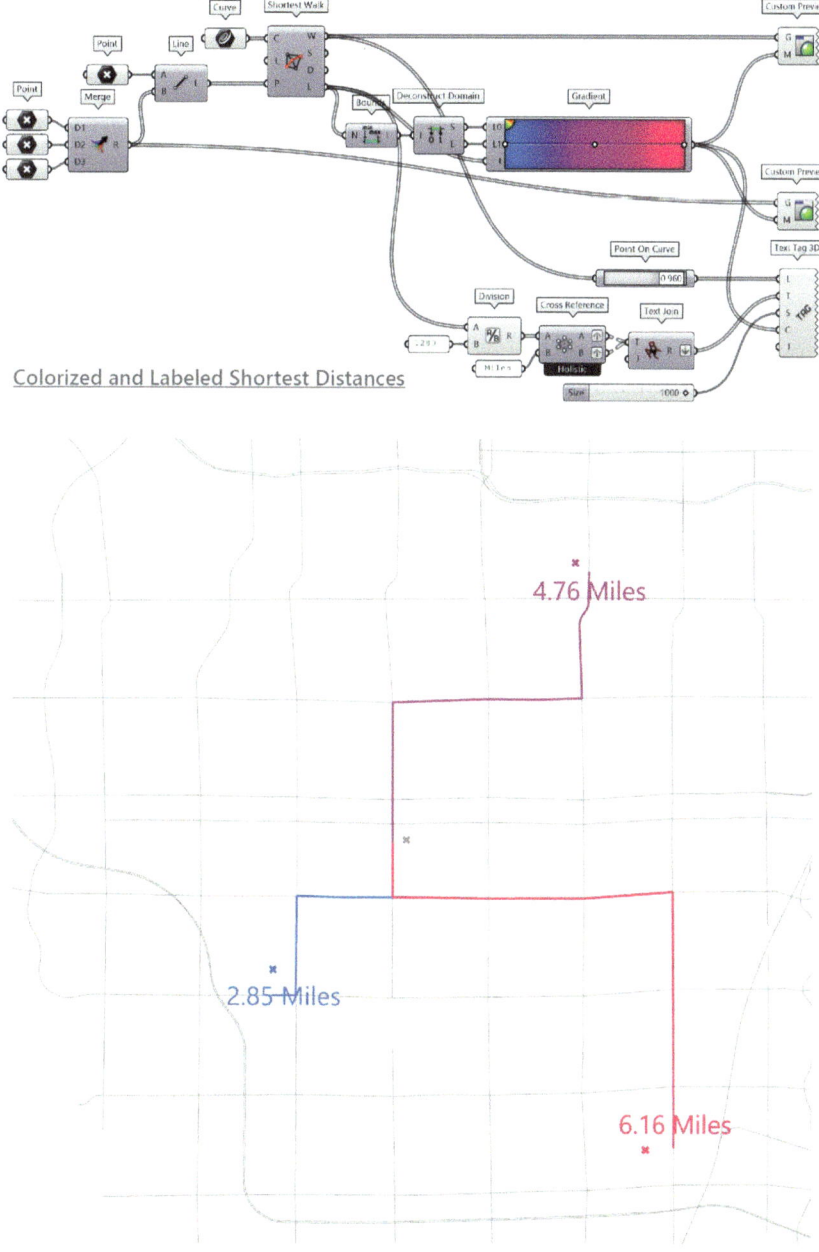

Figure 4.9
Coloring and labeling the shortest network path for different destination points.

the word to separate that from the numeric distance. Because there are multiple list items for the distance value but only one item for the unit *Miles*, the *Cross-Reference* component becomes helpful in repeating that shortest list to match the longer list in the number of values. Both the *A* and *B* outputs need to be *grafted* to ensure that when they are both connected at the *Text Join* input, the first distance value joins with the first *Miles* text value and so on for all the other list items. If these outputs are not grafted, then when they are joined, the resulting output will be one long line of text. After the grafted is joined, the output can then be *flattened* to match the data structure as all the other inputs for the *Text Tag 3D* component.

The size of the text is also contingent on the scale of the model; so for a map of this size, a value of *1,000* is used. To complete the cohesive process of all these different data output, the color values from the *Gradient* component are once again used for the text color to match in all three aspects.

The *Shortest Walk* component processes these network routes quickly and the subsequent operations of colorizing and labeling the distance routes are relatively simple; the task of moving both the start and end points to generate new colorized routes can be quite effortless and efficient in generating multiple distance maps for the analysis amenity proximity. This task will be further analyzed for application to walkability studies in the performance modeling section of this book.

ELEFRONT (BY ELEVELLE)

Rhino geometry is often the starting point of many Grasshopper definitions, and naturally during the analysis and design process of any project, this geometry can change in size, shape, type, and even quantity. Whenever these changes occur, the Grasshopper component referencing that geometry needs to be selected and be re-referenced by selecting the new set of geometry in Rhino. This can be tedious and challenging, as the selection process becomes more complex with many different layers and objects modeled inside of Rhino needing to be turned on/off or locked to make that selection. Most geometry types also need to be referenced within its respective geometry component, for example points are referenced in the *Point* component, so if there are multiple different geometry types used later, then that specific type needs to be selected instead of simply using the *Select Objects* option in the layer submenu.

With the *Elefront* plugin, geometry can be automatically updated within Grasshopper to avoid this re-referencing process. The process can be done one of two ways: starting with either a geometry type or by a layer. The image on the left side of Figure 4.10 shows the process of using a *Panel* to type out the geometry type for referencing into the *Reference by Type* component. If preferred, these labels can be made to not be case-sensitive, so it does not matter whether it is capitalized or not by *right-clicking* the initial referencing component and enabling the *Ignore Case* option. Additionally, this component is not restrictive for some geometry types, for example, typing *Curve* will also reference polyline curves and line-like curves. Now all the surfaces or any geometry type in Rhino can be filtered out into respective layers with the *Filter By Layer* component for each layer needing to be used for that Grasshopper definition.

Figure 4.10
Referencing Rhino
geometry by type
or layer with the
Elefront plugin.

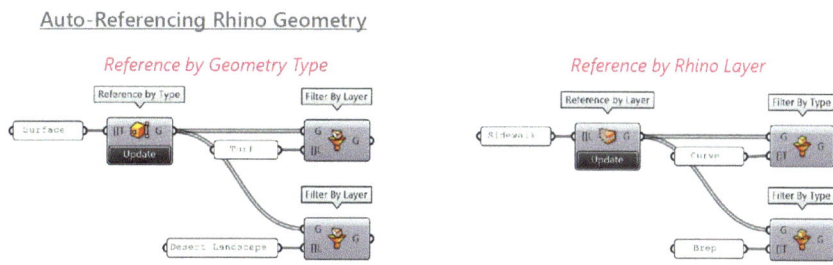

The right-side image shows the process in the opposite order of first referencing by layer and then by type. The *Reference by Layer* and even the *Filter By Layer* components can be performed for both parent and sublayers in Rhino.

Regardless of which order is being used, in order to have any changes to the geometry in Rhino be automatically updated in Grasshopper, the referencing component needs to be *right-clicked* and have the *Disable auto update referenced objects* deselected. By default, these components will not auto update referenced objects or layers.

Although it seems both operations will yield the same results, both have their advantage depending on the purpose of the Grasshopper definition. In some cases, all surface geometries may need to be referenced to compare relative size with each other for the landscape performance modeling of stormwater runoff. An advantage of starting with the *Reference by Layer* component is that there may be a lot of geometry of a particular type modeled in Rhino, so it's easier to manage starting with a specific layer and then filter to a geometry type. Regardless of the order, this plugin can help with time management and create a more seamless and automated workflow of referencing geometry between Rhino and Grasshopper as the modeling process inevitable changes during the extent of a project.

CONDUIT (BY NATHAN MILLER)

Communicating the entire story for site inventory, analysis, and design can be very challenging with parametric modeling alone. It often requires supplemental graphics to provide additional information or re-represent it through another perspective. One of the ways in which this can be accomplished with Grasshopper scripts is the inclusion of charts to visualize data abstractly instead of spatially.

When working in the Rhino workspace, it is important to visualize these charts in a way that is legible and complimentary to the main graphic model. This requires arranging and scaling these charts to fit the Rhino viewports for either exporting as static imagery or even creating dynamic animations.

One of the most effective plugins in doing this is *Conduit*. This plugin includes several presets such as pie, line, and bar charts. In addition to creating the charts from extracted data generated from the parametric model, the plugin also provides extensive layout options to configure the charts to fit within the Rhino viewport along with the model itself. The charts can be stylized and colorized to fit the same graphic style used to visualize the model.

The first step in using this plugin is setting up the size and placement of the charts using the Rhino viewport with the *Get viewport boundary dimensions* component, as shown in Figure 4.11. Depending on the setup of the Rhino workspace, this size can fluctuate significantly. Either turning off panels or having them float in Rhino can widen the viewport and give more room to work with. Using a scale of *1* for this component will also keep the workspace the exact size as the window display in Rhino. With the *RelSize* input of the demonstrated *Horizontal Tiles* component, the number and relative proportion of each tile can be sized across the entire viewport.

The second image of the figure shows a value of *1* being repeated *seven* times to make seven tiles of equal size. The third image shows what would be required to make differing proportions for a set number of tiles. This was a result of inputting the *Panel* component numbers with the list of values going 1,2,3,2,1 on the left-hand side on the image of the script. Because there aren't any dimensions used, the *2* values tiles are twice the size of the *1* value tiles and the *3* value is three times the size in the sequential order of the *Panel* values.

Next, for the layout configuration is the height and padding of the tiles. The *0.40* value being used represents 40% of the height of the viewport and the width and horizontal padding compute likewise. For this example, a *HeightPad* of 0.10 is used to give some spacing between the bottom of the viewport and where the charts will sit at the bottom. Number sliders are used for all these inputs, since the charts and labels have not been created yet and will be subject to their legible sizing and adjusted accordingly. This workflow will most likely require back-and-forth adjustments as settings are applied to the Grasshopper script.

The next step is deciding which tile to use for displaying the chart using a *List Item* component with an index value indicating which item from left to right, as shown in both the script and Figure 4.12. With an index value of *1* used, the second tile from the left is used to display the demonstrated chart design by inputting into the *Bounds* of the *Draw Pie Chart* component. A *Panel* list of labels are then inputted to identify each list of values being connected to the *Data* input. Hypothetical values are used, in this case, but for future demonstrations in this book, specific values will be inputted, as shown in that figure.

For the *PieStyle* input, the *Conduit Pie Style* component is used to configure the inner and outer radius along with the label location and height. As shown in the script, other settings can also be applied that include the text font and color. Next, a *Palette* is inputted with the *Conduit ColorPalette* component using a color gradient operation, demonstrated in the previous chapter. The values and their extends from the *Gene Pool* component are used just like any other parameter intending to be colorized. Finally, the output from the *Draw Pie Chart* component is inputted into the *Heads Up Display (HUD)* component. The *Bounds* output from the initial *Get viewport boundary dimensions* is connected in its respective *HUD* input along with a *Boolean Toggle* set to *True* to activate the display. At this point, with all the settings inputted, the number slider values can be adjusted to alter the tile selection and padding in addition to the label sizes for the chart.

For the creation of a bar or line chart, Figure 4.13 shows the same operations apply as before in using a *List Item* component to determine which tile to place the chart in. Hypothetical values are once again used to simulate what would typically occur in a stormwater runoff model. A value for each *Category* of the bar chart is inputted, which can also be text values instead of numeric. The chart can also be given labels for both axes to annotate the chart further for optimal legibility and rely less on post-process labeling in other software program such as Adobe Illustrator.

To size all the bars in the chart relative to each other, a *Range* input is used that includes a domain from *0* to the highest value being inputted. A *0* value is used

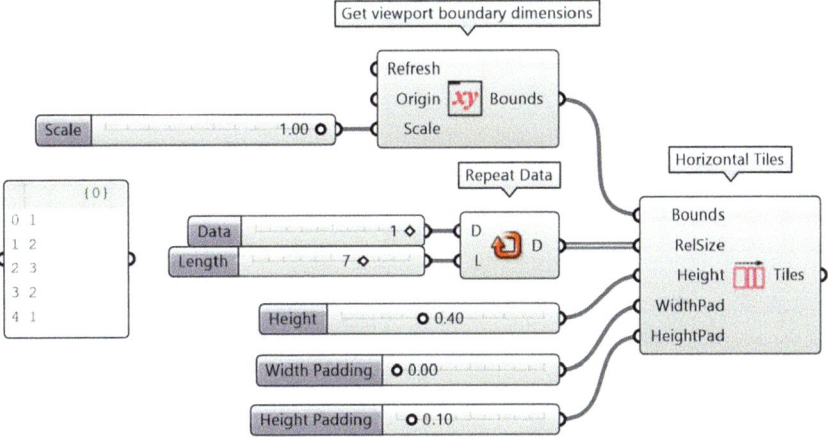

Figure 4.11 Demonstrating different viewport options and settings with the Conduit plugin.

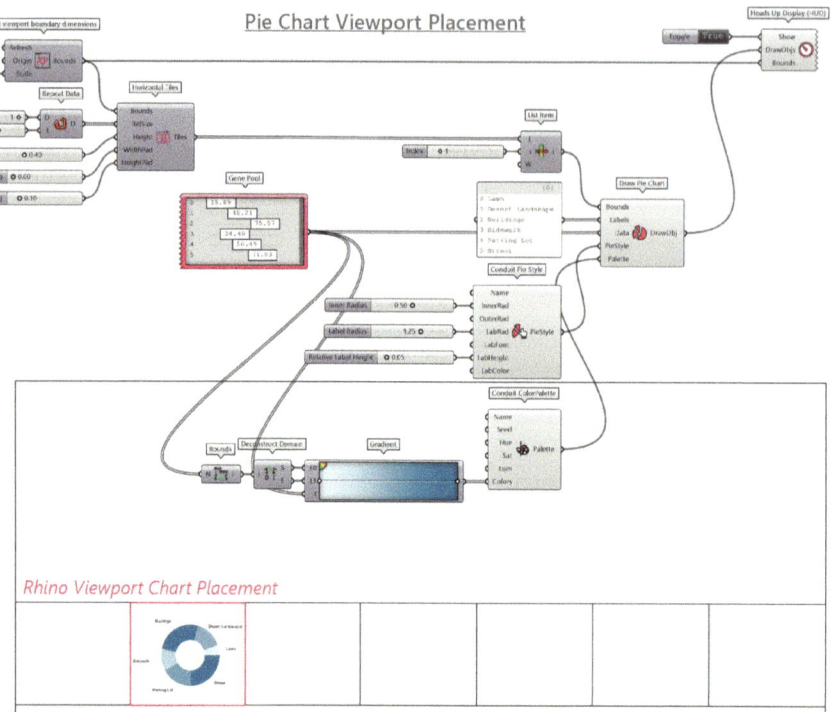

Figure 4.12
Placing and stylizing a pie chart for a select viewport in the Rhino workspace.

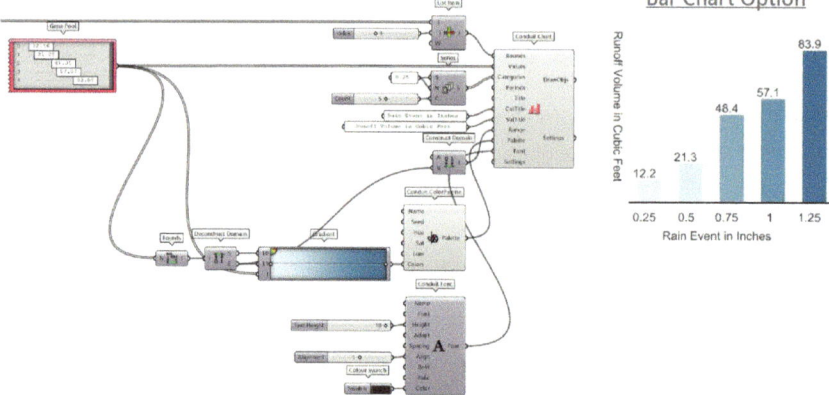

Figure 4.13
Stylizing other chart options with labels and color for the Conduit plugin.

instead of using the lowest value, because then the lowest value would have no height value since they are all relative to each other, so the *0* input serves as a better range instead of the actual value. The *Conduit Font* component can also be used to adjust the sizing accordingly for best legibility. Additionally, the *Conduit Chart* can be *right-clicked* to open additional settings for text orientation and justification.

To show multiple charts simultaneously, connect each chart to the *DrawObjs* input of the *HUD* component as demonstrated earlier and *flatten* that input. The

DrawObjs component will only read the first chart inputted by default, so the function of flattening the merged inputs will read as one chart. Alternatively, the *HUD* operation from Figure 4.12 can be repeated for each chart scripted out.

HORSTER CAMERA CONTROL FOR GRASSHOPPER (BY FRAGUADA)

In addition to utilizing present charts from *Conduit*, it may also be valuable to create custom charts with either Rhino geometry or Grasshopper operations. The use of supplemental data and graphic communication can be taken advantage of even further with the *Horster* plugin to fit these custom elements to the viewport and model workspace similarly to the *Conduit* plugin. The placement, sizing, and types of data fit within the viewport can be more specialized than the previously demonstrated presets; however, it will require more efforts to label and colorize since that is not the primary function of this plugin.

Horster will be used for the aforementioned functionality, but it is also a valuable plugin for animations and camera configurations since that is what its main intention is. The plugin can both extract the current viewport camera settings and create camera settings that include origin and target locations using coordinates.

To begin configuring custom geometry to the Rhino viewport, a plane needs to be created based on the camera location and target using the *Get camera* component. The component first needs to be activated with either a *Button* or a *Boolean Toggle*. The *Button* option can be most effective since it only requires to be *clicked* to activate and update, as the Rhino camera settings change when moving and orbiting around the model space. A *Boolean Toggle* will require the turning off and on action to update. A *Trigger* can be connected to the toggle with a set interval, but that requires significant processing power which is not necessary for a mundane operation.

The construction of a plane will be done using the *Plane Normal* component since that utilizes a vector/normal input at *Z* that can be created with the *Vector 2Pt* component from the *target* output for *A* and the *location* output for *B*. The *O* input of *Plane Normal* will also need to be the *target* output. The reason for this direction and origin choice of the vector is because that will also be the plane to be moved anywhere between the camera and the target locations. If not done this way, the plane will be behind the camera and unviewable or would require additional operations to correct that. The plane constructed from this operation is skewed, as shown in Figure 4.14, and needs to be oriented orthogonally to have the red *x*-axis directed to the right and green *y*-axis directed up.

Figure 4.14
Orientating a plane based on the default outputs from the Horster plugin.

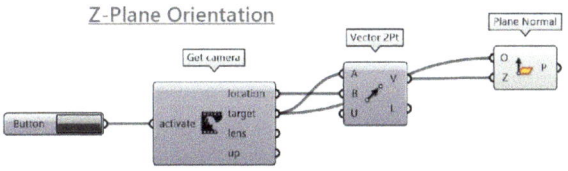
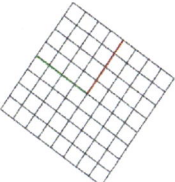

Camera/Viewport Orientation

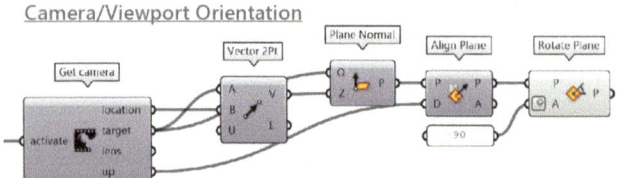
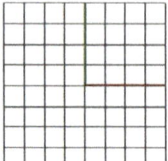

Figure 4.15
Adjusting the
default plane
orientation to
an orthographic
orientation with the
viewport.

Orientating the plane to these settings begins with the *Align Plane* component using the *up* output from the *Get camera* component; however, this utilizes the vector for the *x*-axis which results in the desired *y*-axis pointed to the left instead of up. This can be corrected using the *Rotate Plane* component with an *Angle (A)* input value of −90 set to *Degrees*, as demonstrated in Figure 4.15.

Now that the plane is oriented correctly regardless of where the camera and target locations are, the plane can be moved closer to either the camera (forward) or target (farther) along the *z*-axis. The default location will also begin at the center of the Rhino viewport, so the other settings will be able to move the plane to the left or right (*x*-axis) and up or down (*y*-axis).

Figure 4.16 shows how the orientated plane is deconstructed to get these three axis values so that they can maintain the orthogonal movement using an *Amplitude* component for each direction. This component combines a vector (direction) and distance to move the plane. The number slider values for this component are important to keep in mind as they will serve as a percentage of distance rather than a unit of distance, such as meters or feet. For the *z*-axis, a slider range between 0 and 0.99 is needed, because if the range goes to 1.00, then at that value the plane and geometry will be off-screen. Both the *x*- and *y*-axis sliders should have a value between −1.00 and 1.00 to move the plane in either the positive or negative direction. This range may fluctuate depending on the proximity of the plane to the camera. In this example, a *Panel* with a corresponding label for each axis is used to clarify the function of each slider. It is also important to note that the closer the plane is to the target, farthest away, then the higher the number slider values for the *x*- and *y*-axis need to be to compensate for the distance.

For ease of the plane placement, a *Move* component will be used for each amplitude operation. It is best to begin with the *Z* output from the *Deconstructed Plane* component to determine the best proximity to the camera. This setting will essentially control the size and scale of whatever custom geometry is applied to this plane, so it will most likely need adjusting once that geometry is applied in the last step. This operation can then be repeated for both the *X* and *Y* outputs, but the order in which it is done does not matter. Once these operations are applied to each axis, then the orientated plane can now be moved in the same sequential order as those repeated operations to help organize the movement.

Now that the settings for the plane movement have been configured, custom geometry can now be applied to the plane. Figure 4.17 shows an image for both the custom geometry origin (top right) and where it is oriented within the Rhino viewport (bottom). The process is fairly simple in referencing the custom geometry and inputting it into the *Orient* component. The last *Move* component will be connected as the new reference plane for the input *B*. For this process to orient correctly, when generating the custom geometry for reference, it should be modeled at the Rhino c-plane origin (0,0,0) and centered, as shown in the top right image.

The geometry modeled and referenced is serving as placeholder to simulate a triangular bar chart rather than the previously demonstrated typical bar chart from *Conduit*. Additional data can also be included such as text labels, axis, height values, but that would simply be done in Rhino in the same way in which this surface geometry is applied using their respective reference components. From this example, it is demonstrated how the incorporation of custom content that support a specific graphic style and project narrative can be achieved.

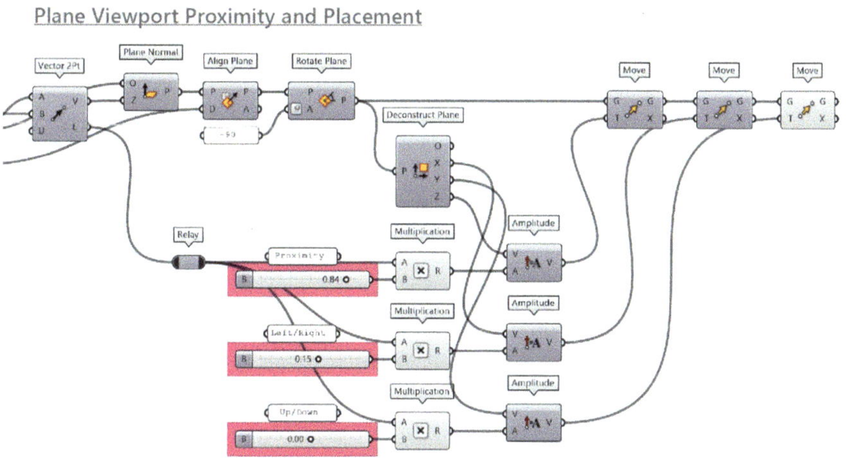

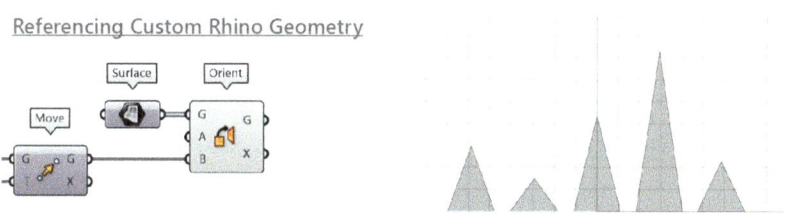

Figure 4.16 Scripting the settings for the plane's proximity to camera and left/right and up/down placement.

Figure 4.17 Applying custom chart typologies from geometry modeled in Rhino.

LADYBUG (BY LADYBUG TOOLS)

Another valuable plugin to have for Grasshopper that is not only critical in measuring landscape performance but in general useful for inventorying and analyzing weather data is *Ladybug*. This plugin can collect weather data from around the world and be charted and modeled to address site-specific climate conditions over a range of different time periods such as hours, days, and months. It is able to communicate different forms of weather data as well that includes temperature, outdoor comfort, wind, and sun positions. Each of these topics will be demonstrated for setup and future use in the upcoming chapters.

The components in this plugin are very robust in computing an extensive amount of data in a variety of ways; so one thing that is helpful to know with them is that by hovering over the component inputs and outputs, additional detailed descriptions and instructions will be explained. Also, this section is intended to just give an introduction to the charting process, but future chapters will go into more detail on the significance and impact they have on outdoor comfort and landscape performance.

The data first needs to be downloaded with *LB EPWmap* component with a *Boolean Toggle* set to *True* to launch the website from an internet browser to find an appropriate station location. From the map, a file can either be downloaded and referenced back into Grasshopper or the url can be copied and pasted into a panel for the *LB Download Weather* component, as demonstrated in Figure 4.18. The *epw_file* output is then connected to the *LB Import EPW* component where all the different forms of weather data can be extracted for the variety of chart models. Once the file or url has been connected to this operation, it is preferred to switch the *Boolean Toggle* to *False* to prevent the map website to be relaunched every time the Grasshopper file is opened.

The first type of weather data to model will be the *dry_bulb_temperature* (air temperature) with the *LB Hourly Plot* component to generate a two- or three-dimensional hourly chart for the entire year. Depending on the appropriate unit of measurement for the project, a *LB To Unit* can be used to convert the default Celsius degrees to Fahrenheit, as shown in Figure 4.19. The data from this component can then be connected to the *data* input of the *LB Hourly Plot* component. By default, this will model a two dimensional chart with a blue, yellow, and red gradient to visualize the range of temperatures for all 8,760 hours of the year (HOY). The location and size of the chart can be changed as an option or a *_z_dim_* (z dimension) can be inputted to convert the chart from two-dimensional to three-dimensional as demonstrated with the number slider set to 250 with the resulting chart displayed at the bottom of the figure.

For the next step of creating a custom legend, a *Panel* is connected from the *legend* output to show the current range of temperatures. This is helpful in the customization of the legend to create specific intervals for the chart.

Before the new *legend_par_* is inputted, the range of temperatures was between *25* and *115*; so for the *LB Legend Parameters* component, those values were inputted for the *min_* and *max_*. The next setting for this component is the number of segments to use between the range of values and because of the extent for this range of values *10* for this input would create even intervals of 10 for the legend.

Figure 4.18
Setting up the
Ladybug plugin to
import climate data
from the available
global weather
stations.

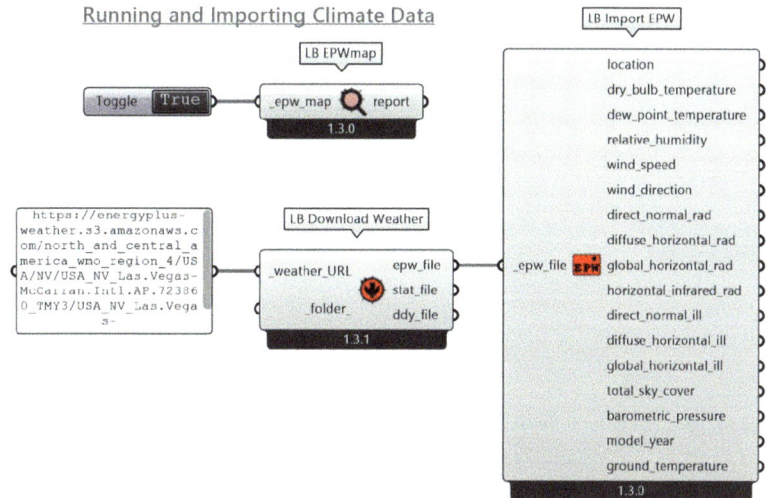

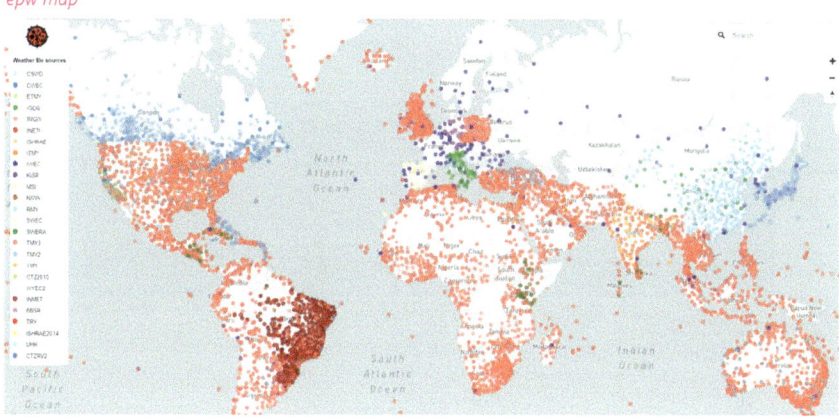

epw map

Each one of the ten intervals needs a color associated with it, so, like the elementary operation in the previous chapter, a range of equal number intervals is created with the *Series* component. The step size does not matter for this operation as long as the number of steps is equal to the number of segments inputted and they are all of equal intervals. The default steps size for the *Series* component is *1*, which is fine for this process. The *L0* input of the *Gradient* component and starting value for the *Series* component are both *0*, so nothing needs to change there; however, the series component's highest value will be *9* instead of count number of *10* because *0* is the first number. Because of this condition, the *L1* input for the *Gradient* component is given an *Expression* of *x–1* so that the *10* value being inputted is changed to *9* to match the high value of *9* from the *Series* component. These are the main settings needing adjustment for the chart; so once this process is complete, the *LB Legend Parameters* component can be connected to the *legend_par_* input.

Other helpful settings that can be applied to this chart for filtering a more concise output are the use of conditional statements and time periods. With a conditional statement, shown in Figure 4.20, temperatures or any other type of data can

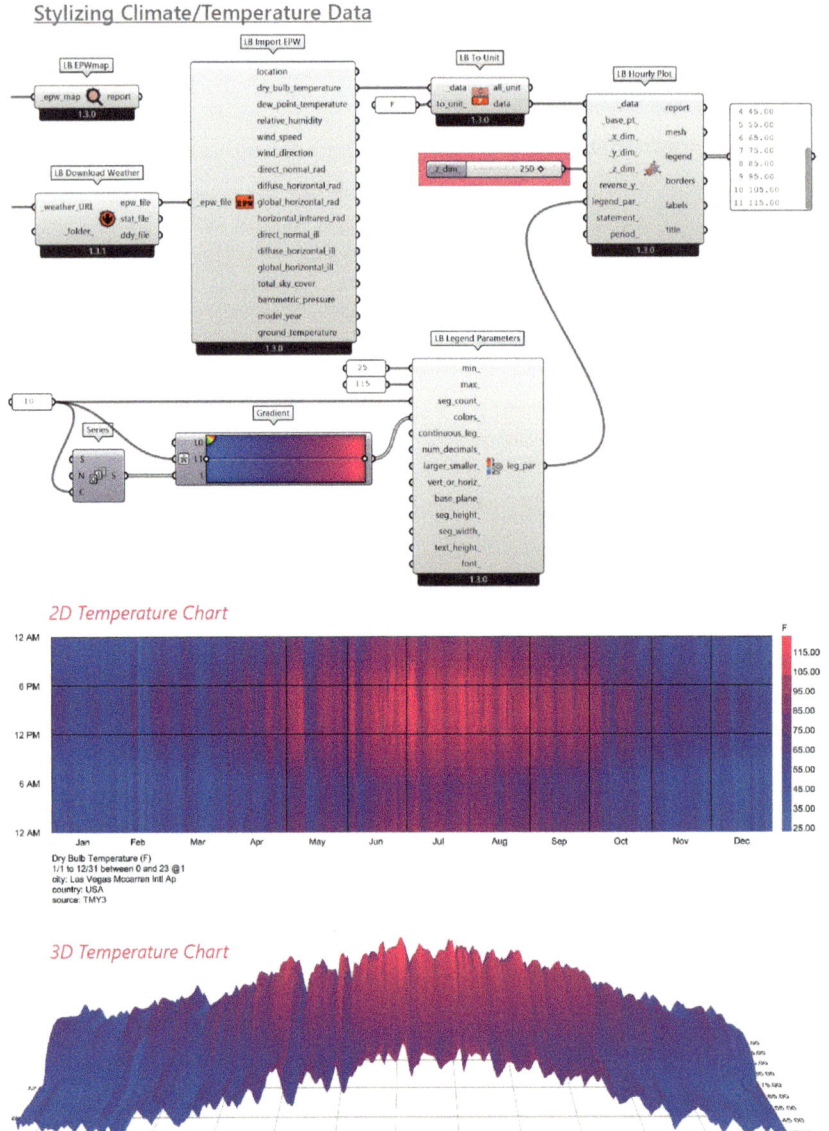

Figure 4.19
Plotting hourly
climate data as
a chart that can
be either two- or
three-dimensional
with custom color
palette.

be filtered to only plot values above, below, or between a range of parameters. For this demonstration, a conditional statement of *a>80* will only plot values above 80 degrees Fahrenheit. When applying these conditions, it's important to disconnect the *legend_par_* input to see what the new range of temperatures are and apply those changes to the *LB Legend Parameters* component. The resulting chart model can be seen as the middle image of the figure.

The other setting is to plot values between specific dates, such as seasons or other appropriate dates. This can be achieved with the *LB Analysis Period* component where a *Panel* is used to symbolize specific months, days, and hours. This can be done as demonstrated with static values or number sliders to allow more fluid adjustments. Like with the conditional statement setting, the legend parameters will need to be adjusted accordingly. The addition of this setting can be seen in the bottom left image of the figure. Lastly, both settings can be applied to look at a temperature condition within a specific period of time as shown in the bottom right image of Figure 4.20.

Another chart type to explore is a wind rose to model wind speed and direction. Similar to the previous charting process, specific outputs from the *LB Import EPW* component are connected to the *LB Wind Rose* component. Once again, *wind_speed* units can be converted, as demonstrated in Figure 4.21, by changing the default unit of meters per second to miles per hour (mph) with the *LB To Unit* component.

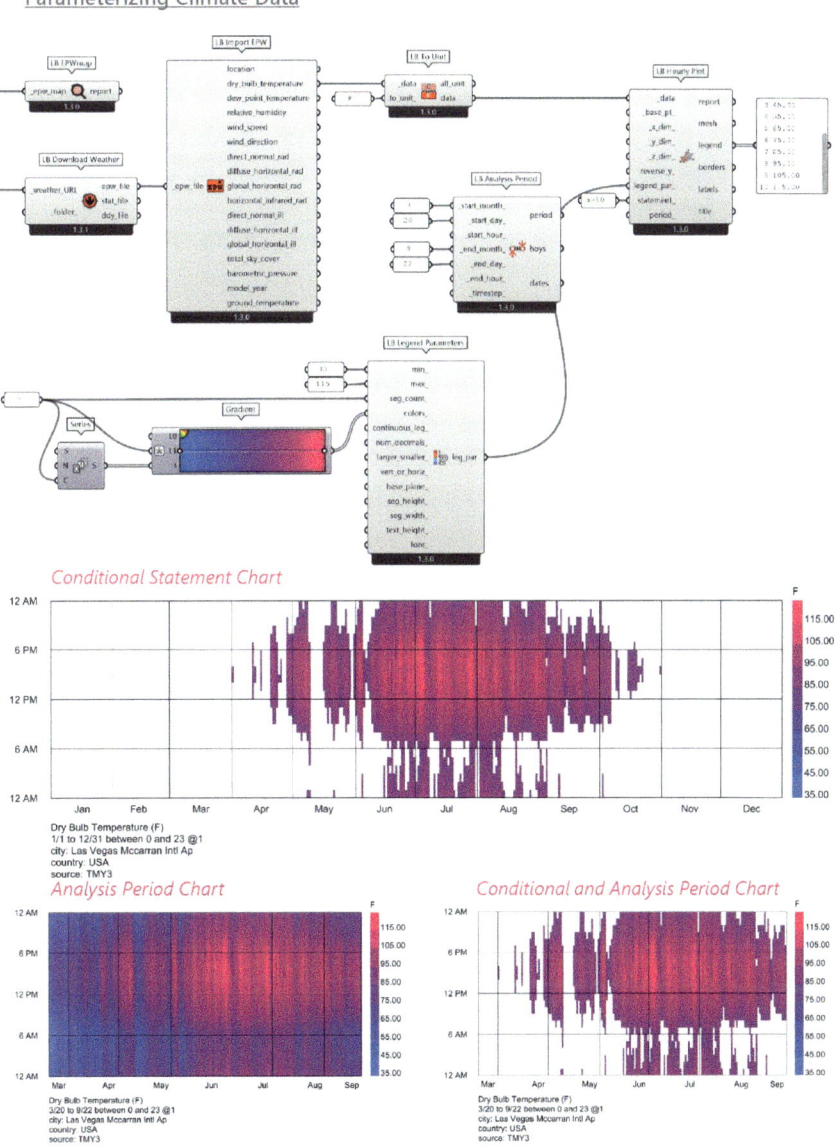

Figure 4.20
Changing analysis periods and conditions of climate data for custom charting.

Next, *wind_direction* is also inputted. The *LB Wind Rose* component also has the ability to change the legend, apply a conditional statement, and isolate the wind rose plots to a specific period of time, as performed in the previous figure.

The last weather chart to demonstrate is the *LB SunPath* component. This chart is extremely powerful in generating sun positions at designated HOY and vectors that can be used to create shadow projections from objects or Rhino geometry. In addition to the instrumental ability, other data such as temperature can also be correlated with the different HOY, like the *LB Hourly Plot* chart, to give a more comprehensive depiction of these daylight hours, as shown in Figure 4.22. To create a sun path for every daylight hour, the *Series* component is used with a count of 8,760 (number of HOY) to connect to the *hoys_* input. This number will be filtered down to omit all the hours outside of that geolocation environment. Specific HOY can also be inputted if it is not desired to look at every hour or a *LB Analysis Period* component can be included to reduce the number of sun positions by confining the time period to a specific month or hours of a day(s).

The sun path chart is three-dimensional, even though it is displayed as a two-dimensional chart in the figure. It is also important to note with this specific chart that its location and size relative to the object being performed, any sun/shade analysis to does not affect the data or shadow projections. In other words, a series of buildings being studied for their shading qualities does not need to reside within the chart but can actually be outside of it without having any discrepancies. Visually,

Charting Wind Speed and Direction

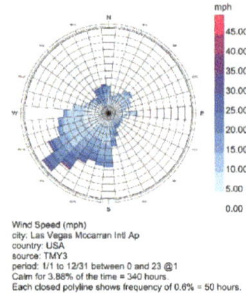

Windrose Charting Speed

Figure 4.21
Charting wind speed and direction with a Wind Rose chart.

Charting Temperature to a Sun Plot

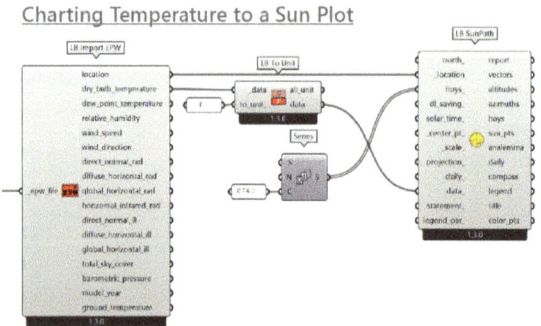

Sun Path (Plot) Chart

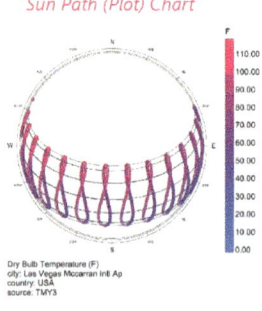

Figure 4.22
Modeling air temperatures at different sun locations with the Sun Path chart.

however, it may be best to try and scale and place them more relative to each other for clear communication of the analysis or landscape performance outcome.

The use of plugins can bring an array of new information (metadata and numeric tables) and access to new forms of modeling (GIS shapefiles and drainage lines) to make the understanding and analysis of terrain characteristics or ecosystem services more comprehensive and robust. They can also provide supporting information through charts and viewport parameters to create more dynamic forms of graphic representation that contain visual and numeric communication of abstract data. It alludes to the fact as well that in combination of all these capabilities, visualizations can be contained within one software program if enough attention is put into the clarity and legibility of these graphic representations. It is still recommended that additional post-process work be done; however, a significant amount of the overall graphic composition can be communicated in this parametric workspace.

SECTION 2

5   Landscape Conditions

Before any landscape performance measurements or analysis can be made, an accurate base model is necessary for extracting various fundamental conditions of a landscape or project site. This can be from a dense urban city center or an undisturbed natural topography. Regardless of the site's characteristics and development, this base model will literally serve as the foundation for all the processes to come. In this chapter, modeling a terrain or topography from contours will initiate this process to assist in the inclusion of surface typologies or ground covers, soil characteristics, and hydrology through drainage lines. With these modeled landscape conditions, the inventorying of additional information can be conducted for future analytical computations with parametrics. The demonstrations throughout this chapter will be of the same project site to ensure a fluid sequential and consistent workflow. The choosen site will also most likely be more complex than a standard landscape to demonstrate the highest degree of challenges that may be required for a landscape performance model. There will be, however, other examples throughout the book that do not have the same level of complexity to showcase the replicability and flexibility of these computational models for any unique project site.

TERRAIN MODEL FROM CONTOURS

As mentioned in the first section of this book, different online resources provide supplemental and foundational data to the grasshopper modeling scripts. Creating accurate terrain models is one of the fundamental elements to any landscape architecture project, regardless of its intention to incorporate landscape performance. For the best results, it's critical to use existing topography data from either contour lines or digital elevation model (DEM).

If you already have a contour shapefile, you can skip these few steps, but regardless, it will be important to still import them into ArcGIS Pro or another comparable GIS software program such as ArcMap or QGIS. This will ensure a consistent coordinate projection for all files used in this process and all other methods relying on the import of shapefiles into Grasshopper.

Chapter Prerequisites

To best prepare for this chapter, please refer to the *@it* plugin in the first section of the book within the *Computational Modeling* chapter. This plugin script will be the first step of the Grasshopper modeling method.

Modeling Methods

A reliable online resource to download DEM data of locations within the United Sates is USGS's TNM Download v2.0 (https://apps.nationalmap.gov/downloader/). At this website, a zipped folder containing all shapefiles can be uploaded for a site location, as shown in Figure 5.1. Once uploaded, the map will zoom to the site location where the *Elevation Products (3DEP)* menu can be selected and have either the *1/3 arc-second DEM* or *1/9 arc-second DEM* selected for download. Although the *1/9 arc-second DEM* is more detailed, there are fewer instances of that file type, whereas the *1/3 arc-second DEM* is more readily available. With those selections made, the *Search Products* can be entered to find available results for the site location under the *Products* tab. For this example, only one product was found and downloaded. If there are multiple products, you can select the *Thumbnail* option for a bounding box to display on the map to determine if it covers the project site.

Once the DEM file is downloaded, it can be imported in ArcGIS Pro as a raster layer as also instructed in the *@it* plugin section and shown below. In the *Contents* panel, the DEM file will display a grayscale color gradient in *meters* where the highest elevation point is in white and the lowest point is in black. All DEM files will be imported in meters. For this modeling method, all conversions to feet, if necessary, will occur in Grasshopper. The previously used site boundary file that was uploaded to TNM Download will also be used to trim the DEM, since that file will most likely be significantly larger than the project site.

In ArcGIS Pro, the *Clip Rater* geoprocessing operation is used with the *Parameter* and *Environments* settings highlighted in the two panels in Figure 5.2. Once these settings are applied, the operation can be run at the bottom.

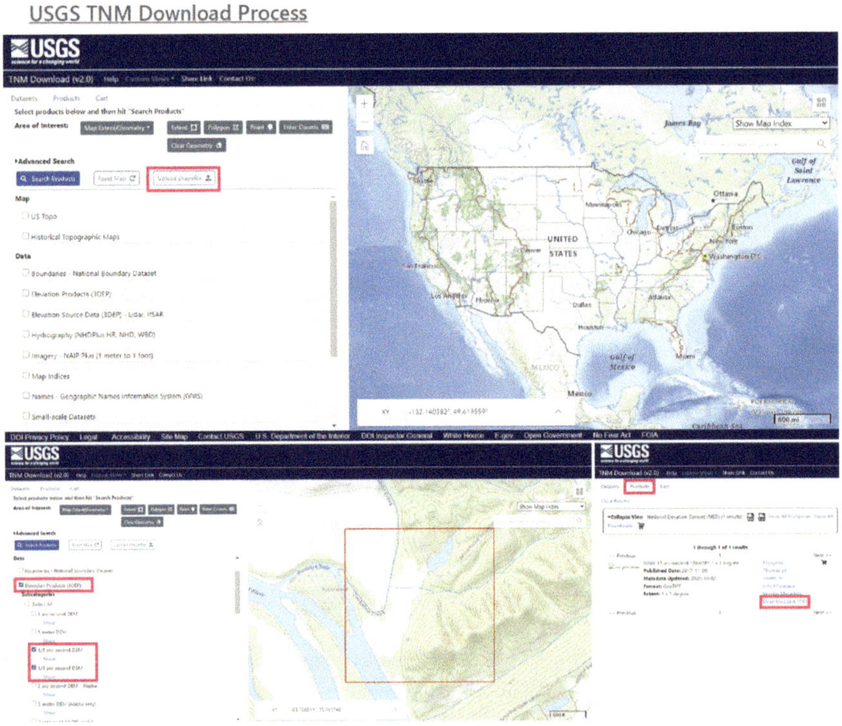

USGS TNM Download Process

Figure 5.1
The search and download process of heightfield data for next steps.

Figure 5.2
Using the project
boundary shape
layer to clip
the previously
downloaded DEM
image along with
configuring the
export settings.

Clipping DEM Image with Site Boundary

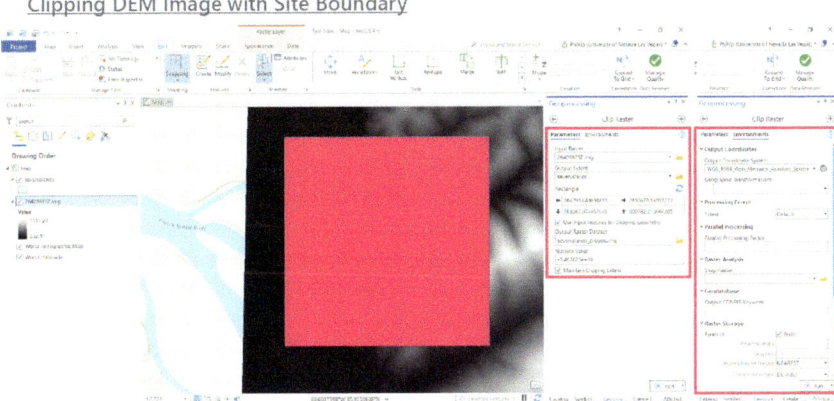

Figure 5.3
Using the
contouring tool
in ArcGIS to
generate a shapefile
for Rhino and
Grasshopper.

Contouring Clipped DEM

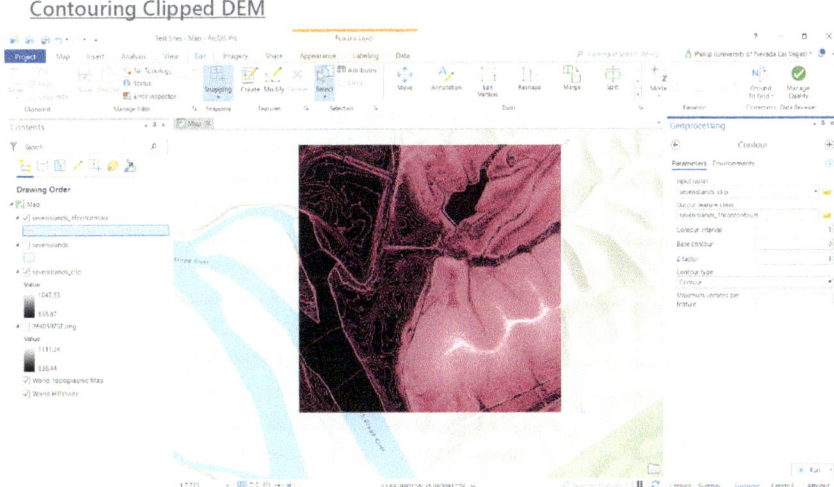

A new DEM file is created from this process and is used to generate contours using the Contour geoprocessing, as shown in the image below. Although the DEM is still in meters, if the *Z factor* is set to 1 the *Contour interval* will read as feet. In order to generate contours in meters, change the *Z factor* to 0.3048. The example below, as shown in Figure 5.3, shows the settings for the contour interval at 1 but for file management purposes, 5 was inputted for demonstrating in Grasshopper. Again, for data and modeling alignment in Grasshopper, use the same *Environments* settings from the previous step.

With the contour shapefile from ArcGIS Pro, it can now be imported into Grasshopper using the previously scripted operation from the *@it* plugin but referred to once again in Figure 5.4. The geometry of the contour lines will remain flat until *z* values are assigned, but the *Scale* component will use a factor of 3.2808 to convert to feet in planar dimensions. As part of the contour geoprocessing in ArcGIS, a features column was created containing all the contour elevation values. In the image below, that data is list item 1 and can be filtered out using the List Item component with 1 set as the index input.

The data from the *List Item* component in Figure 5.5 is used to elevate the contours to their respective height in feet. A *Control Polygon* component is used on the new elevated polylines to extract points which are then *flattened* for the *Delaunay Mesh* component for a terrain model. If this is the extent of needing a terrain model as a mesh, the process can end here, but for the many other performance modeling, it is best to generate a surface model from the mesh as shown in the next steps.

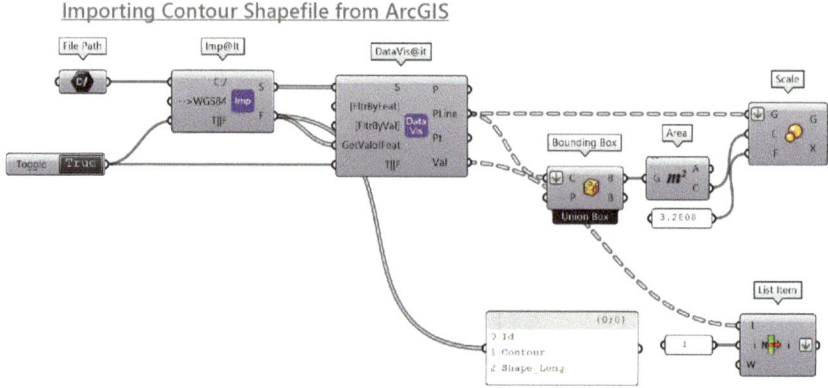

Figure 5.4
Scaling the planar contour curves to the correct unit of measurement for the project, in this case, feet.

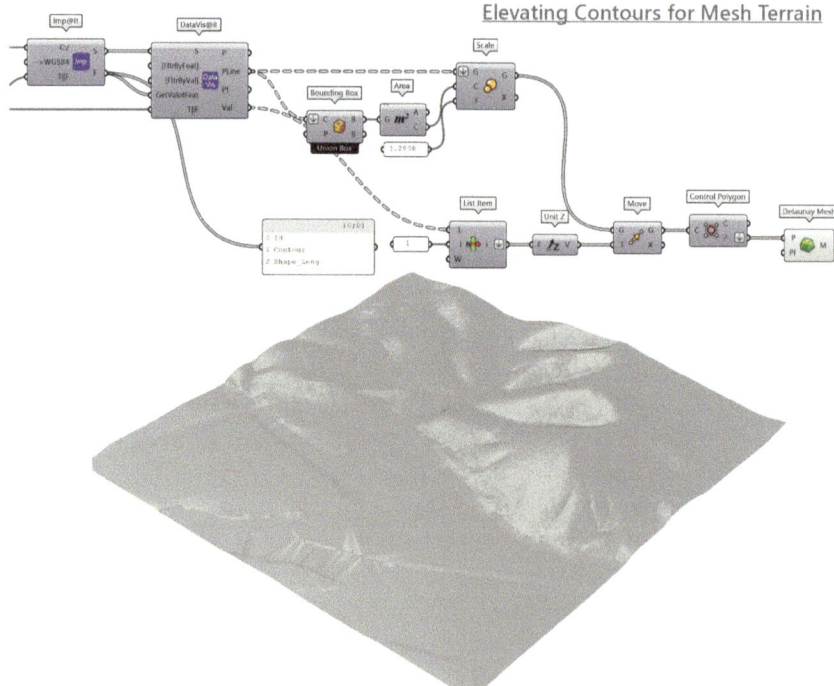

Figure 5.5
Elevating the contours curves to their respective height and converting to a mesh terrain.

Figure 5.6
The mesh
properties are used
to generate a planar
surface for the
conversion process
to a surface terrain.

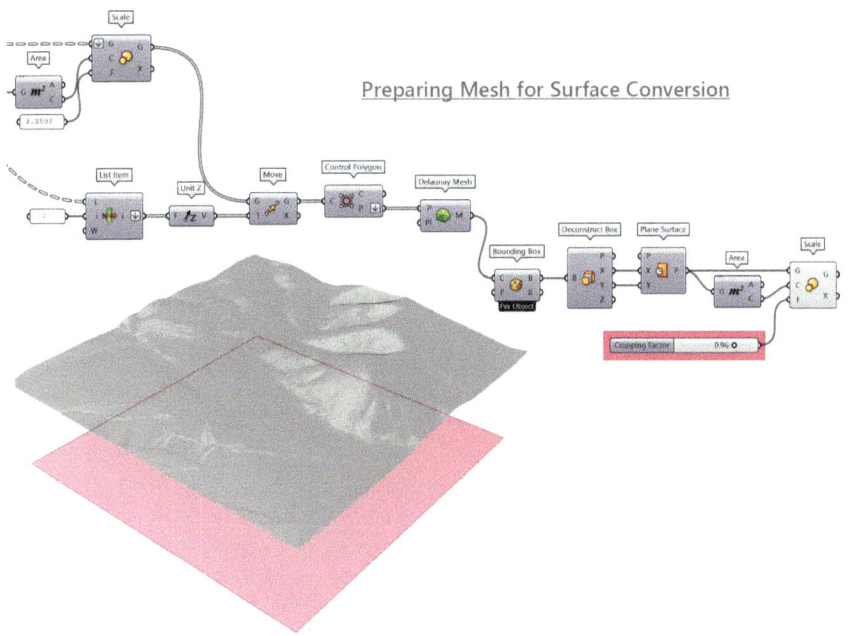

Figure 5.6
The mesh properties are used to generate a planar surface for the conversion process to a surface terrain.

From the *Delaunay Mesh* component in Figure 5.6, the property dimensions can be extracted to generate a planar surface below the mesh model. It is best to make this planar surface marginally smaller for accurate point projection for the surface model. This scale factor will vary but a slider with 0.96 (96%) value was used for this script.

The scaled planar surface, as shown in Figure 5.7, is then inputted in a *Divide Surface* component which can have varying point densities for surface detail using a number slider. The higher value, the more detail and vice versa, depending on the purpose of the terrain model. The points from this component are *flattened* and inputted in a *Project Point* component in the *z* direction on the *Delaunay Mesh* geometry.

If the project site is irregular in shape or doesn't have 90-degree corners, some of the project points will miss the mesh model and create issues with the generation of a surface from points. Therefore, those missed or *null* points need to be identified and replaced with points from the planar surface, demonstrated in Figure 5.8. The scaling or crop factor slide can prevent this in some cases but ultimately this will ensure that a valid surface is generated.

The new set of points coming out of the *Replace Nulls* component can then be used for the *Surface From Points* component. The *U* input for this component needs to have the same value of points used for the *U* input in the *Divide Surface* component but also include an expression of *x + 1*. This was explained in Section 1 of the book, but as a reminder, the *Divide Surface* component creates a number

Defining Surface Terrain Detail with Projected Point Density

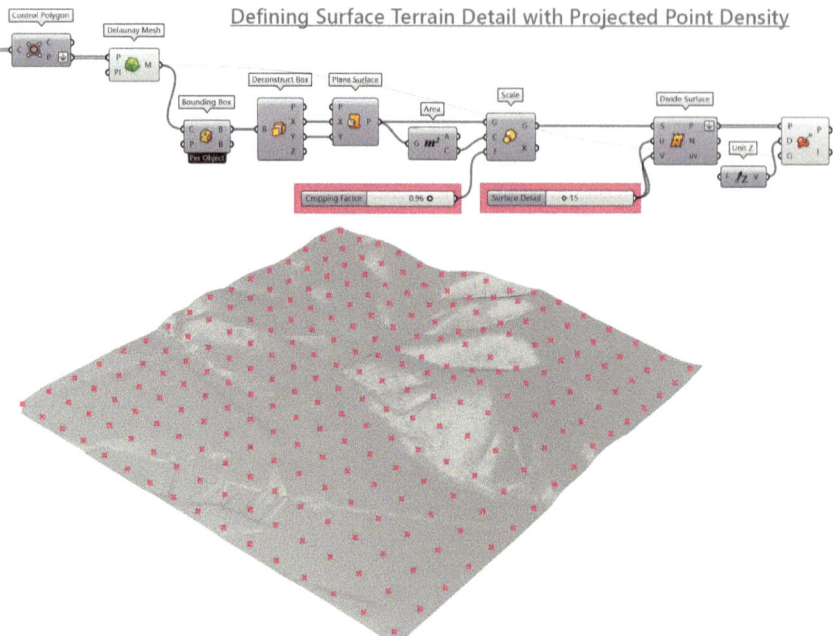

Figure 5.7
Adjusting the point density from the planar surface for projecting onto the mesh terrain.

Removing Missing Points and Contouring Trimmed Surface

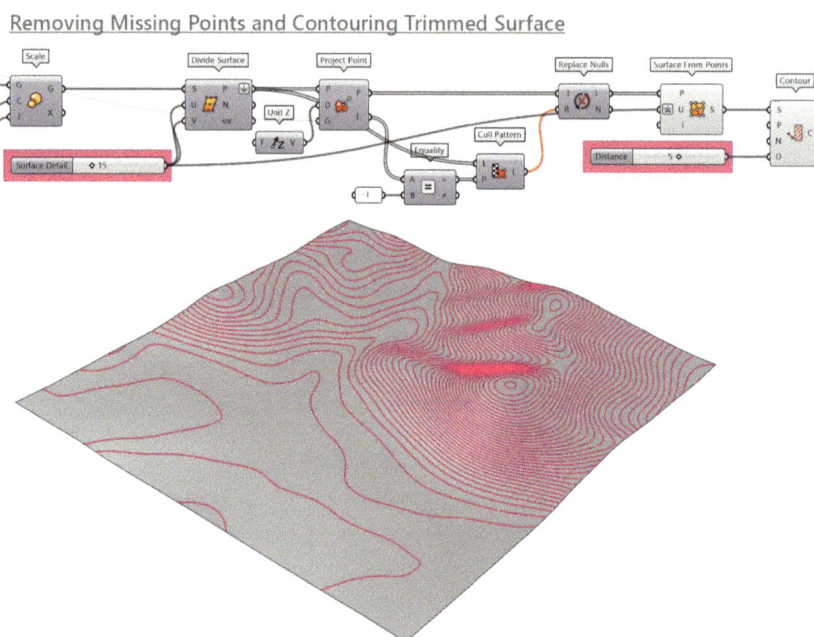

Figure 5.8
The projected points result in a new surface terrain for contouring and future analysis.

(U and V) of spans, whereas the *Surface From Points* component requires number of points which is always going to be one more than the number of spans. This surface can then have contours extracted at a given interval for additional clarity of elevation change within project's topography using the *Contour* component.

SURFACE TYPOLOGIES

Just about every project site will contain different surface typologies or conditions ranging from natural softscape ground covers to impervious hardscape materials on an urbanized site. Regardless of these different conditions, it is important to define these areas in a computational model through either curves or surfaces and sometimes with both. Many of the demonstrations of defining these areas will begin with tracing the different areas with a curve that can then be converted or used to generate a surface. As demonstrated in Chapter 3 in the "Elementary Operations" section, surfaces will yield the most accurate measurements of the area size as well as being a more effective geometry type for more advanced landscape performance operations later on in the book.

Chapter Prerequisites

When it is required to align and reference multiple layers of data (aerial imagery, contour lines, DEMs, etc.) from different external resources, it is important to use a consistent project boundary shape to maintain accurate overlaying in the Rhino and Grasshopper workspace. For doing this, it is best practice to utilize a GIS shapefile since that can be universally applied in many external online resources, as mentioned in Chapter 2. This will require the use of a GIS software application, preferably *ArcGIS* since that will be the demonstrated example. And like the previous section in this chapter, the *@it* plugin will also be required to import the shapefiles to Grasshopper.

Modeling Methods

For the purpose of matching with other terrain model properties, the process will again begin with ArcGIS Pro to ensure the data overlays correctly. The same site boundary, in this case *sevenislands.shp*, will be used to crop an aerial image for tracing in Rhino.

One of the best ways to get accurate delineated areas within a project site is using an aerial image to trace over. As shown in Figure 5.9, the *World Imagery* option in ArcGIS Pro from the *Basemap* menu gives high-resolution imagery for this method. By *right-clicking* on the *Map* layer in the *Contents* panel, the Properties option can be selected to access the *Clip Layers* setting to isolate the aerial image to the project site boundary.

From the drop-down menu in the *Clip Layers* settings, select the *Clip to the outline of features* option. Next, select the site boundary shapefile for clipping. Now that the aerial image has been clipped to the site boundary shapefile, for best resolution, it is important to make sure that the ArcGIS Pro window is zoomed to the shapefile layer by *right-clicking* on it and selecting the *Zoom to Layer* option, as shown in Figure 5.10.

The final step within ArcGIS Pro is to export the aerial basemap as an image that can be imported into Rhino, preferrably a jpeg or png. From the *Share* tab at the top of the menu, select the *Export Map* option to open the panel settings. In the settings, choose the preferred file type and a high *Quality* and *Image Size* setting.

Once those settings are made, the aerial image can be exported from the bottom of the panel. There will most likely still be a bit of a white border around the image, but that can be cropped in either Window's *Photo* program or Adobe Photoshop.

The same scripting process used in the previous chapters will once again be used to establish a reference geometry to import and place the aerial image in the correct location, as demonstrated in Figure 5.11. This is currently one of the better options since there is currently no plugins capable of doing so that can accurately match the currently used coordinate projection settings from ArcGIS. For the shapefile used in the *File Path* component, the *soilmu_a_aoi* file is used. The geometry from the *Scale* component is then inputted into a *Union Bounding Box* component and then converted with the *Surface* component for baking into Rhino.

The surface is then converted into a closed curve using the Rhino command *DupBorder* for ease of process in the next few steps and then deleted. The newly

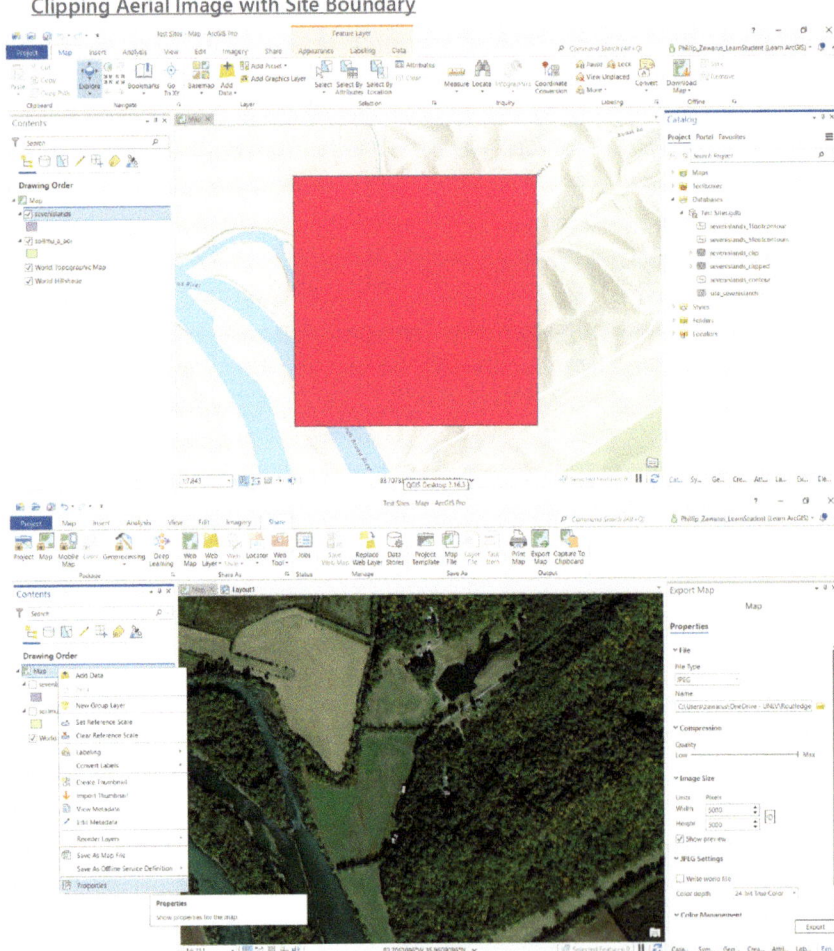

Figure 5.9
Using the same shapefile boundary layer to clip an aerial image.

Figure 5.10
Clipping and
exporting aerial
image for importing
to geo-location in
Rhino.

Exporting Clipped Aerial

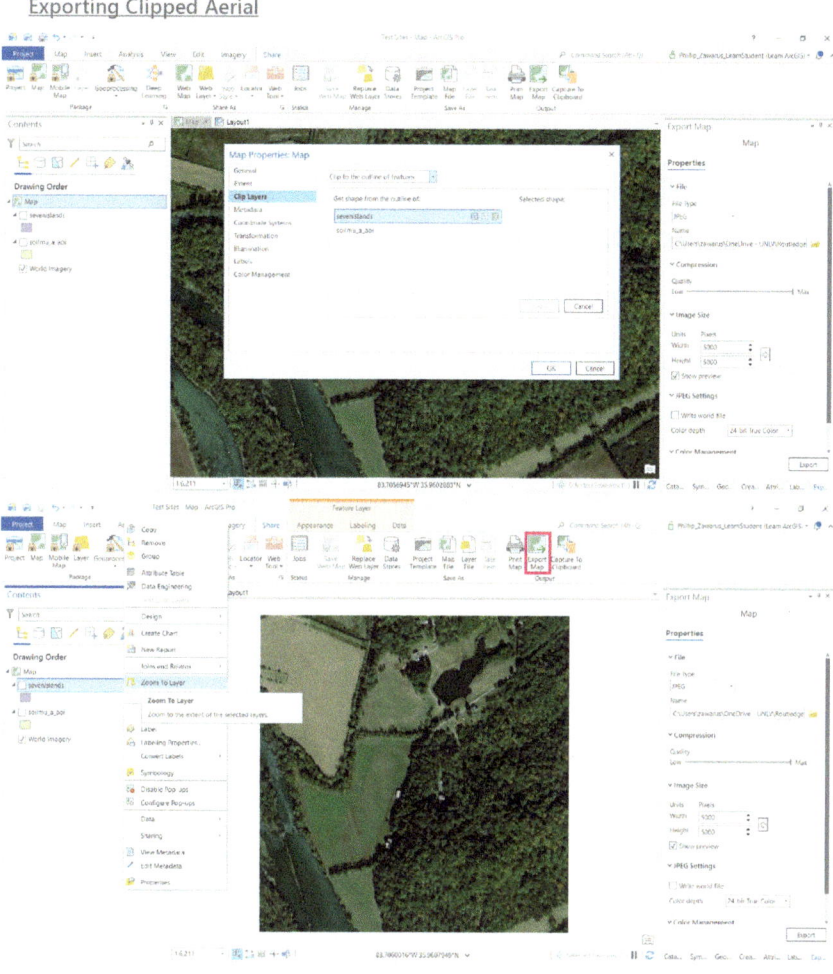

Creating Reference Geometry for Aerial Placement

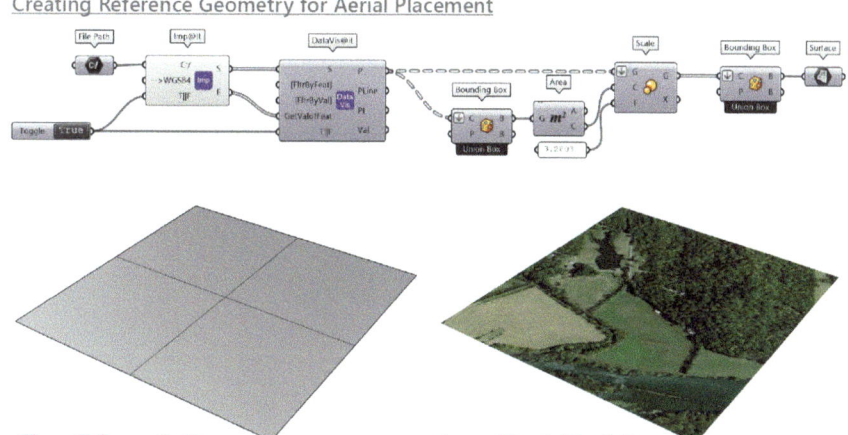

Figure 5.11
Creating reference
surface for accurate
geo-placement of
clipped aeria image.

Planar Reference Surface

Image Material Applied

created closed curve is used as a reference to match with the aerial image exported from ArcGIS Pro. The aerial image can be imported using the Rhino command *Picture*, then snapped to the bottom left corner of the curve geometry and scaled to its extent. Both the closed curve and aerial image are on their own separate layer and labeled.

The aerial image can now be traced using either polyline curves or control point curves from Rhino. For each surface type, there should be a separate layer. For organization purposes, all the surface types are put into sublayers labeled accordingly so that all can either be collapsed or turned off with an empty master layer that has been labeled *Surface Conditions – Curves*.

When beginning the surface tracing process in Figure 5.12, it is important to identify three major surface types and try and trace in this specific order for accurate and efficient delineation. A general rule of thumb is to begin with the less complex surface types, which are identied as *pocket* and *bisecting* types. For this project site, the grassland areas are, for the most part, either surrounded by forests or adjacent to water bodies so they can be considered *pocket* surfaces. The water bodies, specifically the river at the bottom of the site, run through two

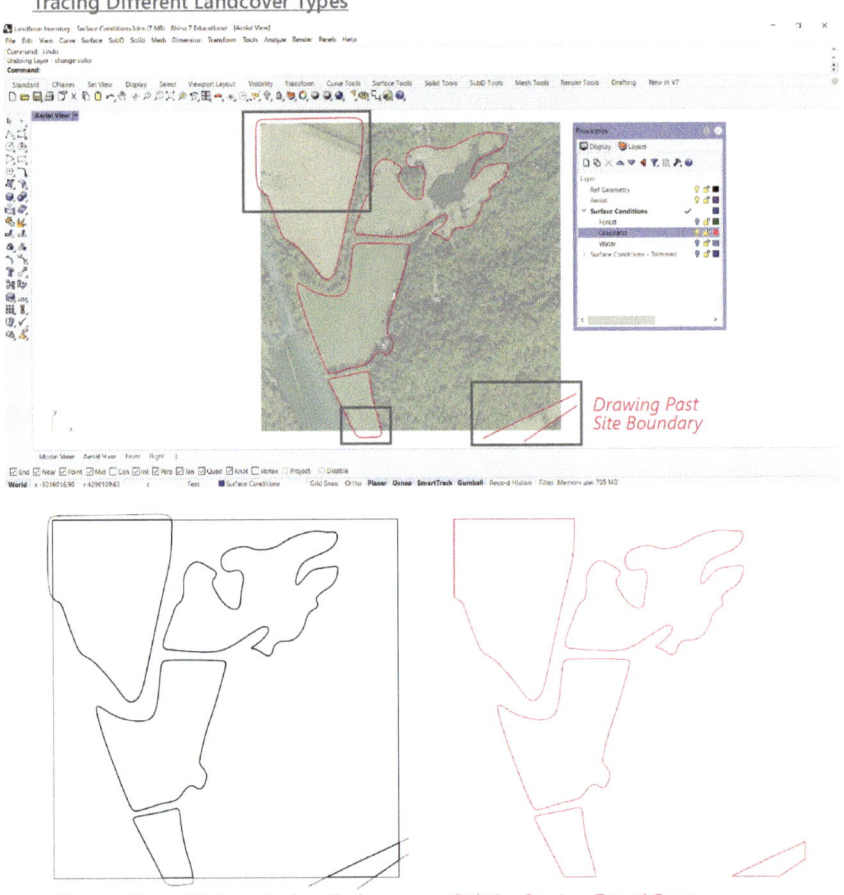

Figure 5.12
Creating effective landcover traces with the Rhino *CurveBoolean* command.

boundary edges so they would be considered a *bisecting* surface type. What remains is the most complex forest surface type, which will be referred to as an *enclosing* surface.

Because the *pocket* grasslands are the simplest type to trace, the process begins there. The Grassland layer is the active layer in Rhino for tracing with a curve. The grassland areas that are contained within the site boundary can simply be traced as accurately as possible but when they excede or end at the site boundary, it is preferred that they be overdrawn for accuracy and efficiency with an additional Rhino command. There are instances of this occuring at the top left corner, bottom middle edge, and bottom right corner.

To cleanup those overlapping areas, the *CurveBoolean* command is used by selecting the *Grassland* curves and the site boundary curve on the *Ref Boundary* layer. An easy way of selecting all the objects on a layer is by *right-clicking* the layer and choosing either *Select Objects* or *Select Sublayer Objects*, depending on where those objects are nested in. Now that the objects are selected with that command, the settings can be used as shown in the command line of the image below.

By selecting within the regions to be trimmed and kept, a bold outline will appear to indicate which areas have been selected. Notice that the selected site boundary curve acts as a clipping edge to form a new flush edge. Once all the intended area are selected in bold, the command can be completed with the new curves. The old overlapping curves can now be deleted leaving the remaining new boolean curves.

The same tracing process is now performed on the *Water* layer, shown in Figure 5.13. Similar to the *Grassland* layer, once the area reaches the edge, the curve continues to ensure a closed intersection for the *CurveBoolean* command. The curves tracing the river on the bottom left are all curve segments, whereas the pond at the top right of the site is a closed curve, so its important to keep in mind that when tracing surfaces, a closed curve is not required. Another technique that can be used is when this surface type is adjacent to an already traced surface that the curves can overlap since it is easier to redraw that area with the *CurveBoolean* command rather than try to make an identical curve sharing that adjacency. This is done at both the river's edge adjacent to the grasslands and along an edge segment of the pond.

<u>Duplicating Curves of Shared Edges</u>

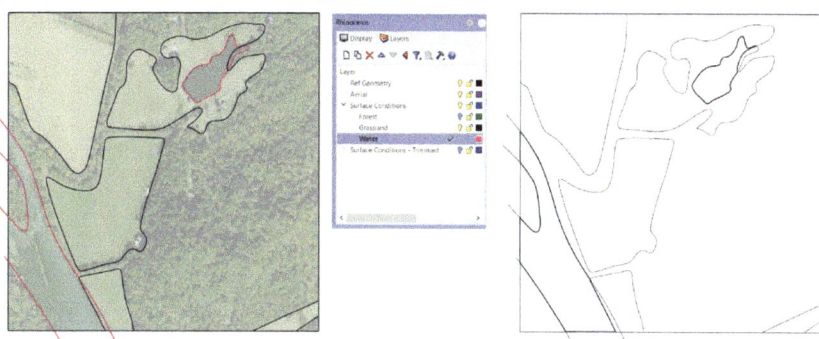

Figure 5.13
Repeating the previous step for the next landcover layer.

Tracing with Overlapping Curves   *Trimming Overlapping Curves*

After all the water features are traced, those curves along with both the site boundary curve and grassland curves are additionally selected for the *CurveBoolean* command. The grassland curves need to be part of the selection so that they perform like the site boundary curve in the previous step and form a new edge for the water features, as shown in the bold selected regions below. The overlapping segment remains as thin line indicating that it will not be part of the newly created curve. From that Rhino operation, the water features and grassland areas now share flush adjacent edges.

Now the final surface type can be traced for the forest, as demonstrated in Figure 5.14. What's great about this technique is that not all surface types need to be manually traced. Because the forest is termed an *enclosing* surface type, which is very complex, all the previously traced surface types serve as the edge for it when running the *CurveBoolean* command. All the remaining interstitial areas are selected forming bold outlines now, as shown below, creating the last remaining surface curves.

When going through this process of tracing different surface typologies, it is valuable to have multiple geometry options, depending on future Grasshopper scripting. It will require some time and effort to trace the areas as curves; however, it is nearly effortless to convert them into surfaces.

Another nested group of layers for each surface typology is created within a master layer labeled *Surface Conditions – Surfaces* to house these new geometries, illustrated in Figure 5.15. This will help keep objects organized and easy to select and reference in Grasshopper. First, the layer for the surface needs to be activated and then by *right-clicking* the sublayer in the curve group, as shown in the image below, all the curves can be selected with the *Select Sublayer Objects*.

"Tracing" Leftover Landcovers

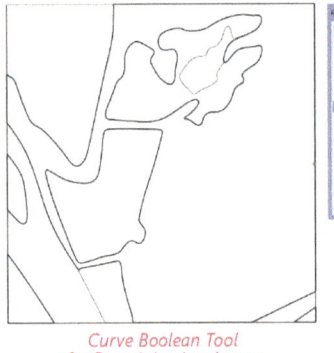
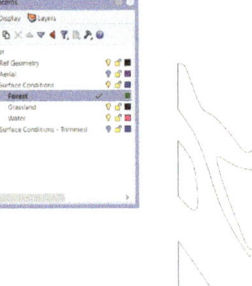

Curve Boolean Tool for Remaining Landcover

Results of Remaining Landcover

Figure 5.14
Using the *CurveBoolean* command for the remaining 'leftover' spaces on the last landcover.

Converting Traced Curves to Surfaces

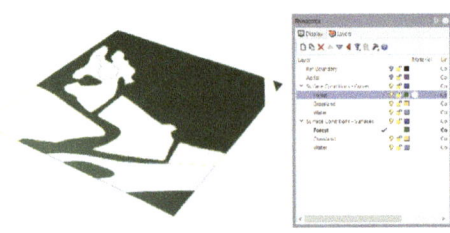

PlanarSrf Command for Conversion

Command Applied to All Layers

Figure 5.15
Converting closed curves to planar surfaces.

With the curves selected, the Rhino command *PlanarSrf* can be executed to make surfaces for the new sublayer within the surface group. This process is repeated for the remaining surface typologies to create a complete set of geometric surfaces.

Now that all the surface types have been modeled out, they can now be referenced in Grasshopper to be used as assets for separating and delineating data. The original surface terrain model from the first chapter of this section will be referenced, shown in Figure 5.16, and divided into a grid of points. The density of the grid is dependant on future intentions, so it is irrelevant what number slider values are used for this operation. The grid of points are then slightly elevated vertically to ensure later operations in the script function correctly.

The surfaces created in Rhino can now be referenced into Grasshopper within their own respective *Surface* components, shown in the script of Figure 5.17. The order in which they are inputted into the *Entwine* component does not matter, but should be noted for future steps. The *Entwine* component will group the three surface typologies while maintaining their data branch structures, an important function for this script and more comprehensive scripts in Section 3 of this book.

Similar to the previous 'Soil Characteristics' chapter, these planar surfaces will also be morphed to the existing Rhino surface terrain model. Since the *Surface Morph* component requires the referenced geometry to have volume, the non-planar surfaces are given a default extrusion of one foot for the *Bounding Box* component. The referenced surface in the *Brep* component is relayed for the other inputs of this component for organization purposes. Just like how the grid of points needed to be elevated slightly, the new morphed surfaces need to be extruded; a *Unit Z* value of 50 is used for the next step that detects wheather points are within a closed Brep.

Between the grid of points and morphed surface, an operation in Figure 5.18 is used to determine not only which individual Brep any givin point is contained within, but also maintains the organization of surface typologies. If this operation is not performed, it would be near impossible to detect which specific surface type a point is contained within, as it relates to many other data characteristics such as soil conditions and slope for a comprehensive analysis of a terrain model. This will become more apparent in Section 3 of the book.

A *Point in Breps* component is used to determine which points are within an individual Brep: a *flattened* input. From this component, indices are outputted that correspond with each referenced surface. In order to first assign an index to each geomeotry, all the geometry needs is to be flatten into one pool of data and

Defining Analysis Detail with Point Densities

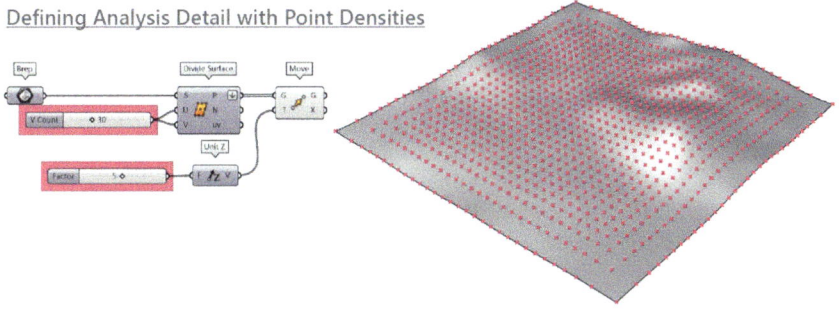

Figure 5.16
Dividing the
referenced surface
terrain in a grid of
points.

then reorganized with a *Partitian List* component to create a set of domains that each point is assigned to be included in. This was elaborated on in the 'Elementary Operations' section of Chapter 3.

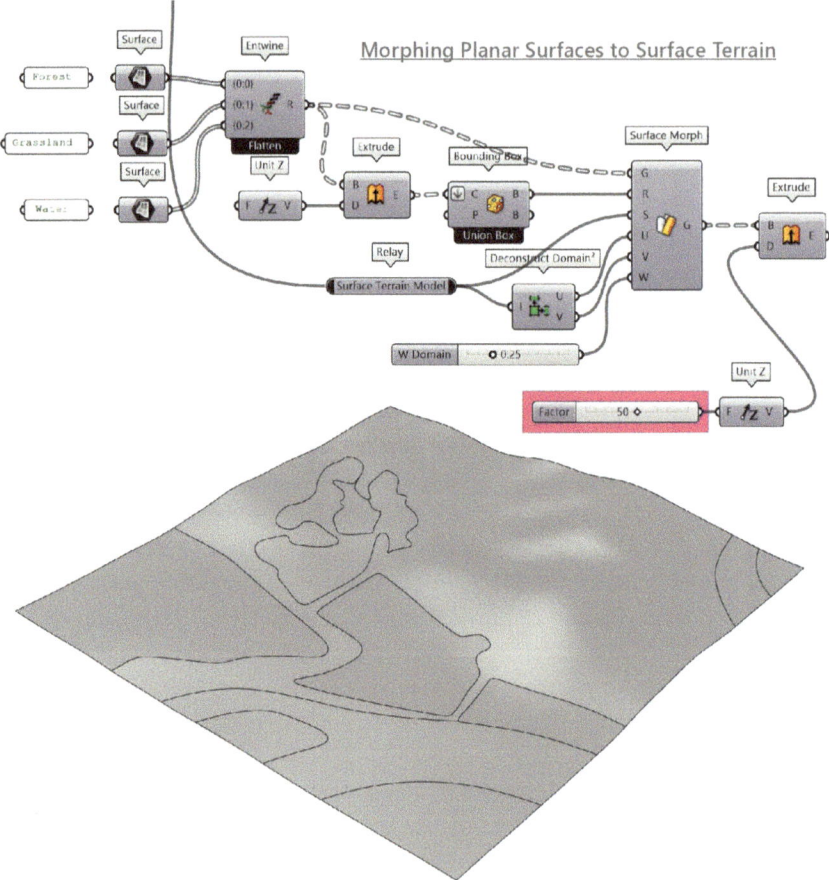

Figure 5.17
Referencing planar Rhino surfaces for morphing to the previously generated surface terrain.

Partitioning Points into Landcover Boundaries

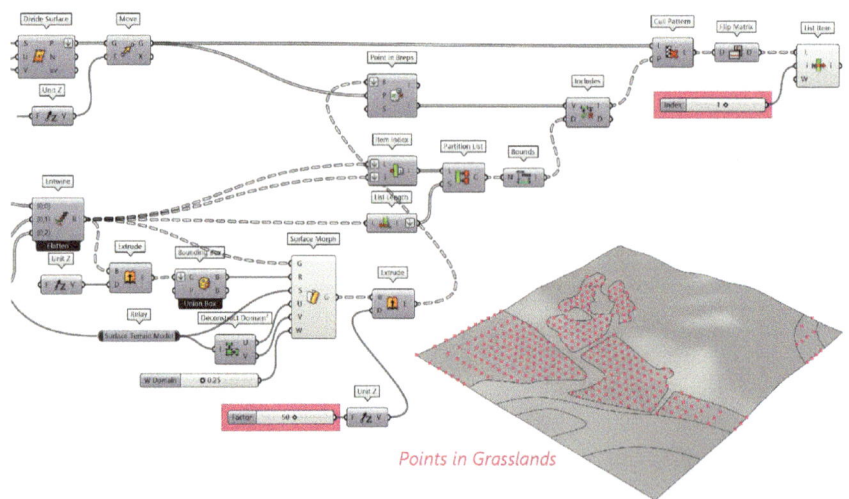

Figure 5.18
Partioning the grid of points by the morphed soil survey Breps.

The *Cull Pattern* component is first inputted with the geometry from the *Move* component at the earlier stages of the script. The *Includes* component generates a list of boolean patterns that can be inputted into the *Cull Pattern* component and flipped into a *List Item* component to visualize which points are included within the different surface types. This process and operation can be used for other geometry, such as surface panels or custom geometric modules. A number slider is used with values between *0* and *2* corresponding with the three different surface typologies. The slider is currently set to value *1*, which is the second list item, and if referred back to the beginning of the script of ordered referenced surfaces, it would be the grasslands.

This last operation can have multiple benefits beyond just visualizing data within a closed shape of a specific surface typology. It will also be valuable to cycle through the different conditions to do a more indepth analysis of a specific surface condition. Also, by omitting the *Filp Matrix* and *List Item* components, the data will still be partioned into all the different surface typologies which is beneficial for more advanced computational analysis.

SOIL CHARACTERISTICS

Modeling soil characteristics within a three-dimensional space is valuable for a variety of reasons. Attaining soil data is necessary to measure and calculate many landscape performance outputs as they relate to stormwater management, erosion control, and planting communities. This coupled with it being transformed from a 'flat' map geometry into a terrain model can now have the ability to compound with other physical properties such as slope, elevation, and drainage patterns for a more comprehensive and robust dynamic model.

Chapter Prerequisites
To best prepare for this chapter and ensure proper overlay of data, continue to utilize the terrain model, either mesh or surface, generated from the 'Terrain from GIS' chapter. Remember to keep the terrain model at its geo-location if baked into Rhino instead of moving it to the c-plane origin (0,0,0).

Modeling Methods
The shapefile for the site project boundary originally used for the 'Terrain Model from Contours' chapter will again be used for this demonstration as it is uploaded to USGS's Web Soil Survey (WSS) website, known as the Area of Interest (AOI) in Figure 5.19. It is preferred that a zipped shapefile with all the additional file extensions is used instead of uploading those files individually.

Once that zipped file is uploaded, select the *Soil Data Explorer* tab at the top, then the *Soil Properties and Qualities* tab in the submenu shown in Figure 5.20. Within that submenu, a side panel will pop-up to select *Soil Physical Properties* where *Surface Texture* can be selected by scrolling down toward the bottom. The default settings for this soil property are fine for this demonstration, but is always worth exploring further for any other specific needs. After all the settings have been made, clicking *View Rating* will process the map and soil survey data.

This will generate a colorized map of all the similar soil textures for this AOI and can be a great reference for the modeling process later in Grasshopper. For this demonstration, I also select the soil pH and K Factor characteristics which can be found in the other property menus.

USDA Web Soil Survey Website

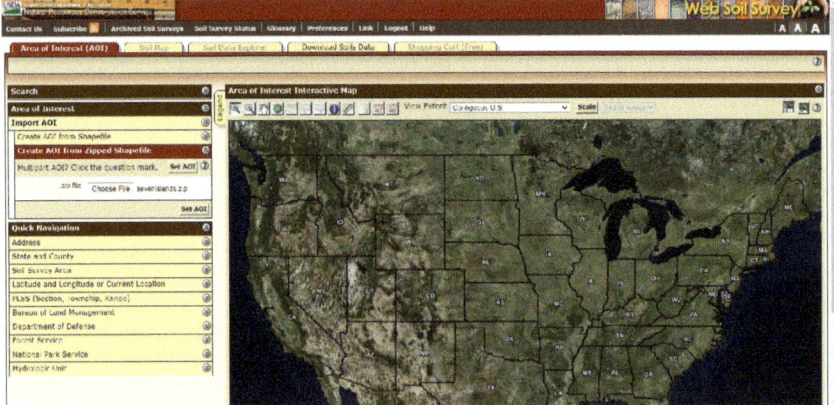

Figure 5.19 Uploading the site boundary shapefile as the area of interest (AOI) on the Web Soil Survey website.

Importing Site Boundary for Soil Properties

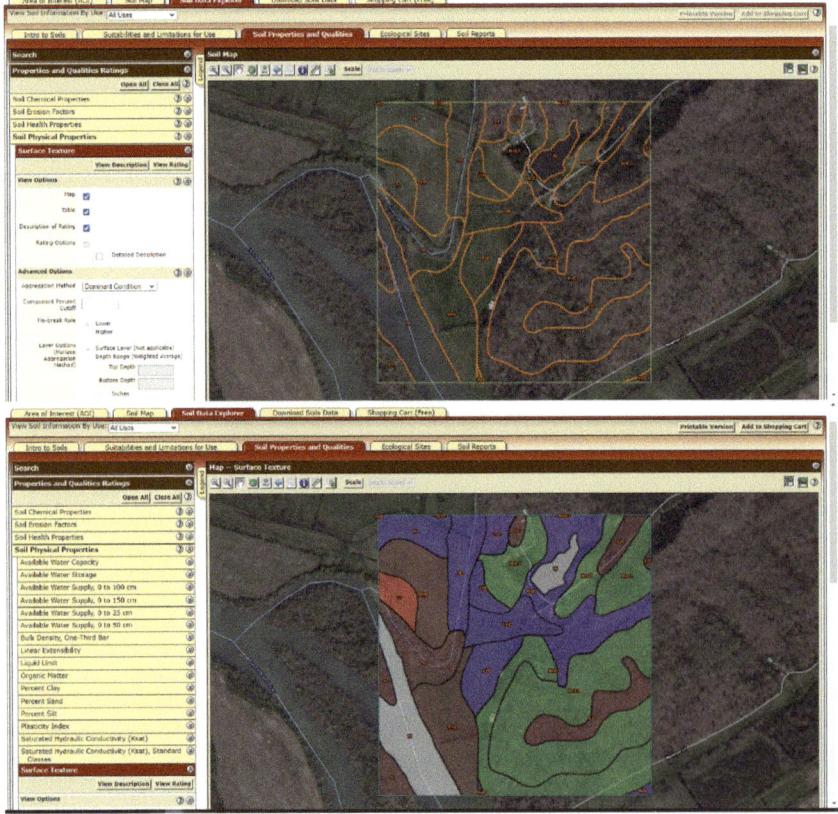

Figure 5.20 Exploring different soil characteristics for export from the defined AOI.

After all the preferred soil properties are selected and viewing for their ratings, the data can be downloaded by going to the *Download Soils Data* in the submenu at the top of the website. At the bottom right of the page, a *Create Download Link* button will generate a zipped *wss_aoi_* file that will be formatted for alignment in ArcGIS Pro and then used in Grasshopper.

The next step is to unzip the downloaded file. Within that folder, the *spatial* folder will contain all the shapefiles for upload to ArcGIS Pro as shown in Figure 5.21. The specific file need for this demonstration is the *soilmu_a_aoi* which uses polygons to map out all the different soil types as shown on WSS. Although the file will match accurately inside of ArcGIS Pro, the coordinate projection is different than the needed WGS 1984 projection for Grasshopper's *@it* plugin. Therefore, the *soilmu_a_aoi* file needs to be exported as a shapefile using the *Feature Class to Shapefile* geoprocessing with the shown settings under the *Environments* tab. The geoprocessing can then be *Run* from the bottom of the panel to generate a new shapefile to overalay with the same contour shapefile created in *Terrain Model from Contours* in Grasshopper.

The last step before switching to Grasshopper is to do some minor formatting and editing of the different thematic data tables from WSS in Microsoft Excel. In addition to the spatial folder downloaded from WSS, a thematic folder is also provided with all the data of the different soil properties selected, which, in this case, include surface texture, pH levels, and K Factor.

The files themselves are not labeled accordingly and can be challenging if multiple properties are selected in WSS; so by openning the *summary* text file, the associated ID numbers will indicate the specific soil property files. The file-types used for formatting are the *ratingXXXXXXX* files, highlighted in the image in Figure 5.22. These files are comma-separated value files which need to be imported into Microsoft Excel. Open the software and go to the *Data* tab at the top and then select the *From Text/CSV* option within the *Get & Transform Data* section. This will prompt the import wizard.

The first file used for this demonstration is the surface texture file (*rating7417367*). The only important setting for the first window pop-up is to ensure *Comma* is selected from the *Delimiter* drop-down menu. This will separate all the data features into separate columns once the *Load* is selected at the bottom of the

Importing WSS Shapefile into ArcGIS

Figure 5.21
Importing the downloaded WSS shapefile of the soil profile curves.

window. This process is repeated for all other soil property files intended on being used for the Grasshopper model. The next step is to begin editing and merging these files together.

To make the process more dynamic with parameters in Grasshopper, a single file is used that combines the multiple Excel files, where the process in demonstrated in Figure 5.23. In the first file, where surface textures were imported to, the *RatingNum* and *RgbString* columns are deleted since they are not needed for the script; this can be done by *right-clicking* on the column letter and selecting *delete* (Columns C and D in this case). Next, the *RatingStr* column from the pH Excel file is highlighted and copied (Ctrl + C or Command + C) and pasted into the now empty C column in the first surface texture file. This is repeated for the K Factor file as well.

Now that all of the Excel files have been combined into one file, the column headers have been relabeled to reflect that soil property. In the case of this specific file, MapUnitKey 757584 does not contain any specific soil data because it is water. Other factors may also lead to there not being in data but because this is known to be a water feature, *Water* is inputted into the *Soil Texture* column and values of 0 are inputted into the other columns. This will reduce errors and other issues in the Grasshopper script.

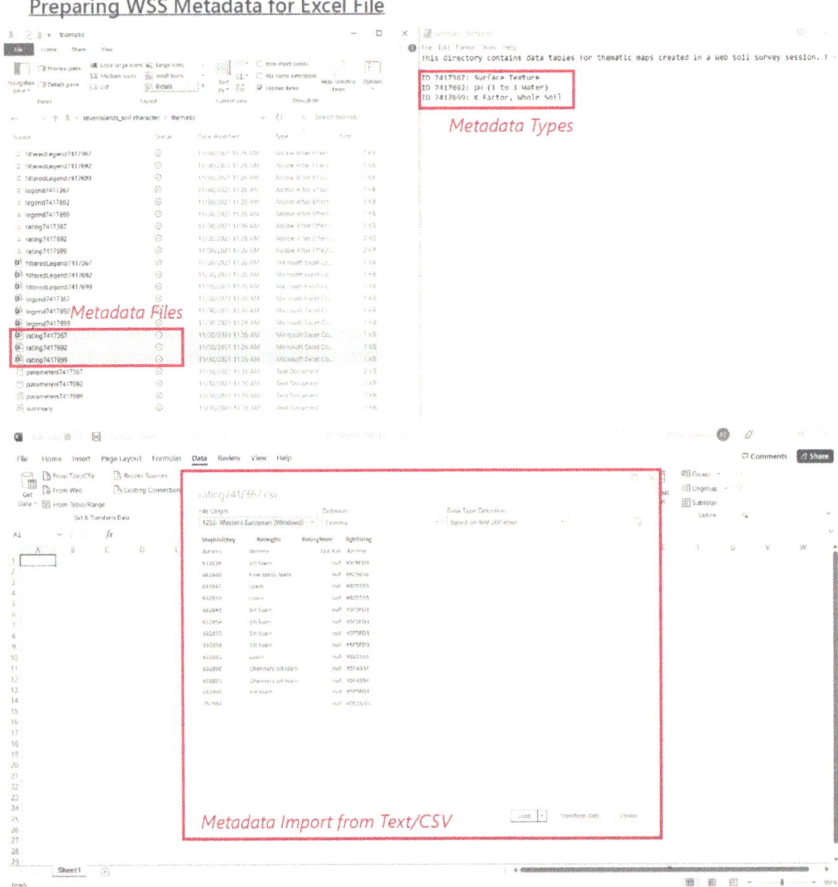

Figure 5.22
Importing the CSV files to Excel from the downloaded WSS thematic folder.

This portion of the script is identical to the process of creating a terrain model from contours, wheras the *soilmu_a_aoi* shapefile created from ArcGIS Pro will be used instead of the contour shapefile, as shown at the beginning of the script in Figure 5.24. From the *F* output on *Imp@It* component, a panel is connected to show the feature columns from that shapefile. List Item 3 *MUKEY* will be used to match with the multiple *MapUnitKey* values in the Excel file. To filter out the *MUKEY* feature values from the shapefile, a *List Item* component is used with a *3* for the *index* input.

These values will then be cross-referrenced with the Excel table using the *Match Text* component to apply the thematic soil property values to the spatial soil

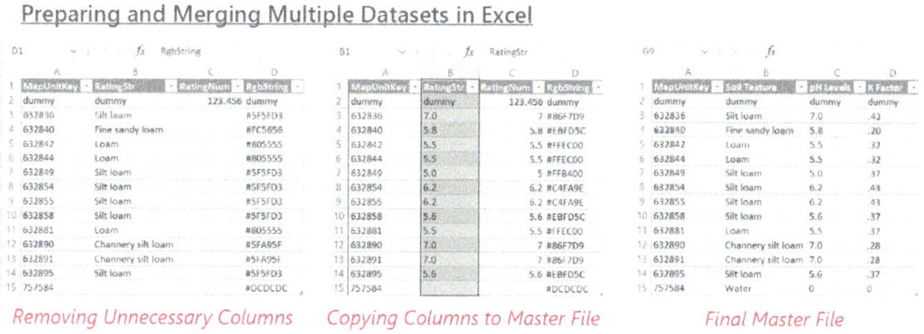

Figure 5.23
Creating a single master Excel file that contains the soil survey metadata.

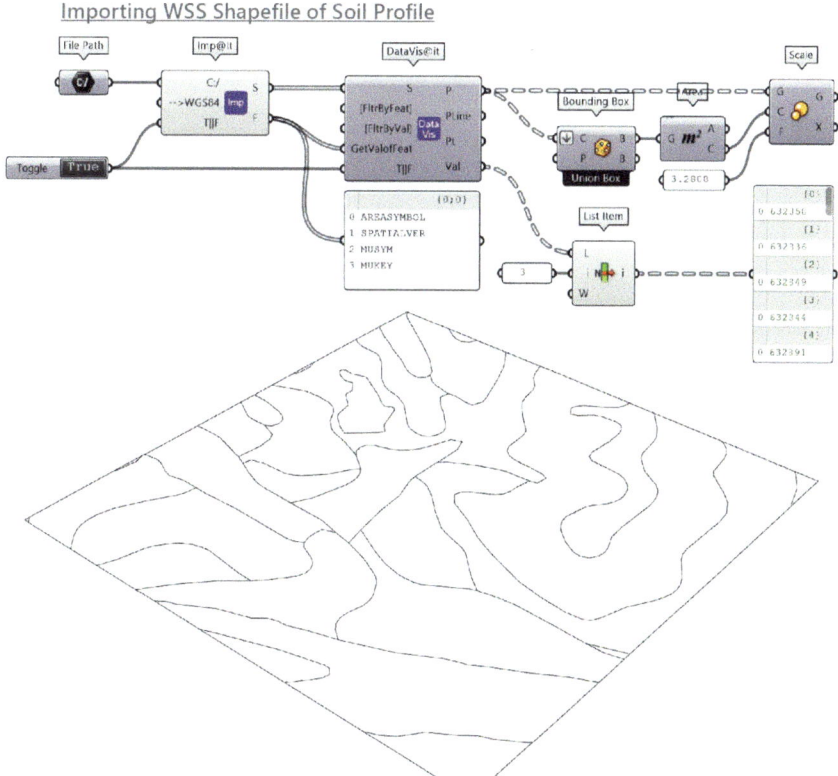

Figure 5.24
Importing soil profile curves with the *@it* plugin and scaling appropriately.

profile geometries. This process can also be achieved within ArcGIS Pro; however, it is important to understand and apply this specific Grasshopper method for future scripts within this book since the ArcGIS Pro method will not be applicable in all cases.

The other part of the script shown in Figure 5.25 is the same operation explained in the *GhExcel* plugin section of this book. This first part of the script operation will use a *0* input for the *List Item* component index to filter the Excel file to *MapUnitKey* column. This operation also includes the addition of a *Split List* component to remove the unnecessary headers in the Excel file (Rows 1 and 2, in this case). Since there are two rows being removed, a *2* is inputted at the *Split List* index. Now that the *MapUnitKey* values are isolated to just the numeric values, they can be inputed into the *R* input of the *Match Text* component.

The *Match Text* component determines which spatial *MUKEY* value corresponds with the *MapUnitKey* value in the Excel file resulting in a True/False boolean

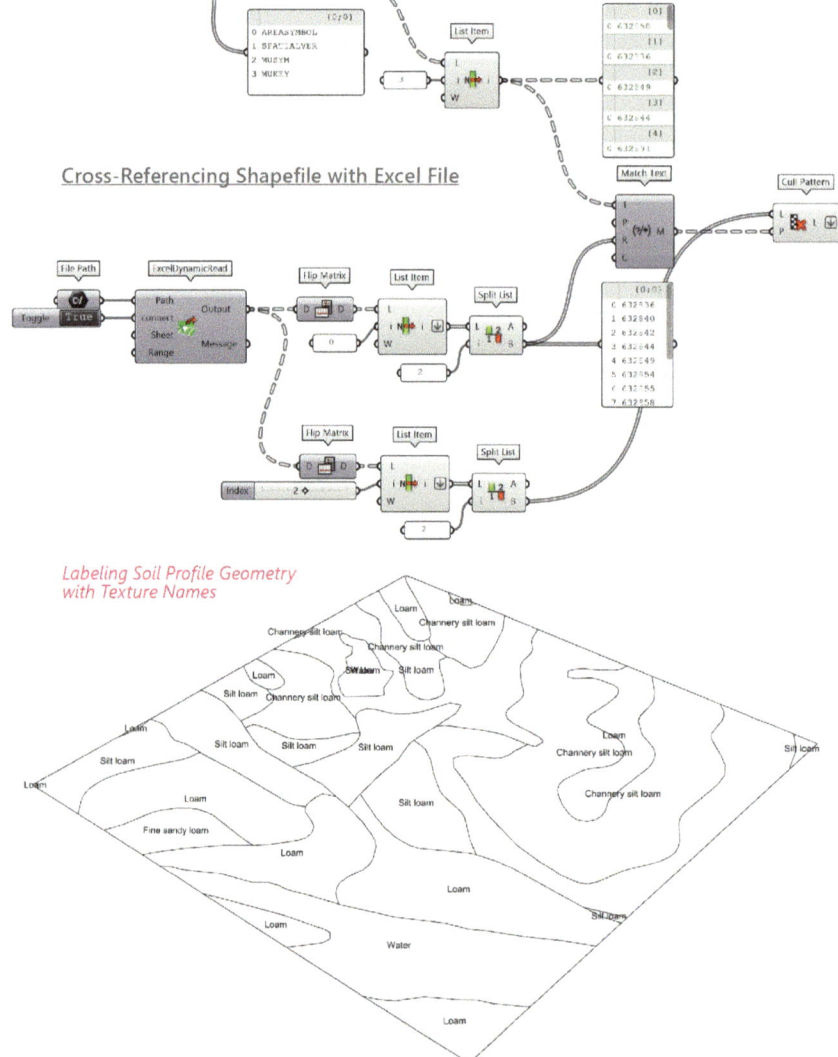

Figure 5.25
Cross-referencing
the soil survey
metadata with the
soil survey curves.

as explained in the *Basic Operations* section of this book. The script operation after the *ExcelDynamicRead* component is copied to isolate the remaining soil property values for the Excel file. Instead of using a panel value for the *List Item* component index input, a number slider from 1 to 3 is used to cycle through the surface texture, pH levels, and K Factor column values. The True/False boolean pattern from the *Match Text* component can then be used to apply those Excel values to the spatial geometry using a *Cull Pattern* component.

For a temporary visual reference only but not shown in the script image, an *Area* component is connected from the *Scale* component to create a central location point for the *Text Tag* component and values from the *Cull Pattern* component for the *T* input. These two components are only serving as a progress check in the model as they are being applied to a planar geometry of the soil profile. The next steps of this script will delete these components and begin the operation of applying them to three-dimensional geometry conformed to the original terrain model from this section of the book. As stated earlier in this chapter, the map generated from WSS can be referred at this point to ensure the data from that website matches the model.

It is also important to note at this point in the process that when there is enclosed geometry, as shown near the top corner of the left-side model in Figure 5.26, there are issues because the enclosing geometry has both the outer and inner curves or two geometries; yet there is only one value for that soil profile. These create a list of data longer than another list which requires the shorter list to have repeated data resulting in errors. The next steps of the script will resolve this issue as it projects this planar geometry onto the terrain model.

Going back to the *Scale* component at the beginning of the scripting process, an operation is applied to remove redundant geometry that creates conflicting data lists, as mentioned earlier. This process will also make the transformation of the planar soil profile geometry into a terrain model more effective. The

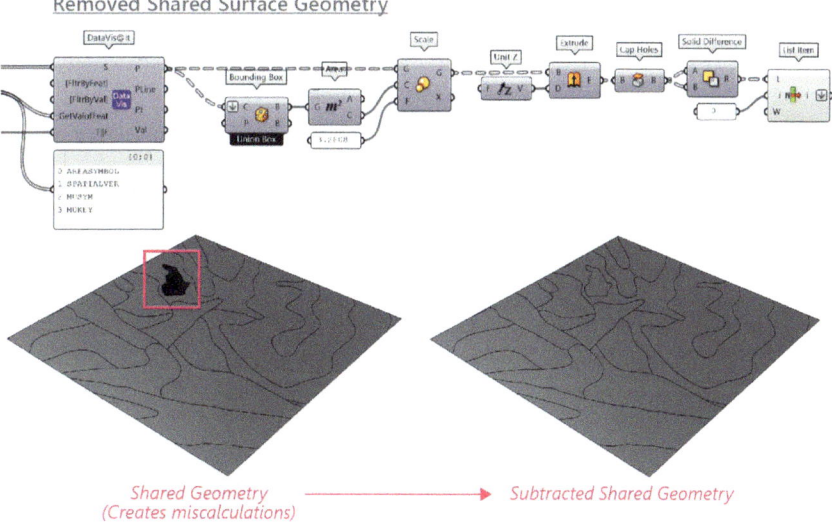

Figure 5.26
Removing duplicate surface from the soil survey geometry.

planar geometry from the *Scale* component is extruded vertically, the value is not relevant but the default value for the *Unit Z* component is *1* and works fine. The extrusion is capped to make solid closed geometries that can have any self-intersecting soil profiles removed using the *List Item* component with a *0* value for the index input.

The *List Item* component will remove the intersecting geometry produced by the larger enclosing geometry since it will have two list items. The *0* index input will select the first item of all the lists, thus culling or removing the second intersecting geometry, as shown in the right side model image.

Now that the soil profile geometry has been refined, it will be projected onto the already existing terrain model previously generated in this section of the book using the *Surface Morph* component in Figure 5.27. A *Bounding Box* component is used with the *Union Box* setting selected by right-clicking the center of the component with *flattened* geometry from the *List Item* component. The same flattened geometry also goes into the *G* input of the *Surface Morph* component.

The Rhino geometry of the terrain model is now referenced within a *Brep* component and connected to the *S* input. The *Surface Morph* component also needs dimensional values of the referenced geometry which can be extracted using the *Deconstruct Domain²* component for the *U* and *V* inputs and a small value, in this case, 0.25, to maintain a 'dimensionless' terrain model.

The newly created soil terrain model can now be colorized and labeled using the various feature values from the previously included *Cull Pattern* component, shown in a variety of ways in Figure 5.28. The process is the same as previous steps in setting up the *Gradient* component for material applied to the *Surface Morph* geometry using a *Custom Preview* component.

The last step to this process is to include a *Text Tag* component to place label within each soil profile geometry using an *Area* component for a centrol point as the *L* input and the values from the *Cull Pattern* for the *T* input, as shown in Figure 5.29.

Modeling out the different soil characteristics serves as a valuable inventory component for any site project as it will impact vegetation, hydrology, erosion, and many other factors related to ecosystem services and landscape performance. And as mentioned with the WSS website, there is an extensive amnout of datasets that can be acquired for many other project objectives beyond what will be covered in this book.

Morphing Planar Soil Profile to Surface Terrain

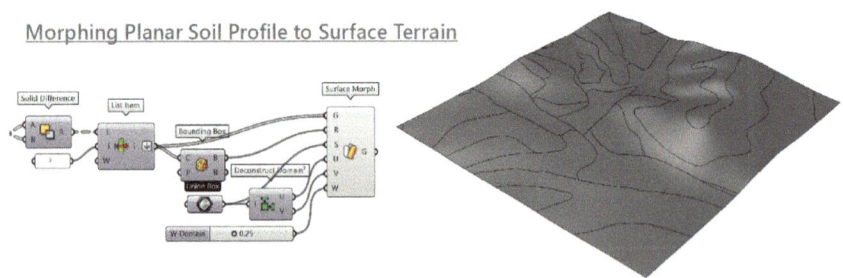

Figure 5.27
Morphing planar soil survey surfaces to surface terrain.

Figure 5.28
Colorizing and
labeling the
different soil survey
characteristics.

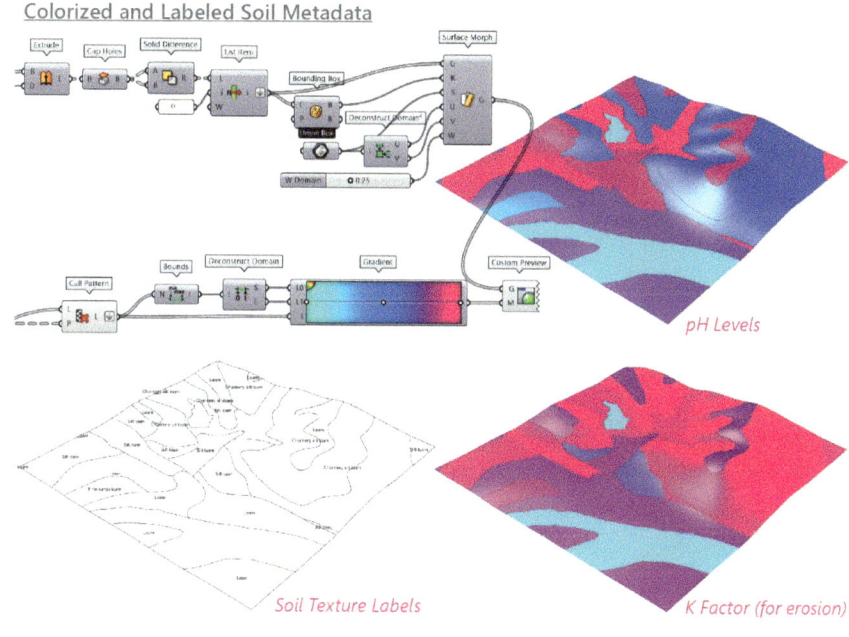

Figure 5.29
Characterizing the
model with color
and labels from
the soil survey
metadata.

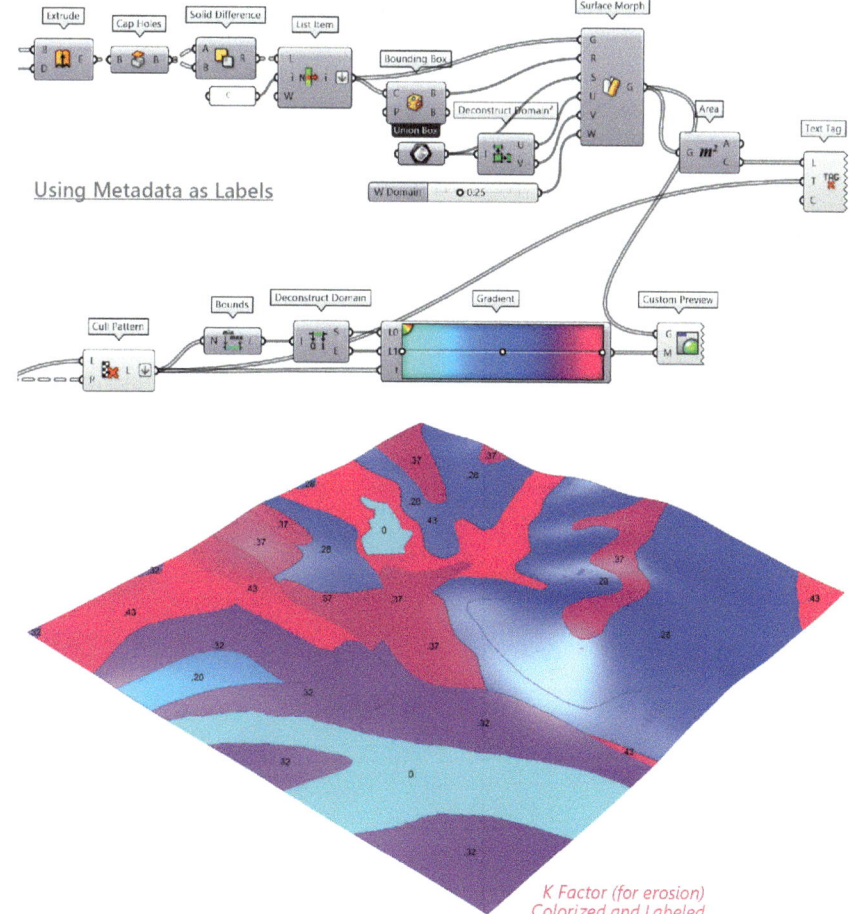

TERRAIN HYDROLOGY

Communicating terrain hydrology will be demonstrated with drainage lines, ridges, and watersheds in this section. These characteristics are not the only elements that make up a site's hydrology; however, they do serve as the fundamental modeling components for future examples as well as other analytical objectives. The drainage lines will dictate the other characteristics and is pretty straightforward, as already demonstrated from the *Mosquito* plugin earlier. With a few additional operations applied to the drainage lines, a robust model can be formulated to provide a more comprehensive understanding of a site's basic hydrological conditions.

Chapter Prerequisites
As already stated, this section will only require the additional *Mosquito* plugin to complete the demonstration beyond the default Grasshopper components.

Modeling Methods
The first step is beginning with the same script operation from the *@it* plugin in which the previously generated surface terrain model will be used. Depending on the size of the model, different densities of grid points can be used for varying amounts and details of drainage lines, shown at the beginning of the script in Figure 5.30. The *Quad Panels* component is used for this demonstration since it works well for the later steps; however, the *Lunchbox* plugin provides a variety of other options for dividing the surface into different forms of panels.

Remember that for large area models like this one, different number slider values will need to be used. The number slider for the *Calc* input can range between 1 and 5, depending on how much is needed/wanting to be shown; however, the lower values (more details) will require more processing time and can begin to slow down the computing as the script progresses.

Not only do drainage lines indicate the directional flow of surface runoff, but they can also be used to determine ridgelines and delineate sub-basins within a terrain model, as the operation in the script of Figure 5.31 demonstrates. The *Point on Curve* component is used to find the end point of the drainage lines where, in this case, aggregate at nine different cluster points in the model shown with the black dots. Although this does provide an important visual of where runoff drains and collects at, it is just as important to partition the data into these same groupings using the *Point Groups* component.

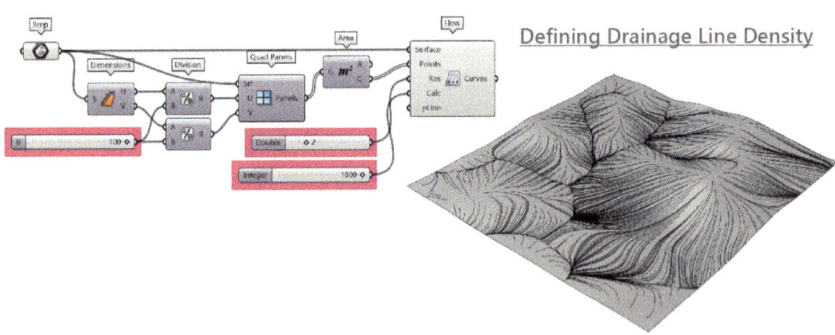

Defining Drainage Line Density

Figure 5.30
Defining different drainage line densities on the surface terrain model.

Figure 5.31
Grouping drainage
line endpoints to
establish future
sub-basins and
ridgeline geometry.

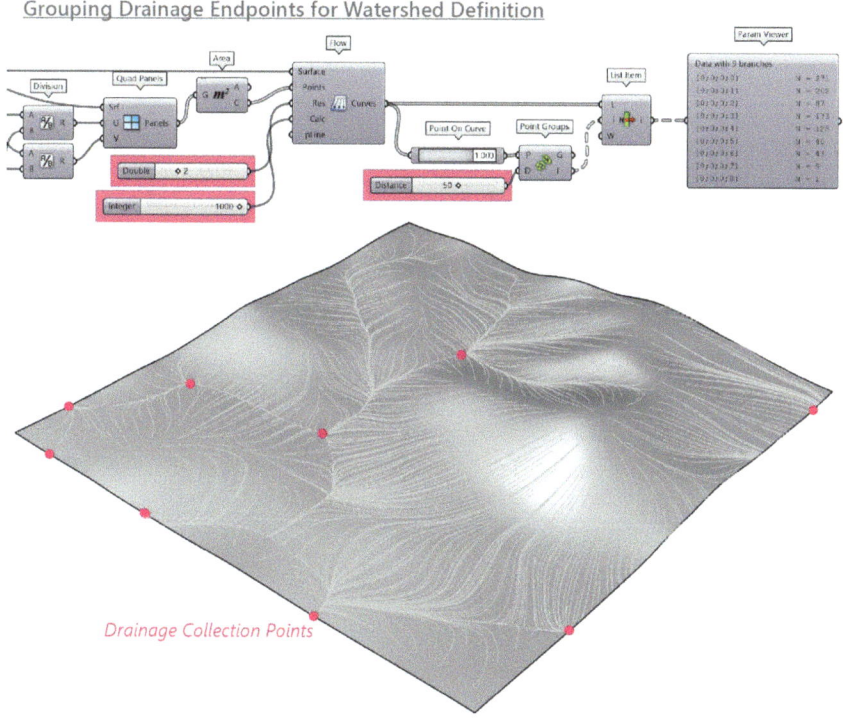

Again, depending on the size of the model, a different value may need to be used for the *Distance (D)* input. From this component, each corresponding index number for the individual drainage lines are grouped based on this distance proximity and used for the *Index (i)* input of the *List Item* component along with the *Curves* output from the *Flow* component for the *List (L)* input. Visually, there will be nothing to indicate this operation; however, with the *Param Viewer* component, the number of drainage lines for each group can be seen.

With the data now partitioned according to the drainage end points, the newly branched data can be used to analyze all the sub-basin areas collectively or individually. Collectively, the drainage lines can be measured and colorized for their length with the *Length* component, shown in Figure 5.32, to input into the color gradient and custom preview operation. The *Gradient* component may turn orange with a warning, because some terrain models may generate a drainage line or lack thereof with a distance of zero. Fortunately, this will not affect the outcome of the script and visualization.

Another way to visualize the sub-basins of this terrain model is to create ridgeline curves and colorize each sub-basin according to the overall size. For this operation, instead of using the *Curves* from the *Flow* component as the input for the *List Item* component, the *Quad Panels* component is inputted to colorize according to the overall size of each sub-basin. This process begins with calculating the entire area of each sub-basin using the *Mass Addition* component to add all the individual panel areas. These mass additions are then *flattened* for the *Bounds* component to compare each sub-basin area.

To create the ridgelines for each sub-basin, Figure 5.33 showcases how the quad panels for each data group are merged with the *Brep Join* component to create a solid surface with the border extracted using the *Curve* component.

The same operation demonstrated previously for labeling data is used to convert the area from feet to acres and display at the center of each sub-basin. Because of the variation in slope throughout the terrain model, a *XZ Plane* component is used to orient the text vertically to minimize clipping.

The next few operations are taken from the previous chapter to analyze other characteristics of the different sub-basins that can be further visualized individually, which include slope percentage and elevation difference. For the slope analysis in Figure 5.34, the *List Item* for the *Quad Panels* is still being used to perform that operation with an *Average* component at the end to calculate the average slope percentage which can be used similarly to the previous step to label at the center of each sub-basin. For the elevation difference, the *Centroid (C)* output from the *Area* component is used for the *Deconstruct* component with a *Bounds* component to show the lowest and highest elevation point of each sub-basin.

The previous steps in the script looked at all the sub-basins within the terrain model collectively. These same data collections can also be isolated and cycled through to focus on a more comprehensive individual sub-basin. For organization purposes, a *Relay* for each of the following was used to extend the data away from the previous script operations.

In Figure 5.35, the first relay is used for the partitioned drainage lines coming out of the *List Item* component connected to the *Flow* plugin component. The second relay is used for the partitioned panels coming out of the *List Item*

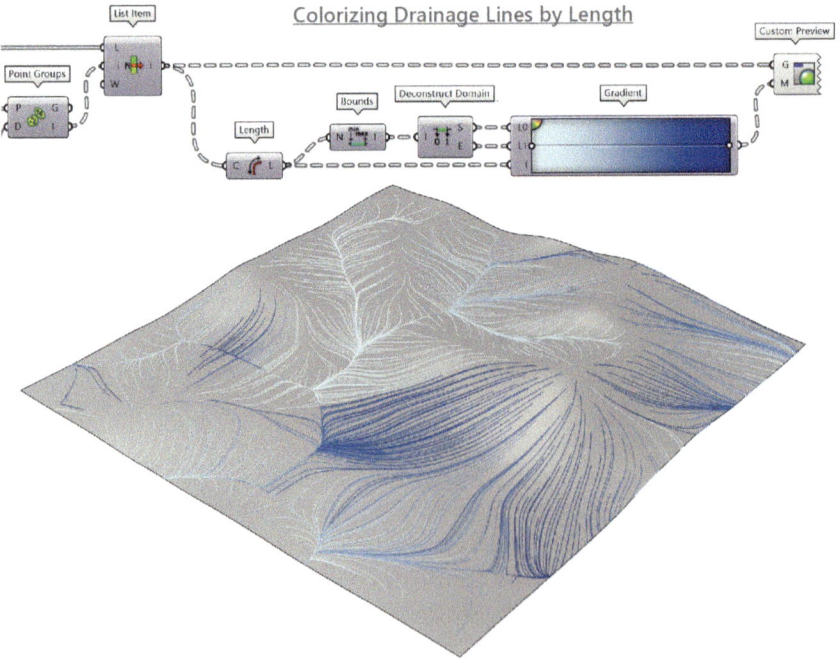

Figure 5.32
Colorizing drainage lines by their length.

Figure 5.33
Defining sub-basins
with a colorization
operation.

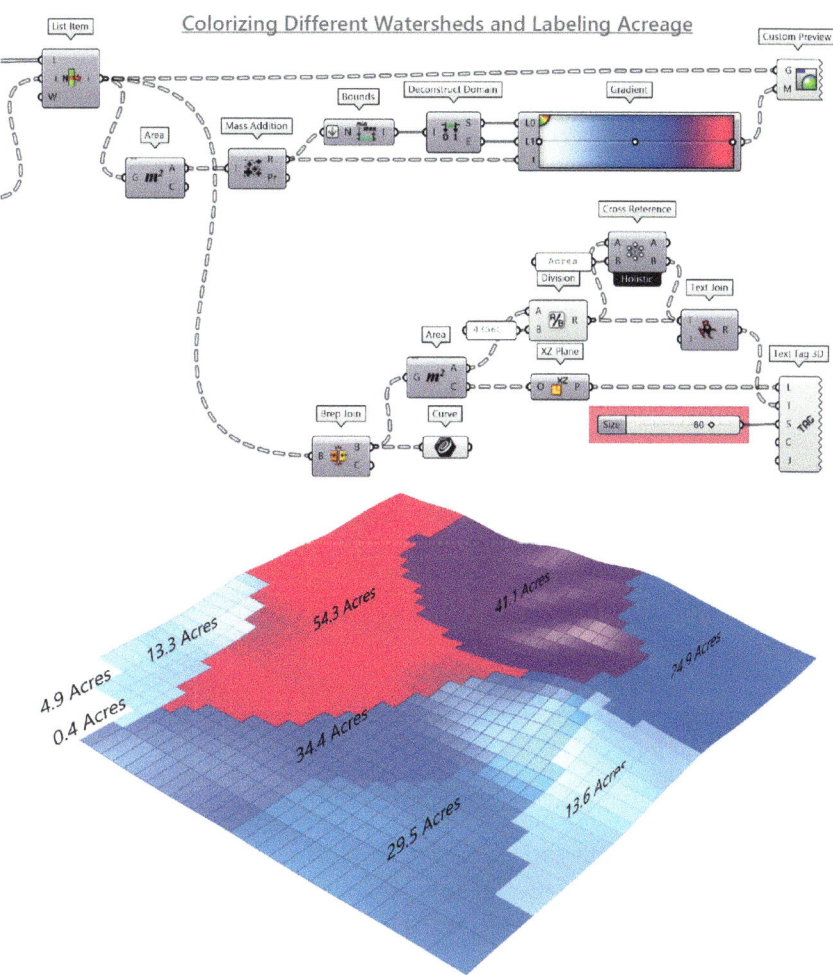

Figure 5.34
Extracting different
analytical data from
the model that
includes ridgelines,
average slope, and
elevation different
for each sub-basin.

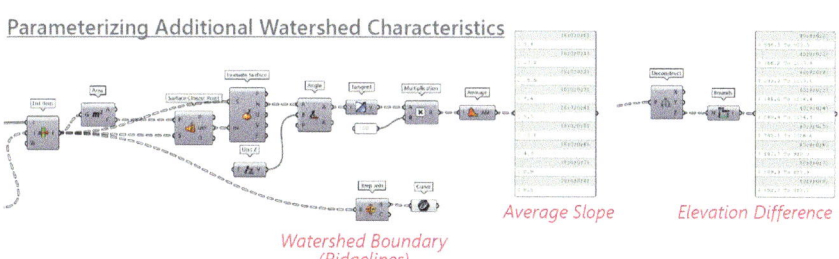

component connected to the *Quad Panels* component. And the last relay is used
from the *Multiplication* component as part of the slope percentage operation from
the last image. The operation of connecting the relays to a *Flip Matrix* and *List Item*
component with a flattened output can be copied and pasted for all the relay connections with an additional one connecting to the *Area* component from the *Quad
Panels* relay.

A number slider is used to cycle through the indices that correspond with all the different sub-basins. The best way to determine the range of the number slider is by referring back to the *Param Viewer* component and counting the number of branches, in this case nice. If the initial grid density of points used to create the drainage lines is ever changed, it can alter the number of sub-basin groups and may need to have the number slider renumbered accordingly. Lastly, a *Custom Preview* component is used with a color swatch to visualize the drainage lines for the selected sub-basin.

The next layer of information to visualize is the size and color of the individual panels within the sub-basin to reflect the slope percentage. The second *List Item* component contains the panel geometry for the *Scale* component, which can also be used to extract center points to scale from. For scaling, a factor larger than *0* and equal to or less than *1* is best to avoid any overlap or clipping of geometry that results in illegible data. Since the slope percentages go well beyond this range, the *Remap Numbers* component is used, as demonstrated in previous chapters, to recalibrate the values to the preferred range. The reason to avoid a lower limit of *0* for the *Domain Start* is because that will create an error of resizing the geometry, so a value of *0.15* is used to significantly scale down the lower values and geometry. If the intention is to use a value less than *1* for the *Domain End*, then a number slider can be used but if not the default value of *1* can be used.

In Figure 5.36, the newly scaled geometry is then colorized using the standard gradient operation for the *Custom Preview* component.

The last procedure in Figure 5.37 shows how to effectively visualize the unique characteristics of an individual sub-basin to display the data values through text for the slope percentage. To reinforce hierarchy and contrast of data in this graphic representation of terrain hydrology, the slope percentage data to be displayed as text is resized accordingly; in other words, the higher the slope percentage, the larger the text is displayed. Similar to the remapping of numbers for scaling the panel geometry, the same is done for the *Size (S)* input of the *Text Tag 3D* component. The number slider should again be set value that provides legible display of text but have enough range to demonstrate hierarchy.

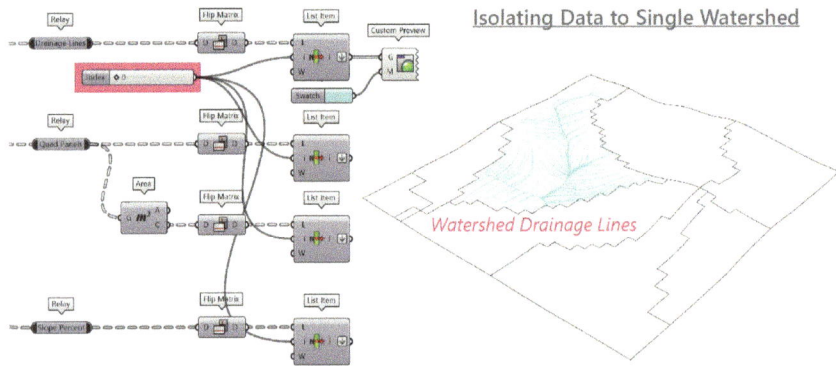

Figure 5.35
Filtering different hydrology characteristics by watershed sub-basins.

Figure 5.36
Colorizing the
scaled quad panels
for the filtered
sub-basin.

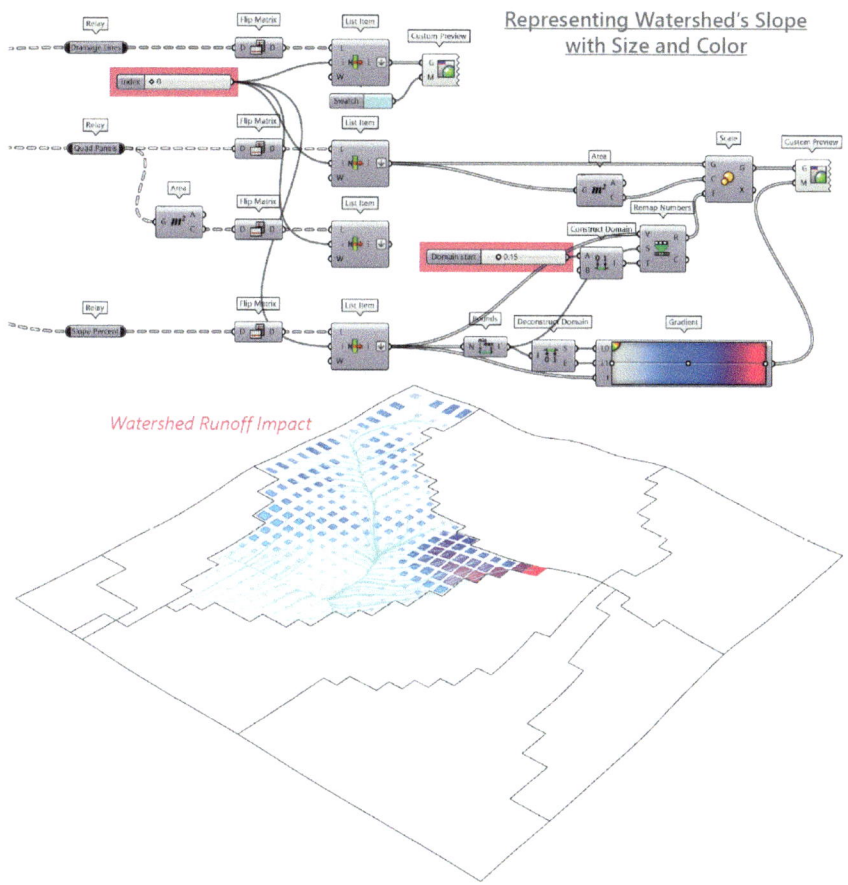

Once this process is completed of visualizing the drainage lines, panel size, and slope percentage text, the different sub-basins can be cycled through using the designated number slider connected to all the corresponding *List Item* components. The different sub-basin visualizations are shown in Figure 5.36.

Additional *List Item* operations can be added to the process from the earlier stages of the script that include the average slope percentage and elevation difference or any other unique characteristics incorporated from previous chapters. The key element for this operation and process to work is to maintain the branching structure of the data within designated groups from the *Point Groups* component.

This chapter has provided several scripts to model some of the basic and fundamental characteristics that will be needed for any inventorying, analysis, and design of a project site. They showcase more complex scenarios than may be typically expected with a project site; however, this will allow for plenty of flexibility and replicability for any project site in the future. The development of the models from this chapter will also serve as the foundation for the modeling procedures in the proceeding chapters to develop additional site inventory, analysis, and landscape performance.

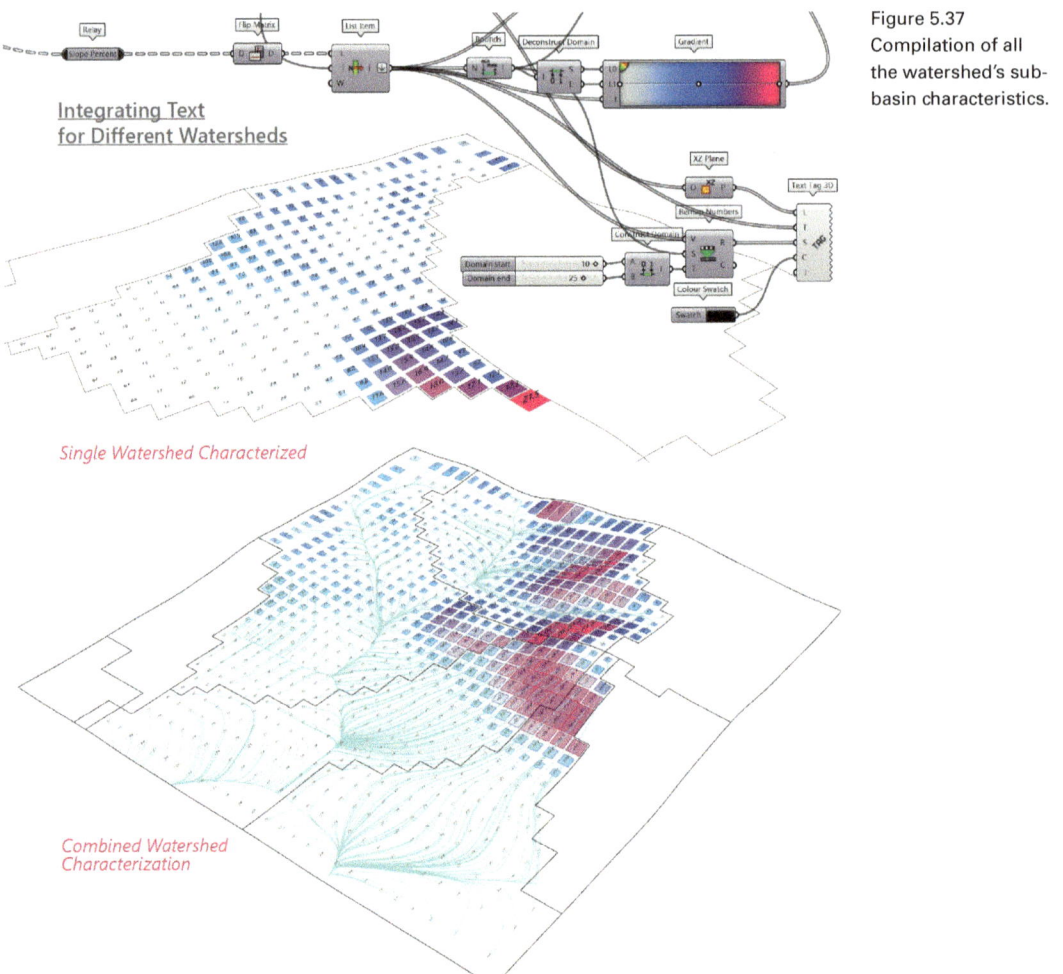

Figure 5.37
Compilation of all
the watershed's sub-
basin characteristics.

Integrating Text
for Different Watersheds

Single Watershed Characterized

*Combined Watershed
Characterization*

6   Landscape Inventory

Modeling the different characteristics of a landscape along with all their unique range of information is helpful in itself but will be even more valuable once those different layers of data become cross-referenced with one another to inform an analysis of landscape performance. But before those robust methods can be deployed in a digital model, their fundamental computations need to first be understood and applied. Like any other method of the design process, modeling the landscape inventory of systems will result in raw data that can be analyzed and synthesized with other sets of data.

Some of the terrain characteristics that will be modeled and inventoried in this chapter will include elevation, slope, aspect, and climate. These fundamental characteristics of a terrain model can be measured and applied in a variety of ways as they relate more specifically to different analytical metrics of ecosystem services and landscape performance. The modeling methods of this chapter will also include basic filtering and highlighting of appropriate datasets to communicate their own significant qualities as an individual system before compiling them into landscape performance methods.

VISUALIZING TERRAIN ELEVATIONS

In the visualization of a terrain model, this chapter will use the generated mesh and terrain from the previous chapter. This terrain model can be managed within the previous script or baked into Rhino and referenced back into Grasshopper. For this example, I will use both a baked mesh and surface terrain model to visualize different elevation graphics. It is important, as mentioned previously with imported shapefiles, that if the terrain model is baked into Rhino, then it maintains its exact location and not moved to the Rhino c-plane origin (0,0,0). That is because if any other analysis is to be done using other GIS shapefiles, it will still be projecting at its geo-location in Rhino not at the c-plane origin.

Chapter Prerequisites

To best prepare for this chapter and ensure proper overlay of data, utilize the terrain model, either mesh or surface, generated from the previous chapter. Remember to keep the terrain model at its geo-location if *baked* into Rhino instead of moving it to the c-plane origin of the model space.

DOI: 10.4324/9781003208020-8

Modeling Methods

This script will begin by visualizing elevations from a mesh terrain model. The second part of the script will then demonstrate the process from a surface terrain model which varies slightly. Referencing different model types (mesh or surface) will yield varying details to the visualization of elevations as a heightfield. The mesh contains significantly more detail than the surface and you will see that in the final images of this chapter. The referenced Brep, or surface, will be converted into a mesh where both options will be inputed into the *Deconstruct Mesh* component to extract the points or vertices for their *z*-value.

In the operations of Figure 6.1, all the vertices are deconstructed to find their *z*-value and then used to find the range of *z*-value or relative elevation of the terrain model. The *Bounds* component finds the lowest and highest values to create a domain that can then be donstructed to output those extents individually.

In Figure 6.2, those values are then inputed into a *Gradient* component where the *S* output (lowest value) and the *E* output (highest value) are inputted into the *L0* (lower limit) and *L1* (upper limit) respectively. All of the *Z* values are inputted into the *t* input to calculate where that values sit within the range or color gradient. The gradient component uses a color range of your chosing and then inputted into the *Construct Mesh* component.

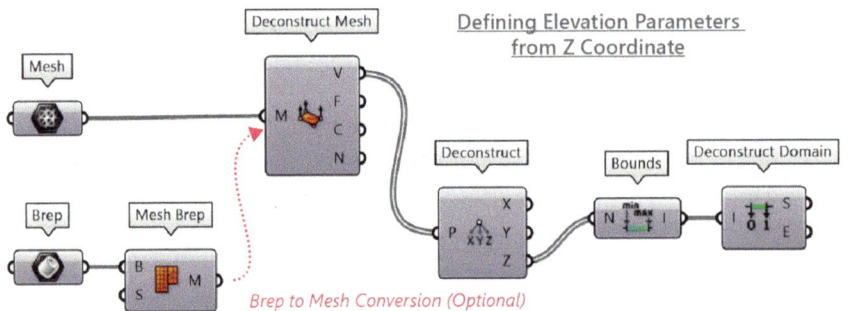

Figure 6.1
Extracting the *Z* coordinates from either a mesh or Brep converted mesh.

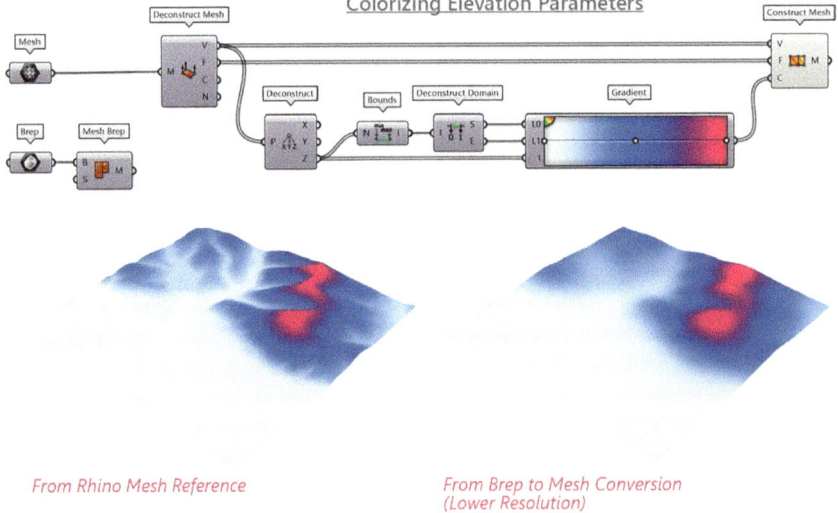

Figure 6.2
Demonstrating the detail different between the two terrain inputs.

Figure 6.3
Colorizing different
panel types
(Triangle and Quad)
on a surface terrain.

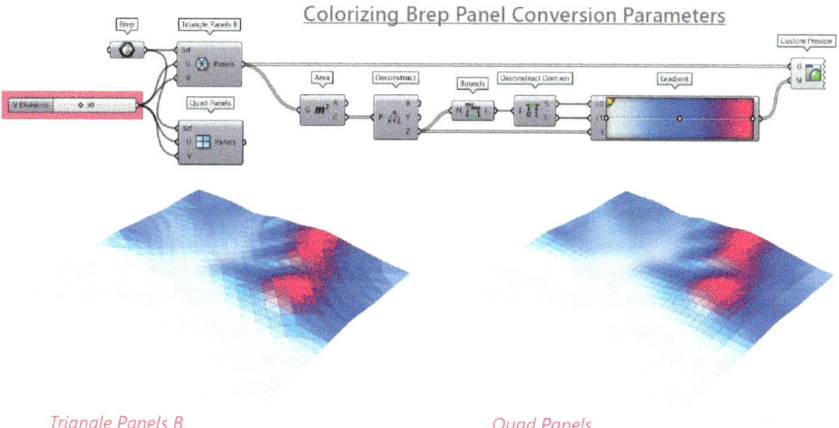

Triangle Panels B   Quad Panels

Unlike the terrain mesh, to visualize a terrain surface, specific geometry needs to be applied for colorization. One of the most effective ways is using the different panel components from the *Lunchbox plugin*, as shown in Figure 6.3. There are numerous panel options, but for this demonstration the triangle and quad panels will be used. Using a number slider for both the *u* and *v* inputs, different densities of the panel type can be applied to the model, as was demonstrated in the *Lunchbox plugin* chapter.

Since these panel components do not create points or vertices like the deconstructed mesh component, the *Area* component is used before utlizing the operation previously scripted. From there, the same process resumes, wheras now the panels are colorized from the *Gradient* component using the *Custom Preview* component.

This terrain model can be further decoded by filtering out high and low points or other elevation points using different parameters. In this example from Figure 6.4, the lowest and highest points are filtered out and annotated. The lowest point can first be displayed with the *Sort List* component that uses the *Deconstruct Z* output to rearrange the elevation values from lowest to highest. The *A* input for this component for sorting optional information synchronously will be the *Quad Panels* center points that were extracted with the *Area* component. A *List Item* component is then used for both outputs to isolate the entire list to only the first list item, which is the lowest point. These component outputs are then used for a *Text Tag* component to display the elevation at that specific location.

The same operation can then be applied to find the highest point, with the only difference being that the two outputs from the *Sort List* component are reversed to rearrange the values to start with the largest values. Another option provided in the script is the ability to also search random indexes (elevation points) for a more thorough depiction of the terrain's topographic variation. One way of doing so is with a *Gene Pool* component to generate a set number of random values for indexes to connect to the same operation of *List Item* components.

The previous visualizations of a terrain's elevation relied more on a colorization approach; however, the more conventional method of extracting labeled contours can also be modeled, as shown in Figure 6.5. After determining the lowest elevation point, a *Series* component can be used to start at that point, in this case, *850* feet. The *Step (N)* size of these contours can then be decided with a number slider where this example uses a value of *10* to signify ten-foot contour intervals. The *Count (C)*

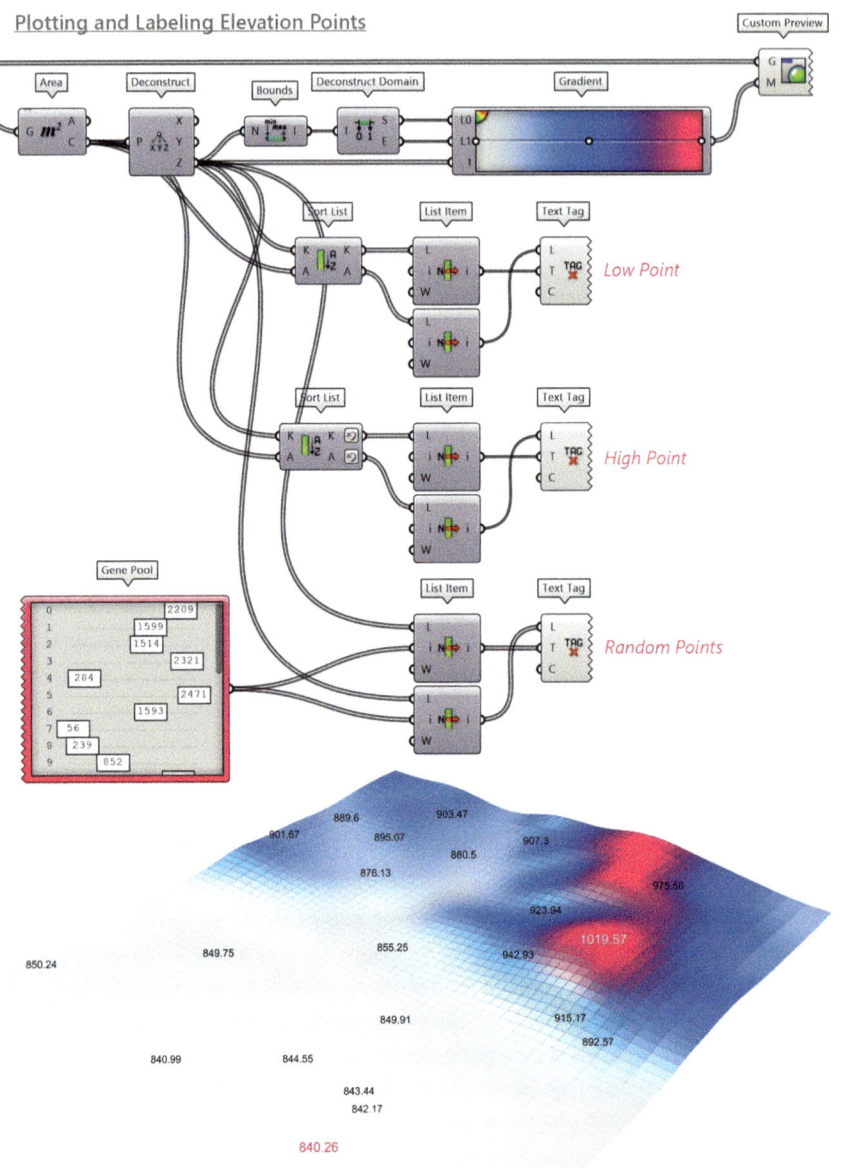

Figure 6.4
Annotating low,
high, and random
elevation points
from a surface
terrain.

should be of a larger value to ensure a contour is exracted at every possible interval. The outputted values from this component will be used to plot the *Z* coordinate of the *Construct Point* component which is then utilized for the *XY Plane* component. These series of planes will be used to generate the contour curves with the *Contour (ex)* component for the terrain surface referenced in a *Brep* component. In order to ensure accurate slicing of the model for contours, the *Distances* (*D*) input needs to be set to *0* to prevent any offsetting of the plane's *z*-value.

The next set of operations will generate the contour labels at the specific elevations for a more comprehensive communication of a topography model. The

Figure 6.5
Generating and
labeling contour
lines on a surface
terrain.

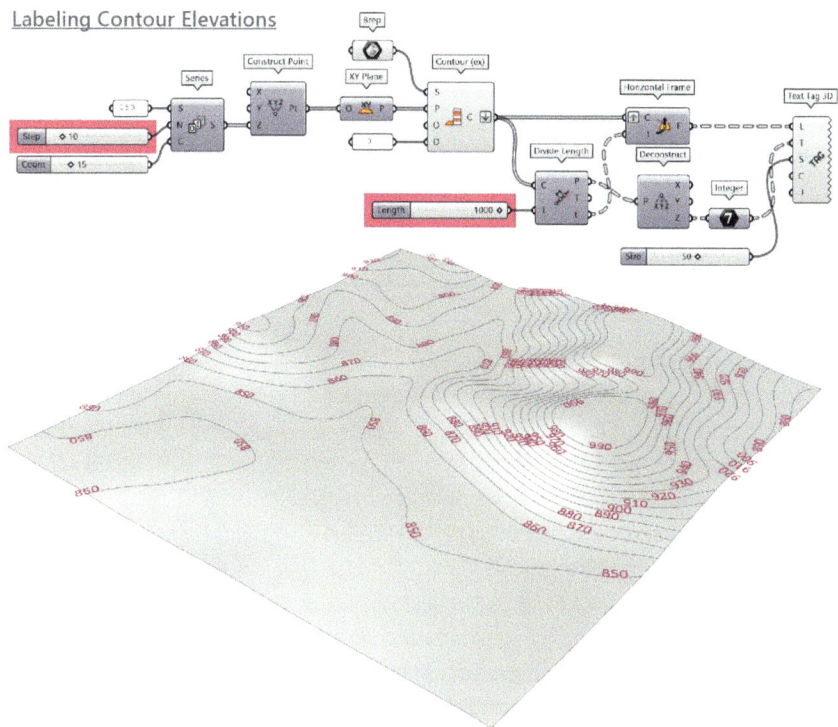

outputted contours (*C*) needs to be *flattened* for the restructuring of the dataset and inputted into both a *Divide Length* and *Horizontal Frame* component, which will require the input to be *grafted*. The first component will dictate the distance between labels, set to a value of *1,000* feet for this scale. This will output the point's *Parameters* (*t*) for the *Horizontal Frame* component which will be used as the *Location* (*L*) of the labels with the *Text Tag 3D* component. The *Horizontal Frame* component is used so that all the text labels will be tangent with the undulating contours curves instead of at a singular fixed orientation. The *Points* (*P*) output can then be extracted for their *Z* coordinate with a *Deconstruct* component and have the output converted to an *Integer* as a whole number instead of decimal. This will then be inputted to the *Text* (*T*) input and have a number slider for the *Size* (*S*) input to determine an appropriate scale of labels.

VISUALIZING TERRAIN SLOPE PERCENTAGE

Indicating different slope conditions on a terrain model is extremely valuable when considering and analyzing a project site, as it will impact erosion, stormwater runoff, tree-planting, accessibility, and other conditions for a project site. This script will initiate the process by visualizing the slope percentage with seemingly arbitrary values. In this demonstration, a range of slope percentages from 0% to 100% and to 30% will be visualized as a foundation for future investigations, and that is why they may be considered arbitrary since slope percentages impact site analysis differently. For example, a 5% slope for stormwater drainage is considered moderate, whereas that same slope is considered extreme for erosion. So the purpose of this script is to demonstrate how these different slope percentages can be visualized but not necessarily their significance which will be covered more extensively in Section 3 of this book.

Chapter Prerequisites

To best prepare for this chapter and ensure proper overlay of data, continue to utilize the terrain model, either mesh or surface, generated from the Terrain from GIS chapter. Remember to keep the terrain model at its geo-location if baked into Rhino instead of moving it to the c-plane origin.

Modeling Methods

Similar to the previous chapter, there are several options of visualizing the slope of a terrain model. The first steps of the script from the 'Visualizing Terrain Elevations' chapter can be applied here, but for this demonstration I will be focusing on only visualizing the slope of a mesh (no conversion from a surface) and one panel option of surface (quad panel). Instead of using the vertices from the mesh, the *normal* or vector direction will be used to measure against a *Unit Z* vector to determine the angle for slope percentage.

Although the terrain model was generated from a GIS shapefile and everything should be oriented correctly, a quick visual check can be performed to make sure the terrain model is facing up, opposed to downward with the normal vectors, as shown in Figure 6.6. This can be quickly done with a simple operation using the *Line SDL* component. The vertices (*V output*) from the *Deconstruct Mesh* component will be used as the starting point for these lines, the normal (*N output*) for the direction of the lines, and then a large slider number for their length to easily see in the model.

If correctly oriented, these lines will be above the terrain model; but if they are below the model, a simple *Reverse* component for the normal output to the *Angle*

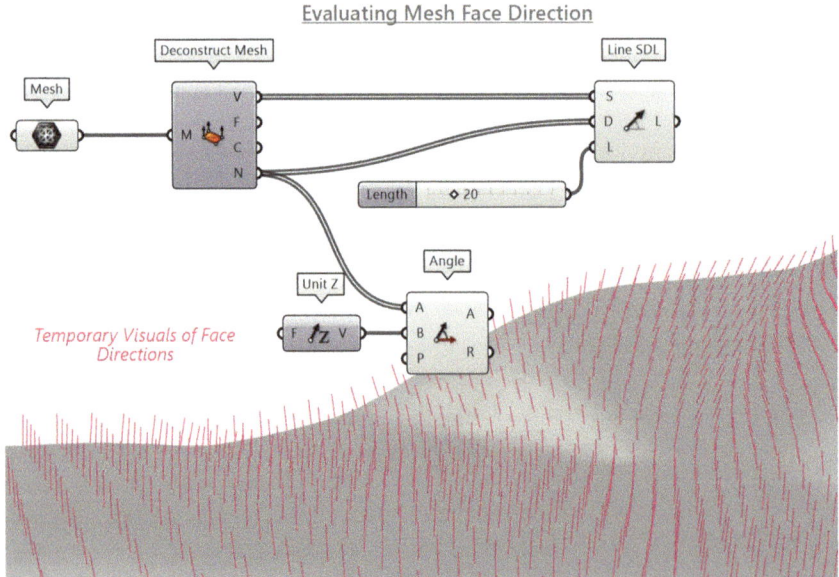

Figure 6.6
Temporarily
visualizing face
directions from
their normal vector
values.

component will correct the model direction. Once that is achieved, the *Line SDL* component and number slider can be deleted.

Although all the normals of the terrain model have been measured, the output from the *Angle* component output *A* is in radians. This next step in Figure 6.7 will use some mathematical components to convert the angle to slope percentage, a much more relatable form of measurement for landscape architecture. The *Tangent* component will automatically convert the radians to slope percentage in decimal format and then a *Multiplication* component is used to convert to percentage values.

With these generated terrain models, it is possible to have obscure elements within the mesh that may skew or distort the results by having a severely high percentage or negative slope values. To correct that possibility, a *Maximum* and *Minimum* component is used to essentially cap those extremes at a specific value. With a *0* value for the *B* input of the *Maximum* component, any negative value – less than *0* – will be converted to *0* or any other value chosen. The opposite is the case for the *Minimum* component where any value above the given *100* input is capped at that value. The order in which these two components are placed will not affect the final output values.

Now that the slope percentage values have been refined, similar to the *Visualizing Terrain Elevations* script, a color gradient operation in Figure 6.8 will be used

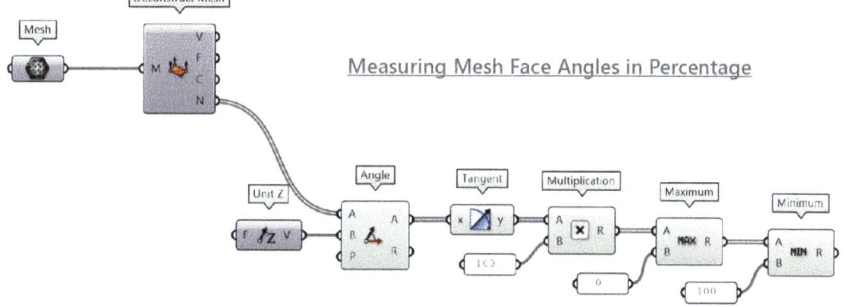

Figure 6.7
Converting surface
slope from radians
to percentages.

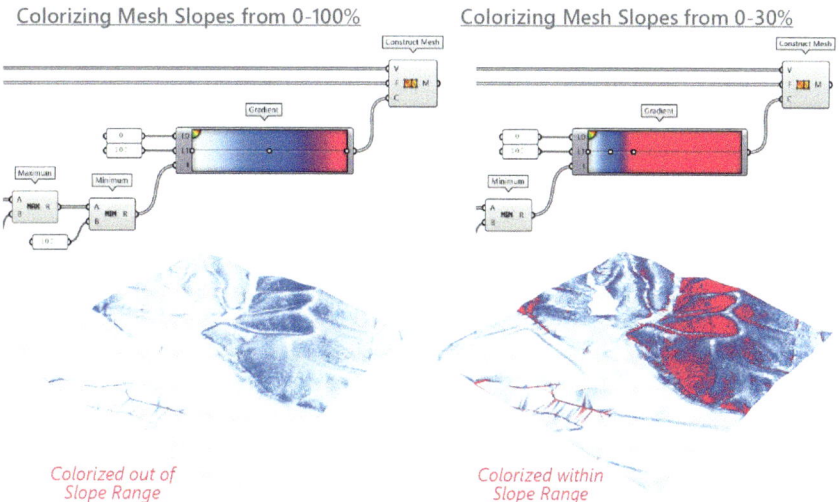

Figure 6.8
Colorizing different
slope ranges.

to visualize all the slope percentages within the terrain model. A *0* value is used for the lower limit *(L0)* and a *100* value is used for the upper limit *(L1)* input of the *Gradient* component.

As shown in the visualization of the slopes for this specific terrain model, the majority of percentage range within the white to blue spectrum or 0%–50%. At this point, these percentages are somewhat arbirtrary and may not be particularly relevant for a specific site project. Because of this, the blue point in the gradient is moved from its current 50% location to 15% and the pink point's 100% location to 30%, resulting in a more legible range of slope conditions. These slope percentages are more applicable to the more comprehensive scripts in Section 3, in which I will go more indepth to the significance of given slope percentages for landscape performance.

Visualizing the slope percentage of a surface terrain model requires a few additional operations since there is not a deconstruction process like that of a mesh geometry, as demonstrated in Figure 6.9. Like the elevation script from the previous chapter, a *Quad Panels* component will be used to analyze the surface at these individual panel. A number slider can be used to adjust the number of panels

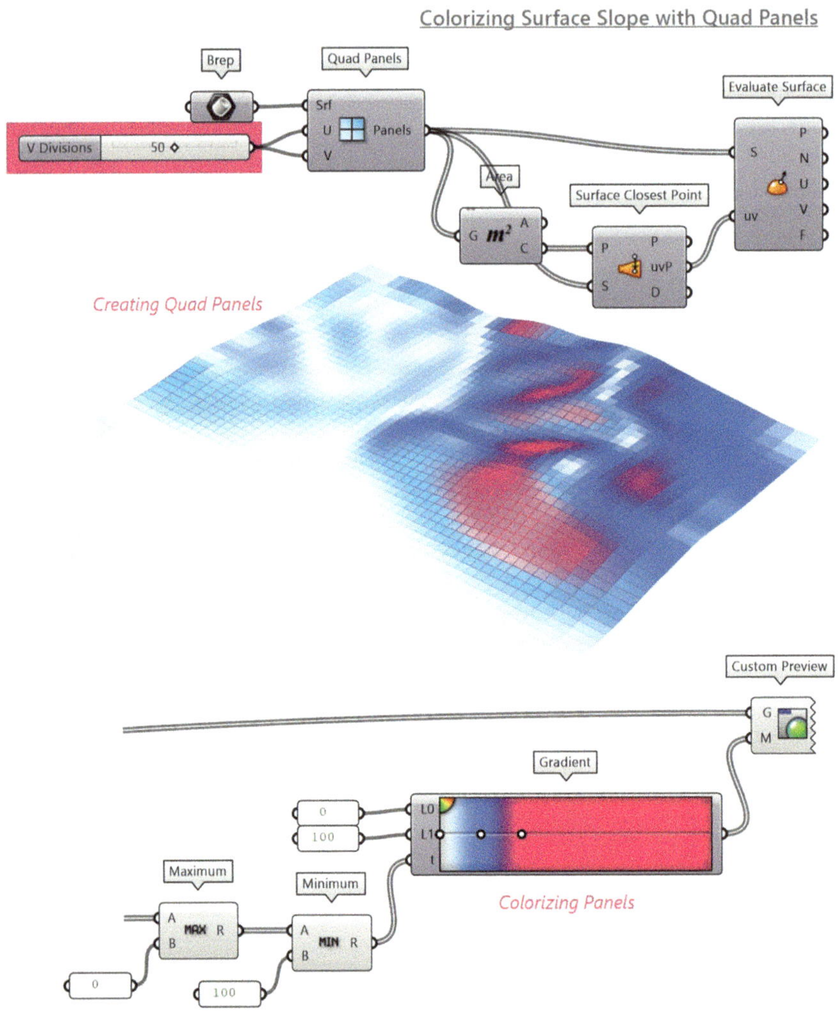

Figure 6.9
Visualizing the slope conditions of surface quad panels.

110 □

in both the *U* and *V* direction of the model, depending on how much detail wants to be shown on the model.

An *Area* component can then be used to extract center points from each panel and inputted into a *Surface Closest Point* component along with its respective panel. The purpose of this component is to get the *UV* parameters which can then be inputted into the *Evaluate Surface* component to find the *normals* just like with the script for the mesh terrain model.

With that operation applied to the surface terrain model, the script can apply the previous steps used to visualize slope percentage from the mesh terrain model, with the only other difference being the use of a *Custom Preview* component to colorize the quad panels from the beginning of the script.

Now that the different slope conditions have been colorized, it is also possible to begin filtering the information for future analysis potential on specific ecosystem services and landscape performance. In Figure 6.10, the script demonstrates how the *Quad Panels* can be used to create boundaries around a specific slope percentage parameter. The first step is establishing what slope percent to filter at by using the *Larger Than* component with a *B* input value to serve as that threshold, in this instance *20*. The resulting Boolean pattern from this operation can then be inputted into a *Cull Pattern* component for the *Quad Panels* created earlier in the script.

The filtered panels can then be merged to create larger surfaces with the *Brep Join* component. These merged surfaces can then have their outside boundary curves extracted with the *Brep Edges* compnent and then have the *Naked (En)* output connected to a *Curve* component to display on those curves.

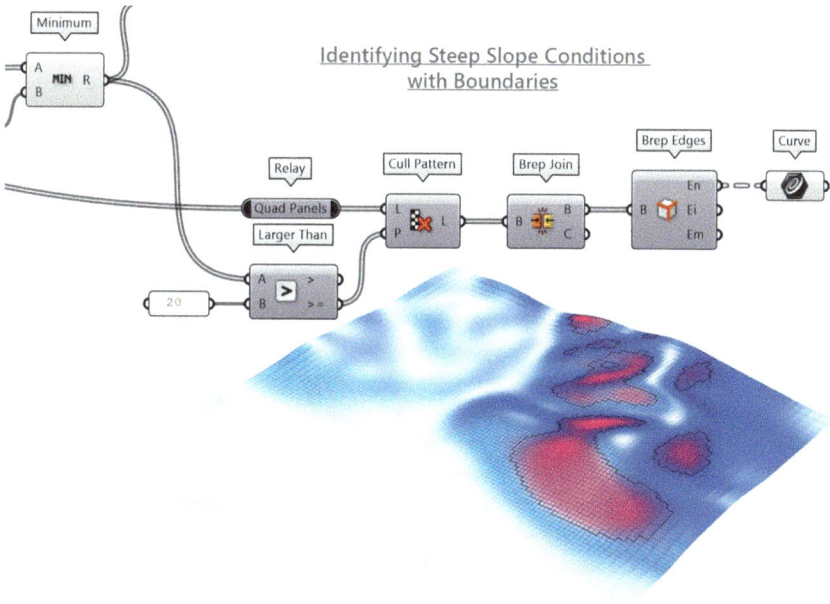

Figure 6.10
Highlighting a specified slope parameter with boundary curves.

VISUALIZING TERRAIN ASPECT

Terrain aspect is useful to understand which direction different faces of a terrain model are facing. This becomes very informative for planting communities, as surfaces that face north and east will be cooler and receive less sunlight than surfaces facing south and west receiving more hours of sunlight in the Northern Hemisphere.

Chapter Prerequisites

To best prepare for this chapter and ensure proper overlay of data, continue to utilize the terrain model, either mesh or surface, generated from the Terrain from GIS chapter. Remember to keep the terrain model at its geo-location if baked into Rhino instead of moving it to the c-plane origin (0,0,0).

Modeling Methods

Once again, this chapter will demonstrate how to visualize the aspect of both a mesh and surface terrain model. Starting with a mesh terrain model, the normals will be used to determine the directional orientation of different parts of the terrain. These normals will be measured against cardinal degrees created by the *Rotate (VRot)* component in Figure 6.11. These angles can then be converted into degrees in preparation for the *Gradient* component. If there are obscure angles in the mesh model, the *Angle* component may turn red, but that will not prevent the script from functioning properly.

The values coming out of the *Degrees* component in Figure 6.12 can then be inputted into the *Gradient* component with a *0* value for the lower limit and *359* value for the upper limit to cover a full *360* degrees of potential terrain face orientations. In order for the colorization to match up with the four cardinal directions (North, East, South, and West), four points will be used at 25-unit intervals

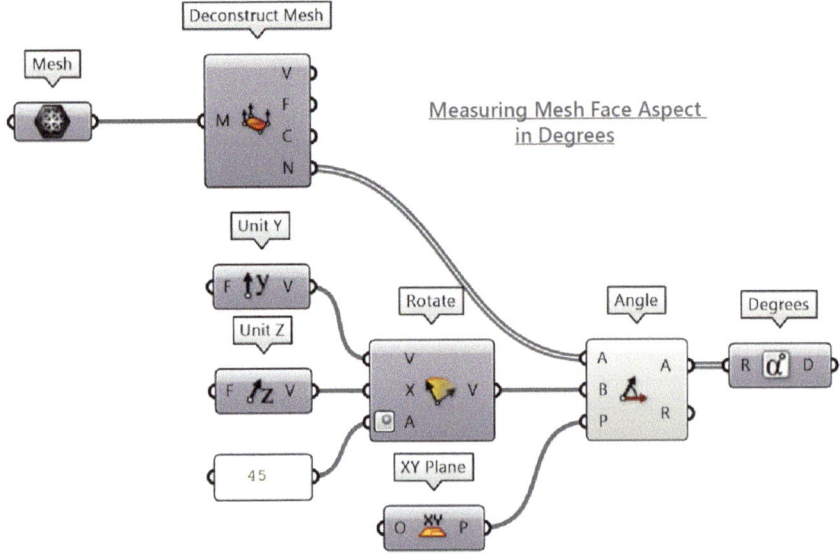

Figure 6.11
Rotating terrain face data to represent a cardinal direction orientation.

Figure 6.12
Colorizing the
cardinal directions
of the mesh faces.

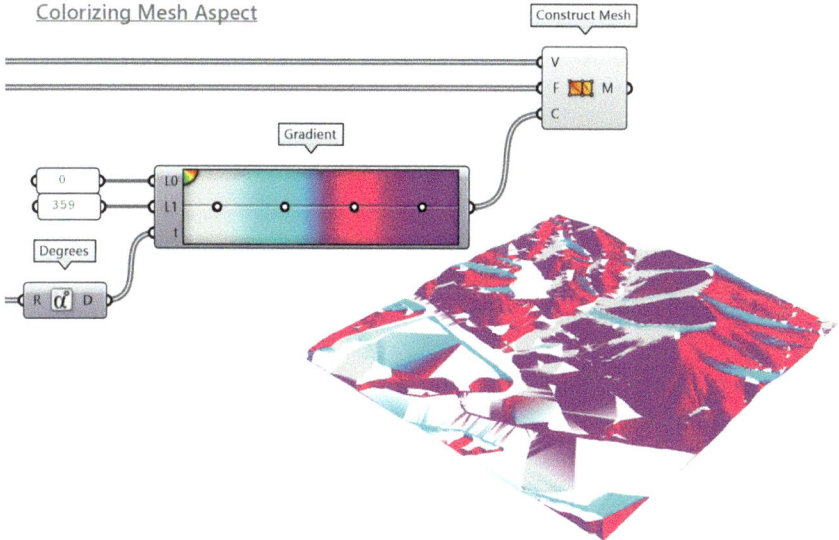

Figure 6.13
Colorizing the
aspect directions
of surface quad
panels.

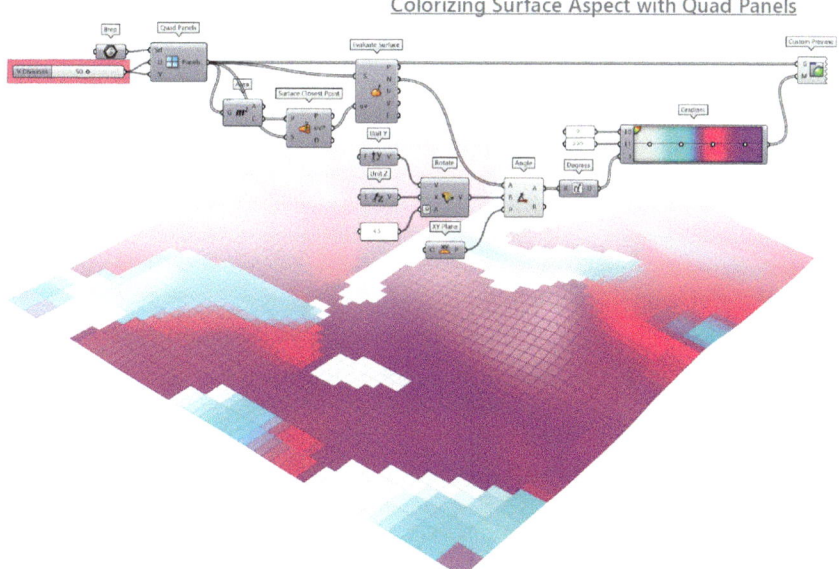

starting at 12. The 12% in the gradient will represent north, 27% for east, 62% for south, and 87% for west.

The visualization of terrain aspect for a surface model is setup the same as the previous demonstrations by turning the surface into a series of panels. The panels in Figure 6.13 will have their center points extraced using the *Area* component, which will be used to evaluate for their normal directions, just like the slope script from the previous chapter. The rest of the script will follow the same process as the previous part of this chapter for the the mesh terrain model with the addition of using the *Custom Preview* component to colorize the panels from the beginning of the script.

As demonstrated in the previous sections, this terrain aspect information can be further elaborated on or filtered to identify parts of the terrain that fall within a specific parameter as it may relate to tree-planting preferences, microclimates, or other relevant analysis of a project site.

CLIMATE DATA

Climate and weather data can be utilized for a variety of environmental, social, and economic scenario models which require unique scripting procedures to convey and substantiate climate analysis for topics specifically demonstrated in this book relating to outdoor comfort. Because of the complex, yet very informative, process this chapter will focus on the impact climate has on outdoor comfort through chart guides, wind patterns, and sun/shade projections. The modeling process will also demonstrate and emphasize the logic of diverting from conventional methods that may be using arbitrary dates in relation to temperatures such as the Winter and Summer Solstice. This is an example of a divergent method that allows for the data to direct the analysis.

Chapter Prerequisites

The *Ladybug* plugin will be needed to import, analyze, and visualize different climate datasets from a defined location from around the world, as demonstrated in Chapter 4. The introduction of this plugin in that chapter provided basic setup instructions, whereas this section will expand on how to analyze the extracted data to inform specific goals related to outdoor comfort from temperature, wind, and sun information.

Modeling Methods – Chart Guides

When modeling any climate data, the process begins by importing an EPW file from the web browser map opened with the *LB EPWmap* component. The EPW file can either be downloaded to your computer or referenced with a copy of the url, which is what is done in this instance. Once that data is referenced, it is best to disable the *LB EPWmap* component so that the browser doesn't open each time the script is loaded.

All the climate data will be coming from the *LB Import EPW* component to make chart guides, wind roses, and sun shade studies. The first chart that is going be generated to demonstrate outdoor comfort will be a psychrometric chart. These types of charts account for dry bulb temperatures (air temperature), relative humidity, and barometric pressure to determine the ideas conditions for outdoor comfort, as shown in Figure 6.14. This becomes a significant chart and process as most traditional analysis for climate usually only considers temperature and humidity as independent sets of data, whereas this process will not only combine those with barometric pressure but will also serve as a guide for future climate charting and modeling in this chapter. The temperature ranges generated from this chart will, however, be the metric for later steps, since it can be impacted by wind and shading strategies.

It is also important to ensure the correct units of measurement are being used for the project site location. By default, all the ladybug components will be inputting

Figure 6.14
Plotting outdoor
comfort data to the
Ladybug plugin's
psychrometric chart.

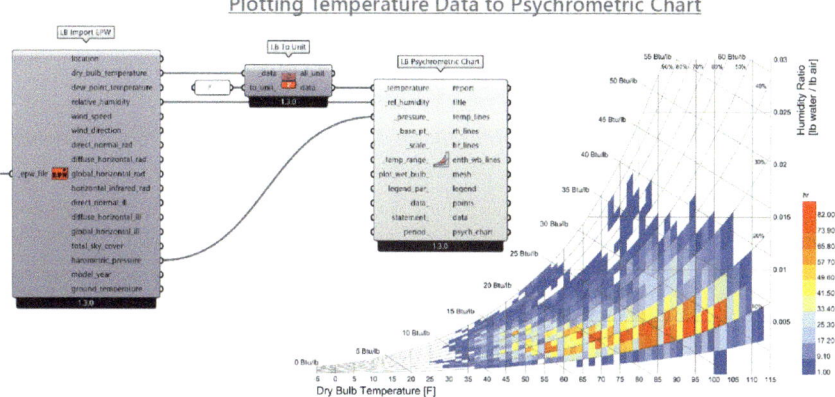

Figure 6.15
Visualizing the
comfort zone of the
psychrometric chart.

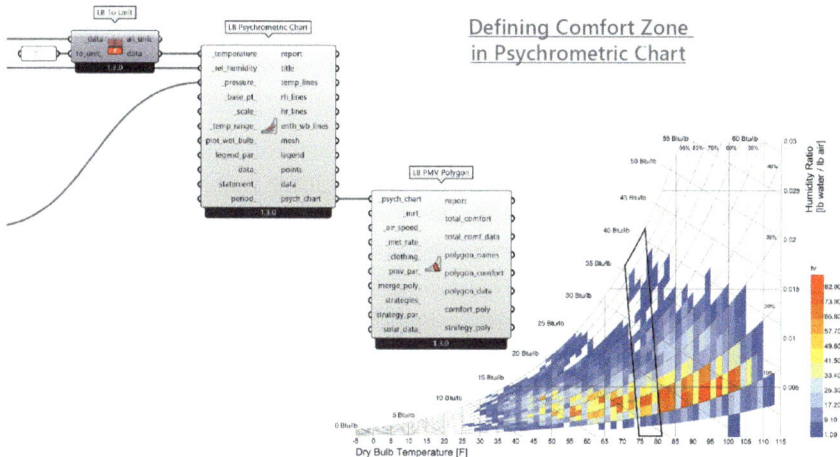

and outputting degrees Celsius and metric units. For this demonstration, degrees is converted to Fahrenheit.

The resulting data from this chart will generate a heat map indicating the number of hours in which conditions match with corresponding air temperature, humidity, and barometric pressure. Although informative, the chart still does not indicate what ideal conditions are depending on these three factors. In order to do so, connecting the *LB PMV Polygon* component from the *psych_chart* output will generate a bold polygon, shown in Figure 6.15, for the range of comfortable outdoor temperatures as humidity and barometric pressure change. According to the psychrometric chart, outdoor comfort for the Las Vegas region ranges between air temperatures of 75–81 degrees Fahrenheit with low humidity (10%) or 70.5–77 degrees Fahrenheit with high humidity (90%). The majority number of hours within this ideal comfort zone are in the low humidity range with slightly higher temperatures.

Another climate factor that can significantly impact outdoor comfort is wind speed. The psychrometric chart does not directly account for this variable; however, the *LB PMV Polygon* component has the option of including those values to adjust the comfort zone. Wind speed data from the *LB Import EPW* component can be extracted using the *LB Deconstruct Data* to separate the values from the metadata (header information) and then used to find the annual average, low and high wind speeds. This input of data for the *LB PMV Polygon* component requires the units to remain in meters, so no conversion is necessary.

With the average annual wind speed (4.5 m/s) inputted in Figure 6.16, the comfort zone now increases to a temperature range between 84 and 91 degrees Fahrenheit. The low wind speed (0 m/s) results in a range from 74 to 81 degrees Fahrenheit and the high wind speed (20.6 m/s) results in a range from 86 to 92.5 degrees Fahrenheit. A number slider that ranges in these low and high values can also be inputted to determine a specific comfort zone for a specific known wind speed. As demonstrated by these different wind variables, these charts do an excellent job in communicating the affect outdoor air circulation has in impacting outdoor comfort.

For locations that are hot and dry, in which outdoor air temperatures are challenging to mitigate, ensuring proper air circulation through the site becomes critical in making outdoor comfort more attainable when temperatures are often high. This fundamental concept can be further compounded with the addition of trees for shade. Chapters in the third section of this book will expand more on outdoor comfort strategies that include evapotranspiration and windbreaks.

A new color legend can also be created to apply a new gradient scheme to communicate these values more effectively, as demonstrated in the script of Figure 6.17. The number slider for the multiple inputs of this operation will be highly contingent on the location values and type of data being charted, so make sure to adjust as needed to provide logical or appropriate intervals. The expression used for the *Upper Limit (L1)* input is $x-1$, since the *Series* component creates a range of values between *0* and *10*.

These comfort models can now be used to guide additional charting procedures to determine the specific dates considered comfortable and having thermal stress.

Another way of modeling outdoor comfort is with the *Ladybug Comfort* components that use the universal thermal climate index (UTCI), an industry standard model, to output a range of comfort considerations. The *LB UTCI Comfort* component requires similar inputs as the previous comfort chart, excluding barometric pressure, but also needs air temperature to remain in Celsius. It is optional to also include the mean radiant temperature (*mrt*), which considers direct solar shortwaves on individuals as well as longwaves with the sky. These components do not create charts, but rather different comfort scenarios with different ranges in values from the *comfort, condition*, and *category* outputs of the *LB UTCI Comfort* component.

The different outputs can then be inputted into the *LB Hourly Plot* chart component, depending on the specific conditions intended to be communicated. The *comfort* output will provide values of *0* (*thermal stress*) or *1* (*no thermal stress*) for all 8,760 hours of the year. The *condition* output will further specify the stress classifications by indicating cold stress with a *−1*, hot stress with a *1*, and *0* indicating comfort. Lastly, the *category* output defines the different stress types with specific levels ranging from *−5* (*Extreme Cold Stress*) to *+5* (*Extreme Heat Stress*), shown in Figure 6.18.

Figure 6.16
Demonstrating the
impact different
wind speeds have
on comfort zones.

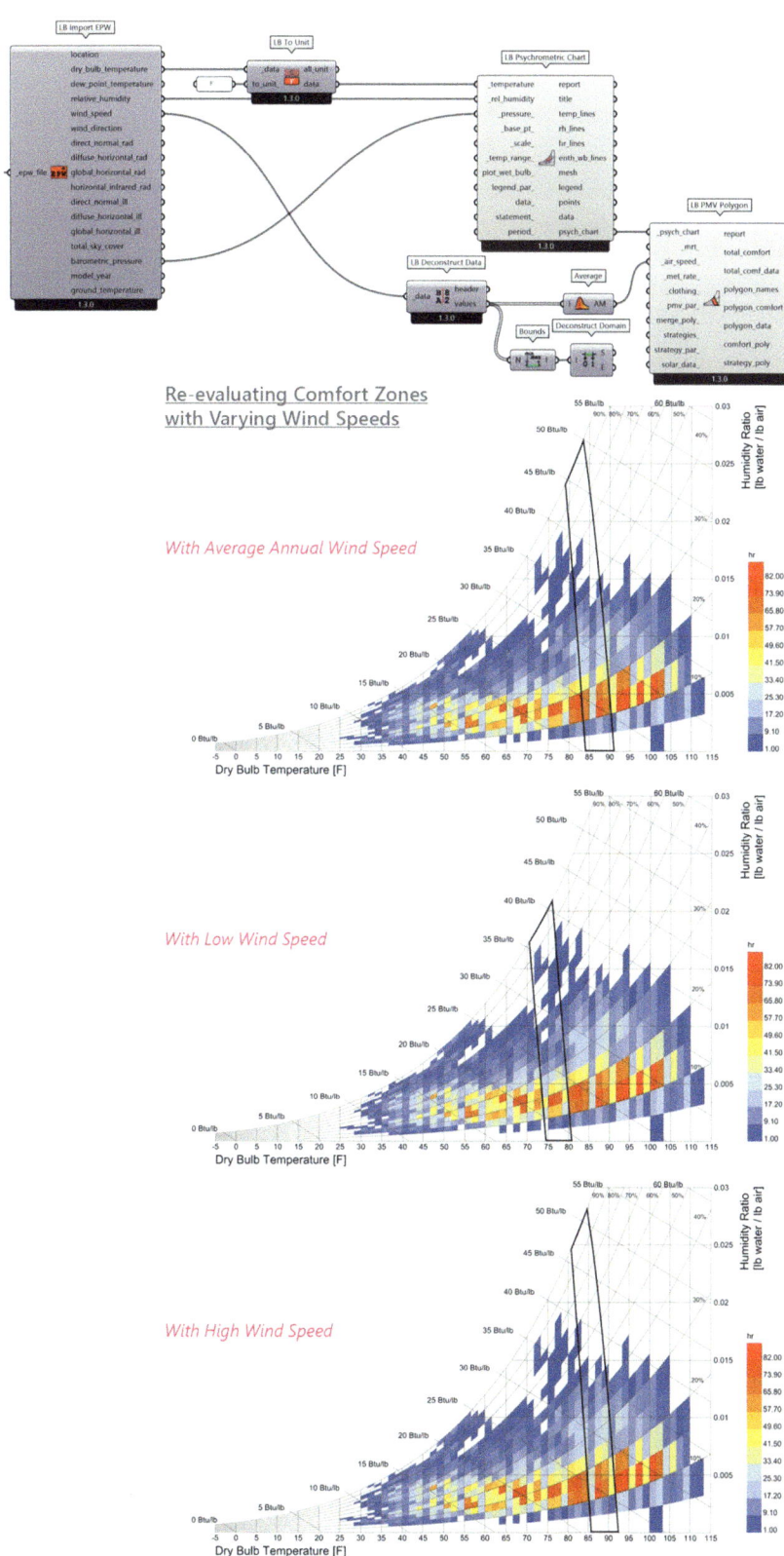

Re-evaluating Comfort Zones
with Varying Wind Speeds

With Average Annual Wind Speed

With Low Wind Speed

With High Wind Speed

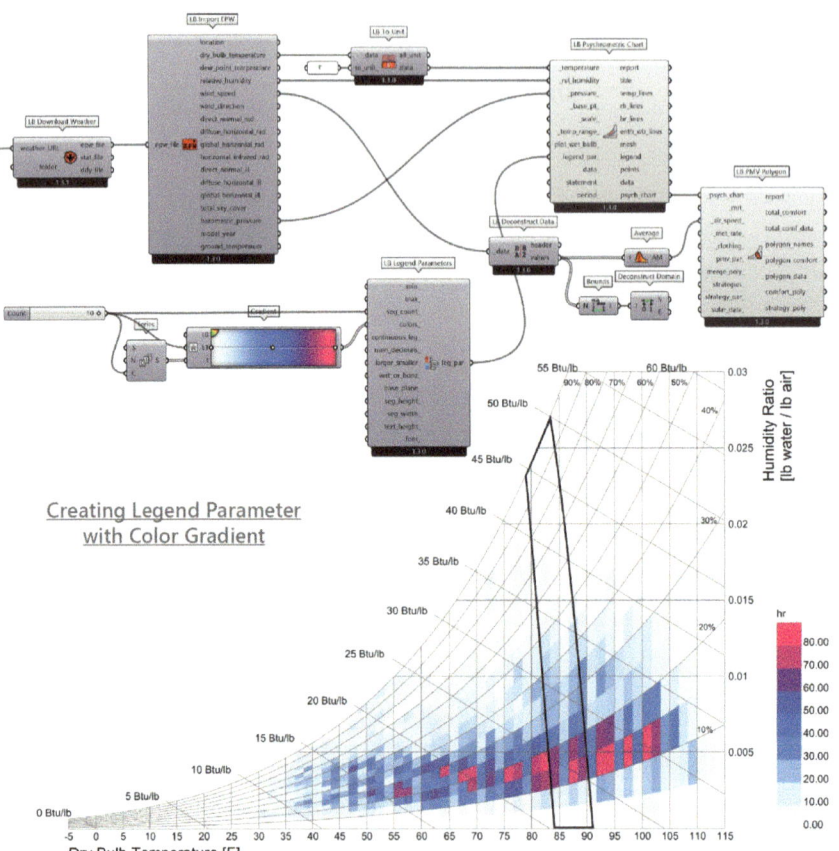

Figure 6.17
Creating a custom
color palette for the
psychrometric chart.

These values also include the corresponding temperatures in Celsius associated with the different stress levels for better understanding of the impact on outdoor comfort. The chart images in the figure show the different colorized charts for these comfort outputs respectively.

As stated earlier, the number slider for the *Legend Parameters* component should change as the number of categories change. Trees have been well documented in reducing air temperatures and making outdoor spaces more comfortable, most notably in LAF's Landscape Performance Series case study briefs. Depending on the region and density of tree canopy, air temperatures can be reduced anywhere between 3 (Martin & Colter, 2014) and 10 degrees Fahrenheit (Chanse & Salazar, 2012).

When referring to the UTCI legend for outdoor comfort, reducing the air temperature by that many degrees can alter the heat stress for levels that fall within the *+1* and *+2* categories. With a UTCI of *0* (*No Thermal Stress*) equating to 78.8 degrees Fahrenheit, any time when air temperature reaches up to nearly 89 degrees can essentially be brought down to a thermal comfort from trees according to the LAF studies. Based on the conversion of Celsius to Fahrenheit, the UTCI *+1* (*Slight Heat Stress*) has an air temperature from 26 to 28 degrees Celsius (78.8°–82.4° F) and UTCI *+2* (*Moderate Heat Stress*) has from 28 to 32 degrees Celsius (82.4°–89.6° F).

Figure 6.18
Charting different
comfort models.

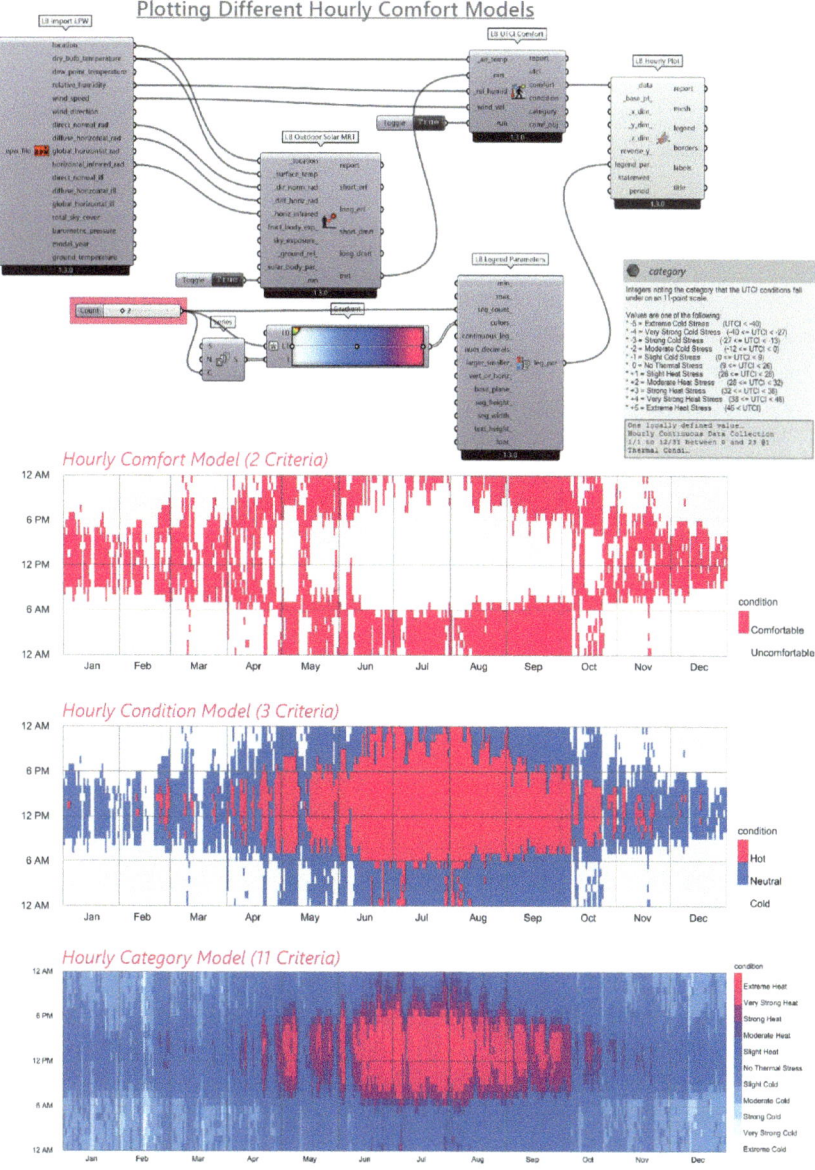

When applying that logic to the comfort model charts, using statements, the number of hours, or percentage of the year can be increased in comfort with the use of trees. In Figure 6.19, different statements are used to represent no trees and a low and high extreme in UTCI reduction. To model out a scenario where no trees are used to increase thermal comfort, a statement of $-1<a<1$ or 78.8 degrees Fahrenheit is used to output only 0 values. The statement $-1<a<2$ will chart out all the hours classified with a 0 and a 1 to model the potential for temperature reduction between 0 and 3.6 degrees Fahrenheit from trees. Then, for more extreme temperature reductions, a state of $-1<a<3$ can be used to look at hours that have temperatures between 78.8 and 89.6 degrees Fahrenheit or values from 0 to 2.

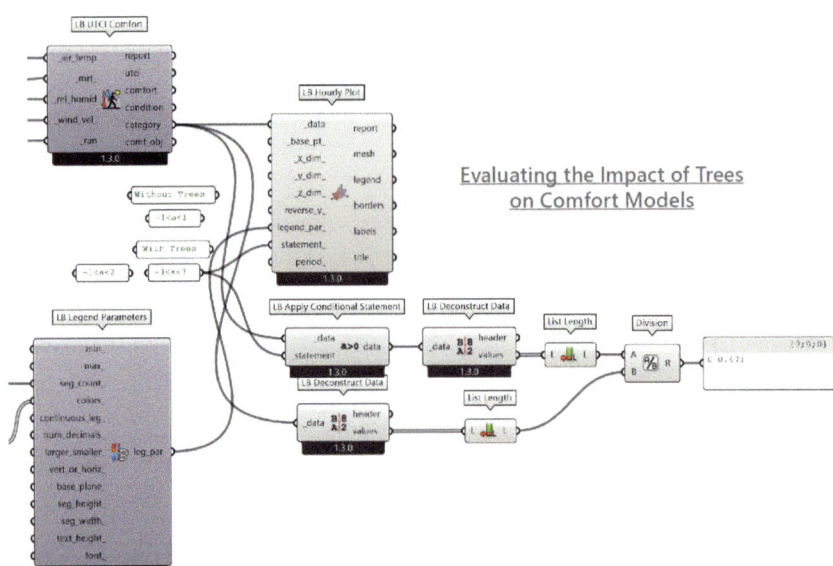

Figure 6.19 Integrating conditional statements to represent the impact of trees on outdoor comfort.

These same statements can also be applied to the *LB Apply Conditional Statement* component to calculate the number of hours since the *LB Hourly Plot* component is unable to do so. The number of values coming out of that operation can then be divided by all 8,760 hours to determine a percentage of thermal comfort throughout the year. According to these models for Las Vegas, the percentage of annual hourly comfort when trees aren't included is 36.8%. For the two statements considering trees and their range of temperature reduction, the percentage of hourly comfort increases to 41.4% and 47.1%, respectively.

For this comfort modeling and the later scripts in this chapter, the use of trees will be abstract and only perceived as data; in other words, there will be no physical modeling of trees in these scenarios. In the later sections of the book, trees models will be applied.

Modeling Methods – Wind Rose

Wind rose charts can be very helpful in determining where and when cold strong winds enter a project site to strategize against with trees as windbreaks. The same UTCI comfort model is also used for this charting process in Figure 6.20; however, other UTCI categories or other comfort models can also be applied using these same principles. The *LB Wind Rose* component will use the *category* output from the *LB UTCI Comfort* component for the *data* input of the *LB Wind Rose* component. This component also requires a wind direction input that can be retrieved from the *LB Import EPW* component at the start of the script.

Since the intention of this demonstration is to determine cold stress values, a statement of *a<0* is used to determine all the hourly values when UTCI is in a range from −1 to −5. Unlike before, when a *LB Hourly Plot* component was used to filter out data, a different operation is used since the data will already be charted with the *LB Wind Rose* component.

Figure 6.20
Charting cold
stress factors to the
Ladybug plugin's
wind rose chart.

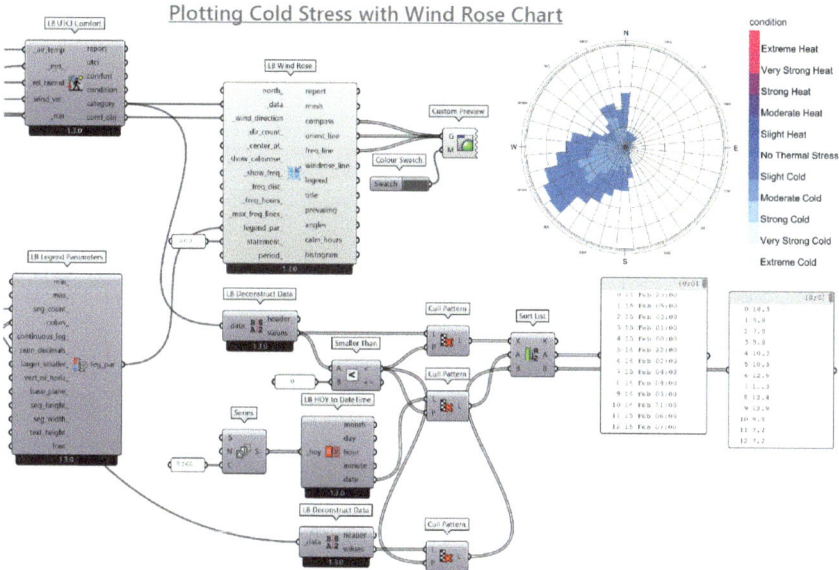

The data is first parsed out with the *LB Deconstruct Data* component to separate the values from the metadata, as previously demonstrated, to input into a *Small Than* component with a *B* value of *0*. This operation mimics the *a<0* statement for the wind rose chart component. This will create a Boolean pattern that identifies which values coincide with that statement that can be used to parse out values that are only considered cold stress along with their respective date, hour, and wind speed (m/s). To get the HOYs, a series component is used to generate sequential numeric values with increments of one for a total count of 8,760. These intervals are then converted to the corresponding date and hour using the *LB HOY to DateTime* component. From the *Smaller Than* component, the Boolean pattern is used for the UTCI values, dates, and wind speed (same input for the *wind_direction* input on the *LB Wind Rose* component).

Next, the script will sort the data by the most extreme cold stress values along with the other corresponding data using the *Sort List* component. This component will sort from the lowest values (*–5 UTCI – Extreme Cold Stress*) to the highest values (*–1 UTCI – Slight Cold Stress*).

The resulting chart and data can then be used to strategically place tress as windbreaks and which species would work most effectively, depending on the majority of dates for these extremes now that the direction from which wind is coming from is known. Later modeling will demonstrate this process more thoroughly.

Modeling Methods – Sun Path
The sun path chart is a useful tool to not only shown which times of the year heat conditions are extreme but can also generate shadow projects from physical elements in the environment (buildings and trees). These shadow projections can be very informative and influential for directing outdoor comfort. Because case studies from LAF have not been able to demonstrate the impact trees have on relative humidity and barometric pressure, it would be inconclusive to use the comfort

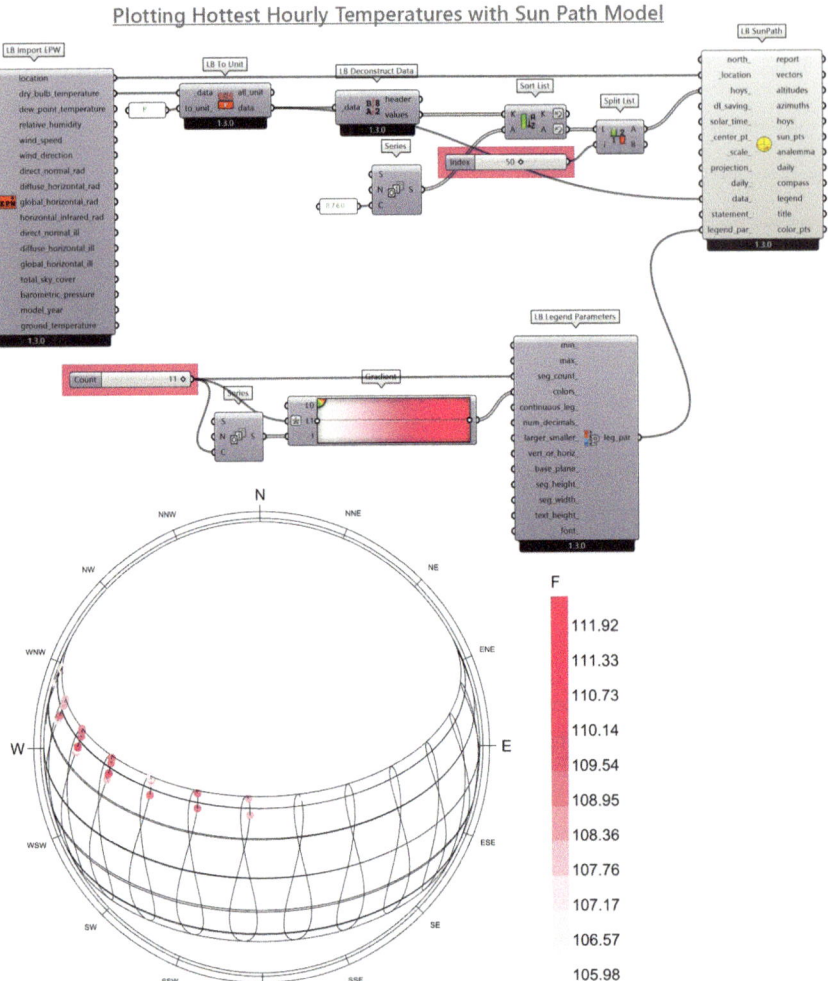

Figure 6.21
Plotting air
temperature data to
the sun path model.

model as the data input for the *LB SunPath* component, as shown in Figure 6.21. Instead, air temperature or the *dry_bulb_temperature* will be used as the data input once converted to Fahrenheit and put through an operation to identify only the hottest times of the year.

Like the previous script for wind rose charts, the converted temperature can be parsed and sorted along with the corresponding HOYs. Since this script will be looking at the hottest temperatures of the year, the outputs of the *Sort List* component need to be *reversed* so that the data begins with the highest values instead of the lowest. The next aspect to consider is how many times of the year is necessary to model the sun path for. For this demonstration, a number slider with a value of *50* is used to split the sorted list into the top 50 hottest times of the year, but can be adjusted at any time to model less or more values. The top 50 hottest times (HOY) is then inputted into the *hoys_* of the *LB SunPath* component to generate solar positions and sun vectors for those specific times. Another important reason for this order of operations is because, in some instances, an extremely hot time can be after the sun sets and therefore not modeled out since this chart will only show data for times between sunrise and sunset.

Figure 6.22
Projecting shadows
from trees and
buildings with the
sun path model.

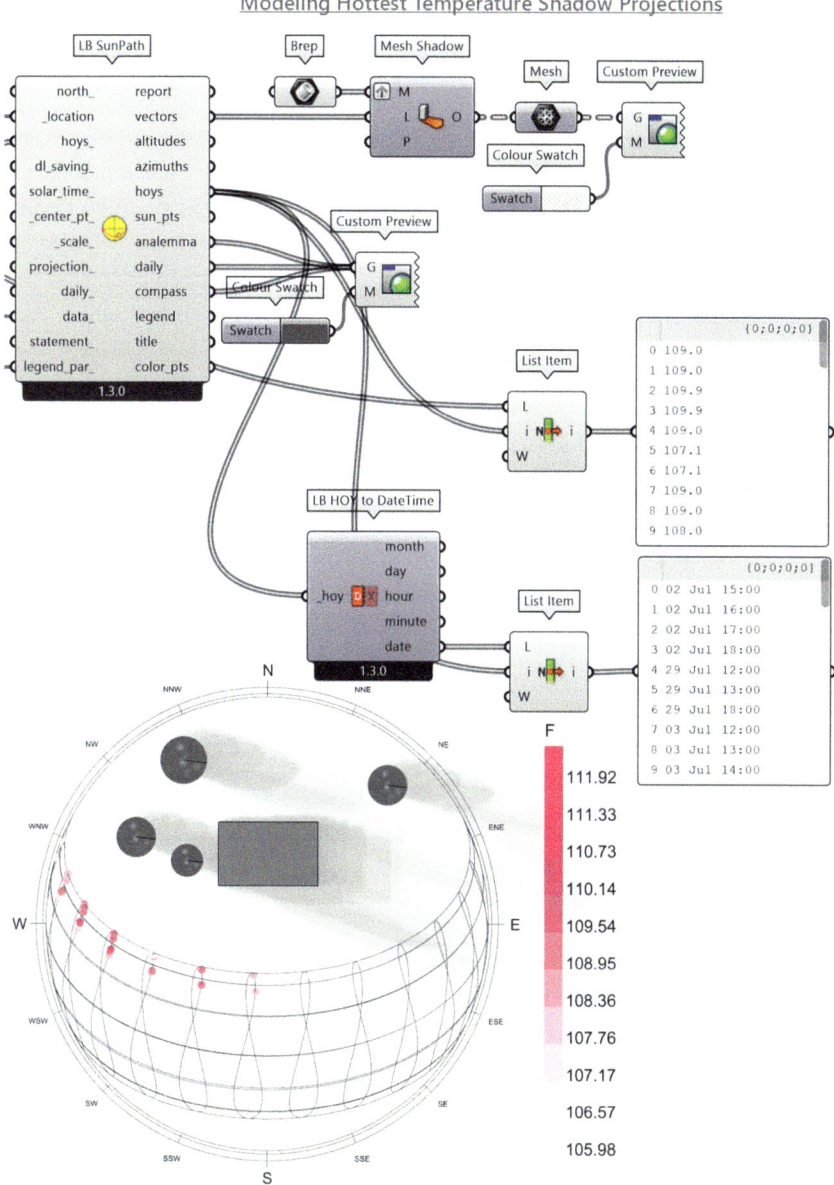

Now that sun positions and vectors have been generated, they can be used to cast shadow projections from elements within a project site, as demonstrated in Figure 6.22. This demonstration uses hypothetical buildings and trees to simulate this information. With the *Mesh Shadow* component, referenced geometry is connected as *grafted* inputs so that each geometry projects a shadow from each sun position. If this is not done, only the last referenced geometry will project a shadow from all the sun positions. For graphic clarity, the output of curves from this component is converted into meshes so that a transparent grey swatch can be applied to visualize overlapping shadow projections.

The model is made more comprehensive by using the *hoys* output (an implied index number) to filter out both the air temperatures, from the *values* output of the *LB Deconstruct Data* component and from the conversion of *hoys* to a date.

This sun path chart once again reinforces the goals of this chapter to critically analyze data and use it as a discovery or divergent process to determine when temperature extremes occur to best strategize for outdoor comfort. Although the Summer Solstice (June 21) does have the least amount of shadow with it being the longest day of the year, in many cases the highest temperatures which affect outdoor comfort more significantly occur in July and August.

With these effective inventorying methods of a project site and terrain characteristics and in combination with the other landscape conditions that have been modeled, different landscape performance calculations and metrics can be applied to further evaluate not only the relationship between these systems but also for their ability to produce ecosystem services. In the following chapters, various topics of landscape performance will be modeled to output robust datasets that can be used in reference to environmental, social, and economic goals and outcomes. It is through that process that makes these basic operations fundamental to any procedure of analyzing a project site through parametric modeling and incorporation of quantified information with spatial characteristics of any area.

REFERENCES

Chanse, V., & Salazar, J. (2012). *Central Wharf Plaza*. Landscape Performance Series. Landscape Architecture Foundation. https://doi.org/10.31353/cs0270

Martin, C. A., & Colter, K. R. (2014). *George "Doc" Cavalliere Park*. Landscape Performance Series. Landscape Architecture Foundation. https://doi.org/10.31353/cs0730

SECTION 3

7   Runoff Rates – Rational Method

One of the methods to evaluate hydrology for landscape performance is through the measured calculation of surface runoff rates and volume that occur during and after a rain event known as the Rational Method. With the computation of these two runoff factors, areas can be analyzed for the impact a terrain's characteristics of land cover, slope, and soil profile have throughout a project site. As a result, areas can be identified as high contributors to runoff where green infrastructure, nature-based solutions, or other stormwater management interventions can be implemented to reduce, mitigate, filter, and convey surface runoff.

Runoff rates, measured in cubic feet per second, guide the sizing of weirs, pipes, and swales with the intention of slowing down the velocity of flow so runoff can be intercepted and infiltrate into the landscape. With the calculation of runoff rates, volume can also be established to dictate the size of detention basins, constructed wetlands, and rain gardens to collect and temporarily hold the runoff so that the same functions may occur with the landscape. When deciding which of the hydrology methods to use for a project site, it is important to consider both the size of the drainage area – recommended less than 200 acres – and the significance of topography. The other method covered in the next chapter can be applied to larger project sites; however, it does not consider slope conditions of the topography as a significant factor.

GOALS

The main goals for modeling the rational method on a project site is to measure stormwater runoff for the analysis of contributing factors and how they may compare to a baseline condition. This analysis can then inform stormwater management practices and design interventions. Below is an outline of how these goals will be achieved:

1. Aggregate the different land cover surfaces, slope percentages, and soil textures
2. Cross-reference the unique terrain characteristics with the calculator's metrics
3. Calculate both the peak flow rate and runoff volume from an assigned rain event
4. Evaluate the primary contributors to high volumes of runoff
5. Delineate the project site into different watersheds with their associated hydrologic features

DOI: 10.4324/9781003208020-10

PERFORMANCE METRICS

Computing the peak runoff rate with the rational method is generally for project drainage areas of around 200 acres or less. The calculation for this method assumes that the peak runoff rate is equivalent to the rainfall intensity multiplied by a land cover characteristic, known as a coefficient, and the total drainage area. The equation reads as $q = CiA$.

The peak runoff rate (q) is measured in cubic feet per second. The coefficient (C) is a dimensionless value that ranges between zero and one and can be seen in Figure 7.1. The rainfall intensity (i) is measured in inches per hour (iph) that represents the duration, frequency, and time of concentration for the rain event. Municipalities will usually have a specific rainfall intensity and rain event frequency (2–50-year rain event) that this equation needs to use for management and design practices. Then, the area (A) of the drainage area is measured in acres.

The drainage area can vary from a space within a project site that is only a few hundred square feet to several square miles of a watershed. These drainage areas will rarely be contained within arbitrary project boundaries, so it is still crucial to consider the entirety of drainage area. The project site will need to either accommodate the additional runoff that is entering the site or consider how any alterations to the project site's runoff potential may influence water leaving the site to downstream areas. This can influence the ecology and access to water resources in both a positive or negative way.

The Arizona State University Orange Mall Green Infrastructure Project (Cheng & Trakas, 2020) is a great example of how the rational method was used to measure the discharge rates of the different bioretention basis on the ASU campus, as shown in Figure 7.2.

Measurements were calculated for multiple rain events and evaluated how different rainfall intensities inundated the basins relative to their performance.

Urban Areas	Coefficients (C)
Downtown Business	0.70-0.95
Neighborhood Business	0.50-0.70
Single-Family Residential	0.30-0.50
Detached Multi-Unit Residential	0.40-0.60
Attached Multi-Unit Residential	0.60-0.75
Suburban Residential	0.25-0.40
Apartment	0.50-0.70
Light Industry	0.50-0.80
Heavy Industry	0.60-0.90
Parks/Cemeteries	0.10-0.25
Playgrounds	0.20-0.35
Railroad Yards	0.20-0.35
Unimproved	0.10-0.30

Urban Surfaces	
Roofs	0.80-0.95
Asphalt/Concrete Pavement	0.75-0.95
Gravel	0.35-0.70

Rural and Suburban Areas	Sandy loam	Clay and silt loam	Clay
Woodland			
Flat (0-5% slope)	0.10	0.30	0.40
Rolling (5-10% slope)	0.25	0.35	0.50
Hilly (10-30%)	0.30	0.50	0.60
Pasture and Lawns			
Flat	0.10	0.30	0.40
Rolling	0.16	0.36	0.40
Hilly	0.22	0.42	0.60
Cultivated or No Plant Cover			
Flat	0.30	0.52	0.62
Rolling	0.40	0.60	0.70
Hilly	0.52	0.72	0.82

Figure 7.1
Matrix of rational method coefficient values for different land cover typologies and soils.
Source: Strom, 2013.

Figure 7.2
Photograph of the
campus basins that
were measured
with the rational
method equation.
Source: Cheng &
Trakas, 2020.

MODELING METHOD

Many of the initial steps with this chapter and some of the future ones will be taking script operations from the Section 2 of this book, so although it may seem redundant, the previous inventory scripts can now be compiled for the analysis of their ecosystem services and landscape performance. Because of this, some operations may not be as detailed in instruction; however, if there are any difficulties with the script, those chapters can be referred to for more elaboration. There will also be references to where these operations are repeated from as well.

Figure 7.3 begins the process of importing the site boundary's soil profile geometry from Web Soil Survey (WSS) using the *@it* plugin. This operation will require the referencing of the soil profile shapefile that was refined from a GIS software program. The activated import and visualization components from this plugin can then be scaled to the correct unit since it was initially measured in meters, but this model is in feet. A quick conversion *Factor* (*F*) of *3.2808* for the *Scale* component will put it at the correct size.

Since there are some enclosed curve geometries on this project site, a List Item component is used to essentially remove the second branch (duplicate geometry) of data. The geometry can then be flattened at the output of this component since its current branching structure is not necessary for the future steps.

Data from the WSS download contains both the spatial geometry of the soil profile as well as the metadata associated with the profile curves. This data needs to be linked to that geometry in Grasshopper as shown in Figure 7.4. Another plugin component *ExcelDynamicRead* is used to import Microsoft Excel file from WSS that contains both a referencing mukey (*MapUnitKey*) and soil texture name. Both these columns of data will follow a similar operation of using a *Flip Matrix* and *List Item* component; however, the mukey column will require a *0* value for the *Index* (*i*) input and the soil texture column will require a *1* value input for their respective *List Item* component. The Excel table also contains two header rows that can be separated from the data to the *B* output from a *Split List* component. A number slider of *2* is used because of those two header rows; however, that is contingent on the layout of the Excel file or can be removed altogether to avoid this additional set of components in the script. The Excel table is also referenced in this figure to demonstrate the use of those index values for the *List Item* and *Split List* components, and panels for each B output visualizes the two columns of data.

The MapUnitKey *B* output is now used to cross-reference with the shapefile mukey values by using the *Match Text* component. From the *DataViz@it* component's *Val* output, a *List Item* is connected with an *Index* (*i*) input set to *3* since that is the index column that contains that data. Again, this needs to be verified first from the *Imp@it* component's *Features* (*F*) output. The *Text* (*T*) input for the *Match Text* component needs to be *flattened* to remove the current branching structure, while its *RegEx* (*R*) input needs to be *grafted* to treat each mukey value separately when cross-referencing. This component will generate a Boolean pattern to parse the soil data and any other set of data in relation to the geometric soil profile curves.

This pattern is first used to restructure the curve geometry from the previous figure's *List Item* component with a *Cull Pattern* component that has a *flattened* output. Another *flattened Cull Pattern* component is used for the *B* output of the soil texture names. Even though each culled pattern is flattened, the intent with it is to reorder and structure both the geometry and metadata in a matching sequence. This culled list is inputted into tow *Match Text* component since silt loams and loams are classified the same in the calculator, but the WSS profile will classify them separately. The first *grafted* panel input needs to list the data in the order of *Sandy, Silt Loam*, and *Clay*. The second *grafted* panel input needs placeholder text (*xxxx*) for the first and third list item but have *Loam* as the second one.

Importing WSS Shapefile of Soil Profile

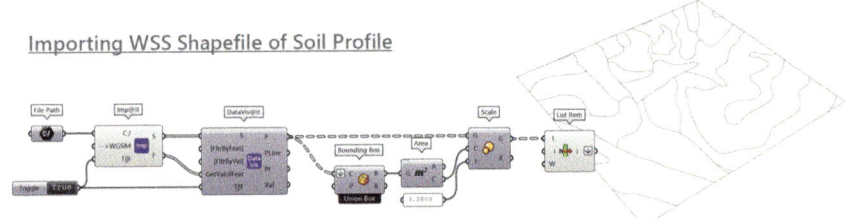

Figure 7.3
Importing soil profile shapefile to Rhino with the *@it* plugin.

Soil Texture Metadata

Cross-Referencing Soil Texture Types with Soil Profile

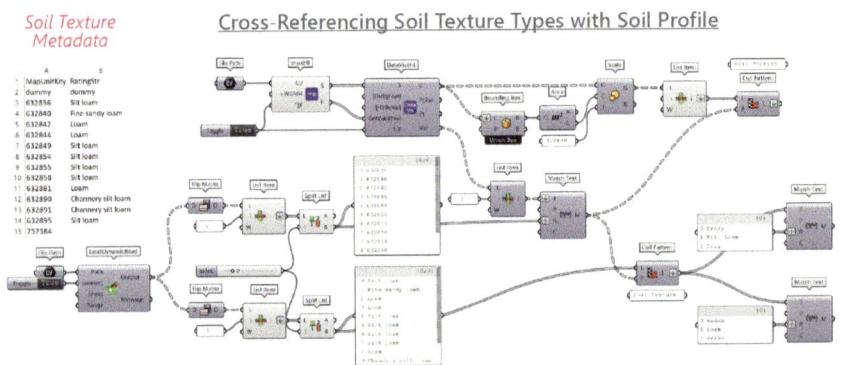

Figure 7.4
Cross-referencing soil survey textures from an Excel file with the soil profile curves.

This series of operations, Figure 7.5, comes from Chapter 5's section on creating a landform from contour lines. The contours are once again scaled to feet, raised to their corresponding elevation, and then extracts *flattened* points with the *Control Polygon* component to generate a *Delaunay Mesh*.

Since it is more effective to calculate the runoff from the project site from a surface geometry, the Delaunay mesh is then converted into a surface geometry, shown in the bottom image of Figure 7.5. After extracting the mesh terrain's attributes, they are used to create a planar surface with a set density of points shown with the *Divide Surface* component and a number slider to adjust accordingly. These flattened points are then projected back on to the mesh terrain with any null points removed, to then connect to a *Surface From Points* component. As stated in that section of Chapter 5, the *U* input for this component needs an *expression* of *x+1* while connecting the same number slider used for the *Divide Surface* component.

This next operation in Figure 7.6 has two main benefits. The first is the subtraction of intersecting or overlapping surfaces (Breps) and the second is for a more effective evaluation of points in these soil profiles. The subtraction process begins from the culled list of flattened Soil Profile curves. They can be *extruded* and *capped* for the *Solid Difference* component to remove any duplicate geometries. The *Unit Z* value for the *Extrude* component is irrelevant since they will all be the same height. These extruded surface volumes can now be 'projected' onto the recently created surface terrain by using the *Surface Morph* component.

Importing Site Topography for Terrain Creation

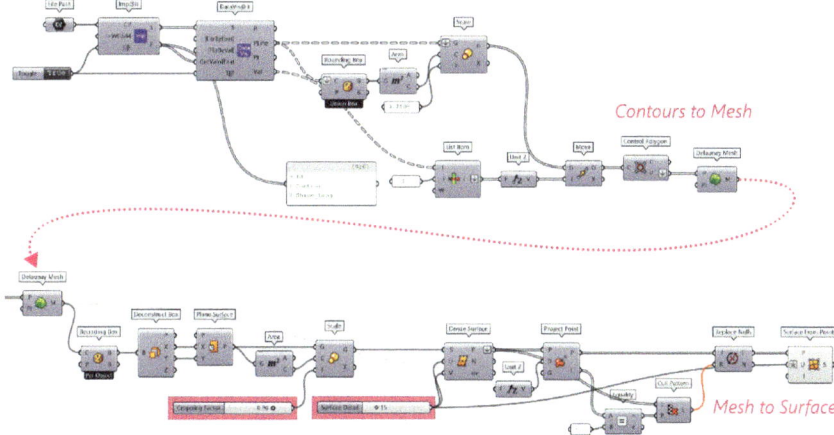

Figure 7.5
Creating an accurate and geolocated surface terrain from an imported shapefile of contour lines.

Morphing Planar Soil Profile to Surface Terrain

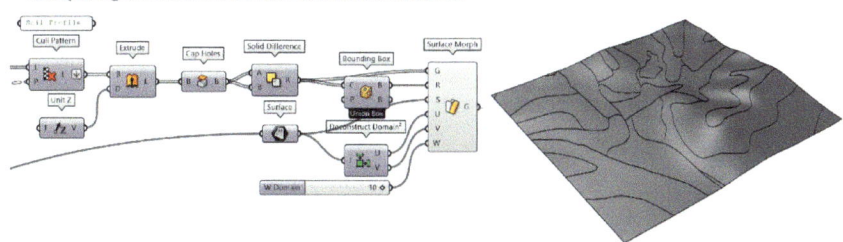

Figure 7.6
Morphing soil profile curves to the previously generated surface terrain model.

For the surface morphing transformation, a *Bounding Box* component, set to a *Union Box*, is used as the *Reference* (*R*) input while the previously created terrain surface geometry is connected to the *Surface* (*S*) input. The *U* and *V* inputs are the same as the referenced surface by utilizing a *Deconstruct Domain2* component to extract those values. The last W input value can come from a number slider, as long as it is of substantial size to ensure that in the later steps all the points of this surface are included inside it. In some cases, it may be more advantageous to construct a domain value in both the negative and positive directions to ensure those points are included. This will be elaborated on further later in the chapter.

This next step will ultimately be a visual and detail preference for the future outputs. In Figure 7.7, the surface terrain model can be converted in a variety of different panel types provided by the *LunchBox* plugin. For this demonstration, however, the *Quad Panels* component option will be used to effectively communicate this ecosystem service. Just like how the detail of the surface terrain model can be iterated from Figure 7.5, the same can be done with the number and density of panels. By extracting the *U* and *V* values of the surface (length and width) and *dividing* them by dimensional value (*100*), a set number of panels in both directions will be generated.

Another set of data that will be important for this process, as it will dictate most of the later calculations, is the center or *Centroid* (*C*) from an *Area* component. These points and their associated panel surfaces will act as the visual output of all the stormwater data, whether it be the runoff volume, watersheds, proximity to drainage lines, or the terrain characteristics.

The following images will now cross-reference the different terrain characteristics of the model (soil, land cover, and slope) with the rational method's metrics to determine the unique coefficient values throughout the site.

The first process of linking the data with these associated geometries is with the soil profile volumes in order to parse them into the different texture types, as instructed in Figure 7.8. Two *Cull Pattern* components will be connected to the *Surface Morph* component, which contains those soil profile volumes, and use the two *Match Text* component outputs, from Figure 7.4, to serve as the *Pattern* (*P*) inputs. These values can finally be *Merged* for the next step.

Now that the two soil texture groups have been merged, then they can now be *flattened* as they are inputted into the *Point in Breps* component, since this

Converting Surface Terrain to Quad Panels

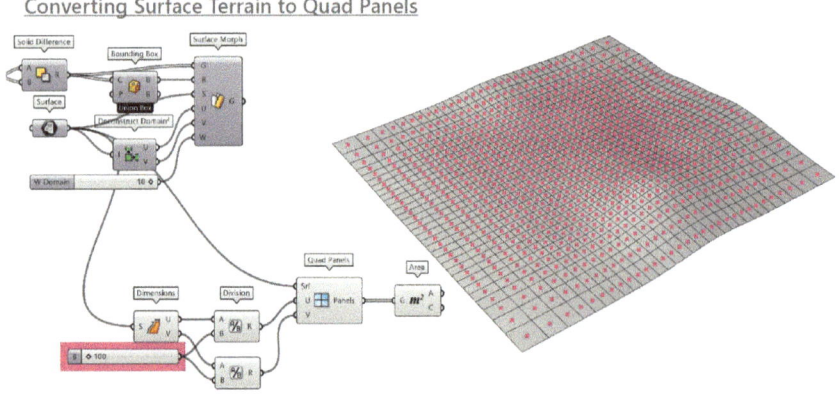

Figure 7.7
Converting surface terrain to a grid of quad panels and points.

Figure 7.8
Partitioning the
point grid into the
soil profile groups.

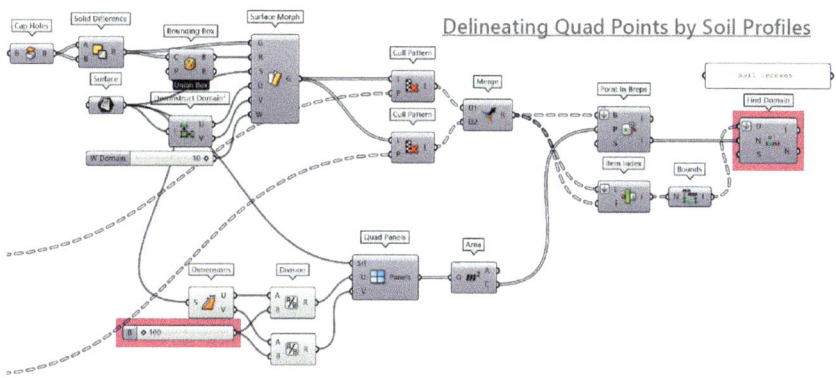

operation will formulate the new data branching structure going forward. The points are tested for inclusion or the centroids from the *Area* component. By using the *Merge* component as a *flattened* input for the *List* (*L*) input and the non-flattened *Index* (*i*) input for *Item Index* component, the soil profile groups can be converted into a series of domains with the *Bounds* component. These domains will be tested for the outputted soil volume indexes (*i*) for which the points are inside. The *Bounds* output needs to be *flattened* when connected to the *Domains* (*D*) input of the *Find Domain* component. This component has been highlighted and labeled *Soil Indexes* for easy reference later in the script.

Since there are three different soil groups (sandy, silty loam/loam, and clay) there are only three potential *Index* (*I*) outputs from this final component. For this example, the only index value output is *1*, which is representative of the second soil group (silty loam/loam). There are, however, − *1* values, meaning those points don't sit within any soil group or in this case the water areas.

Just like how the soil indexes were parsed out in the previous step, the same will be done for the land cover indexes once the land cover types are morphed to the terrain surface, as shown in Figure 7.9. The Rhino modeling of the land covers follows the same process as the 'Surface Typologies' section in Chapter 5. But what is most important in the referencing of these Rhino geometries is the type and order. Even though the land cover types were initially curve outlines of their respective classifications, those curves need to first be extruded into closed Breps before referencing into the script. The three-dimensional volume of this geometry is what will be evaluated for point inclusion as well, just like the previous step.

The other important factor when referencing these geometries is the specific order and maintaining that order consistently throughout the script. Although it is not necessary to utilize this specific order of land cover classifications, the order being used is the same as the land cover matrix in Figure 7.2 as well as the Excel sheet that pulls in that data for Grasshopper. So if the order of land covers happen to be different, make sure to organize the data coefficients the same. This example includes labels for the different *Brep* components going in the order of *Forest, Grassland*, and then *Water*.

The *Brep* geometry is then *Entwined* to maintain their branching structure after the morphing process to create their range of associated indexes. The *Surface*

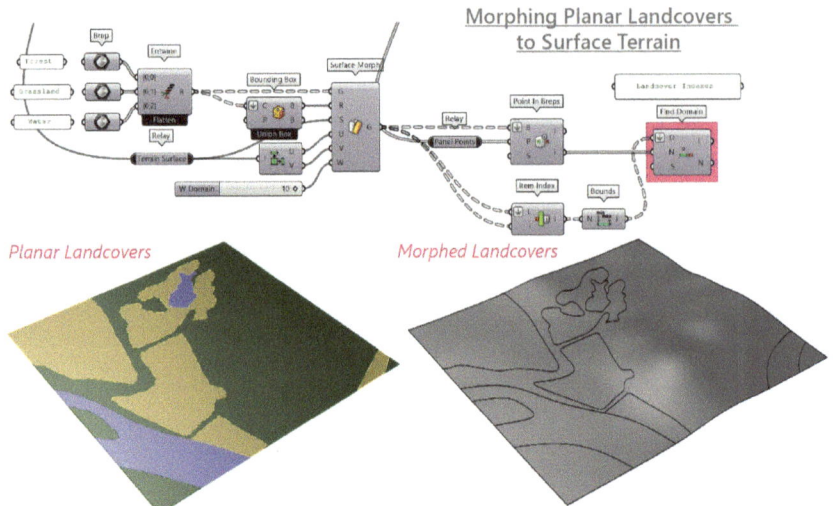

Morphing Planar Landcovers
to Surface Terrain

Planar Landcovers

Morphed Landcovers

Figure 7.9
Referencing and
morphing traced
land cover surface
types to the surface
terrain.

Morph component will once again use the entwined geometry and their *Bounding Box*, set to *Union*, as a *Reference* (*R*). The terrain surface model will target surface along with its *U* and *V* parameters. Another number slider with the same value of *10* will give the morphed surface an extrusion height of ten feet.

The remainder of this operation will follow with the same set of component connections from the *Surface Morph* component's output where the data branching structure is *flattened* at the inputs for both the *Point In Breps* and *Item Index* components. Maintaining the branch structure for the *Item Index* component's *Item* (*i*) input is key in generating a set of index domain values for the *Find Domain* component. Once those domain values have been created with the *Bounds* component, they can be *flattened* at the input of the *Find Domain* component, since the current data structure is no longer necessary. This final component has also been highlighted and labeled *Land Cover Indexes* for easy reference later in the script.

The last set of indexes for this aggregation of metrics is calculated from a slope analysis of the terrain surface, as shown in Figure 7.10. The process of analyzing and extracting slope percentages of the terrain model is the same as the process from Chapter 6's 'Terrain Slope' section but from LunchBox's *Quad Panels* component and their *Centroid* (*C*) data, shown as *Relays* from earlier in the script. The quad panels will first be analyzed for their *Normal* (*N*) direction from the *Evaluate Surface* component. The required *uv* input for this component can be extracted from the *Surface Closest Point* component when connecting both the panels and points to their respective inputs.

The normal values can then be measured with an *Angle* component using a *Unit Z* value for the *B* input and converted from radians to decimal percentage values from the *Tangent* component. Since the rational method has a peek evaluation of slopes percentages to 30%, a *Minimum* component with *0.30* is used to cap all the decimal values that go beyond that degree to 30%. These decimal values can then be converted to percentages with the *Multiplication* component using *100* for conversion.

To begin the classification process of these indexes to the ration method's metrics, these percentage values are parsed into the slope categories. This is achieved

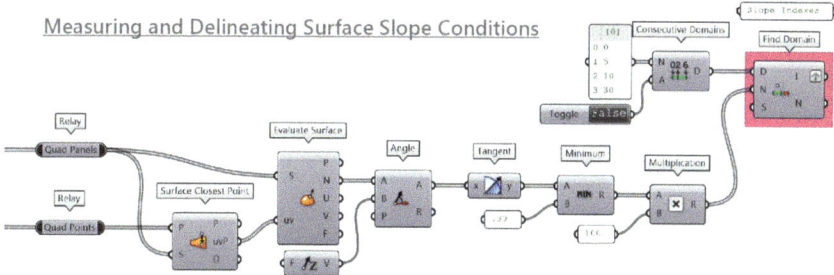

by using the *Consecutive Domains* component with a panel of values listed as *0*,
5, *10*, and *30*. This will create a series of domains that use these as their limits
consecutively once the *Additive* (A) input is set to *False* with a *Boolean Toggle*. Now
these slope percentages can be evaluated, for which range they fit within by using
the *Find Domain* component. The *Index* (*I*) output of this component needs to be
grafted for individual evaluation in the next step.

Now that the data indexes for soil, land cover, and slope have been determined, they can now be cross-referenced with the rational method's coefficient
matrix, as shown previously in Figure 7.2. The calculator's matrix can be translated
into an Excel file for import into Grasshopper with the ExcelDynamicRead component, Figure 7.11, as part of the GhExcel plugin. The Excel file should look like the
one being shown in the image below, matching the same order and values of the
matrix example. Headers and row names are not provided in the Excel file since that
will require unnecessary sorting of the data in Grasshopper. The Excel file already
requires substantial restructuring to match the data logic of Grasshopper so this
avoids the addition of more components as well.

The first restructuring of the data requires the *Flip Matrix* component to be connected to a *flattened* input for the *Partition List* component. This will sort the data
lists from the three columns of soil texture types to lists of the land cover and slope
rows. The *Flip Matrix* component will maintain each row as a separate list but these
need to be parsed into the three different slope categories instead, so that is why
the input is flattened. To create those parsed groupings of slope conditions for each
land cover type, the *Partition List* needs a *Size* (*S*) input of *9*. That is because each
land cover type as three soil types for each of the three slope categories resulting
in nine possible metric values. The partitioned lists can now be restructured again
with a *Flip Matrix* component to make the cross-referencing with the *Land Cover
Indexes Relay* compatible in the *List Item* component.

Now that the land cover indexes have been formatted to match matrix values,
they now need to match the soil indexes by starting off with a *Flip Matrix* component once again. This will allow for the grouping of nine values for each land
cover to be subdivided into the three soil groups using the *Partition List* with a *3*
for the *Size* (*S*) input value. That same value will also be used to repeat all the *Soil
Indexes* values when using the *Duplicate Data* component with the *Order* (*O*) set
to *False*. This will serve as the *grafted List Item* component's *Index* (*i*) input that is
connecting the *Partition List* component as the *List* (*L*) input.

The final step of this process will be to take that *List Item* component and inputting it into another *Partition List* component that has the values flattened. The *Size (S)* of these partitions will once again be set to the same size as before, *3*. This newest partitioned list can now be inputted into one final *List Item* component that uses the *grafted Slope Indexes* from the labeled *Find Domain* component as the *Index (i)* input. This component needs to have a *flattened* output for the next step in calculating the peak flow rate of runoff.

Now that a coefficient value for each of the points and quad panels has been determined as part of the rational method's calculation, they can now be applied to the remaining part of the equation. For this example, in Figure 7.12, the rainfall intensity will be set to *1* for the *Multiplication* component; however, the use of weather data records or municipal guidelines for this equation may be more appropriate in assigning this value.

Lastly, the *Quad Size Relay* connects the values for the panel sizes, *100*, to a *Multiplication* component's *A* and *B* input for the total area of each panel. This value can now be divided by *43,560* to convert the square footage to acres for the necessary unit of the equation and added to the cumulative *Multiplication* component for *Peak Flow Rate (Q)*. The result of this equation will be in flow rate (cubic feet per second), so if it is also valuable to know what the runoff volume from this storm event is, then it can be *multiplied* by the duration of the storm event in seconds.

Cross-Referencing Terrain Coefficient Metrics with Model

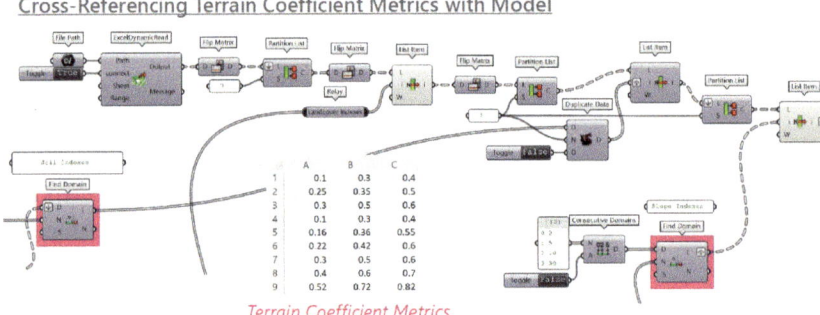

Terrain Coefficient Metrics

Figure 7.11
Applying the coefficient metrics from an Excel file to the partitioned points.

Applying Rational Method Equation to Model

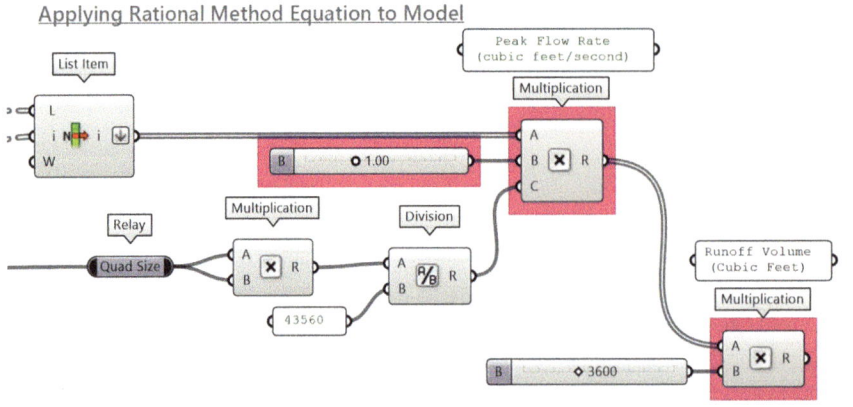

Figure 7.12
Calculating all the rational method variables to the equation.

This demonstration uses a value of 3,600, or 60 minutes or one hour, storm event. This can be adjusted to match the average duration of storm events on a project site or be used to simulate different scenarios of frequency or intensity of rain events. Either value can be connected to the rest of the script as shown in later steps.

These operations determined the peak flow rates and runoff volume data for all the different surface panels throughout the project site. The next series of steps will demonstrate the parsing of this data to watershed regions and drainage lines with different visual analyses. The combination of the two will help guide strategic design interventions from this analysis.

As shown in the 'Terrain Hydrology' section of Chapter 5, the use of drainage lines can not only show where runoff will flow to but also how they can be used to inform different watershed regions and ridgelines. Figure 7.13 will demonstrate how the data used to generate watersheds can also be utilized to aggregate the calculated runoff data along with highlighting specific attributes or contributors to high runoff volume. So, like Chapter 5, the process begins with creating a set density of drainage lines from the *Mosquito* plugin's *Flow* component. The density of points, from the *Divide Surface* component, that act as the starting point of these drainage lines should not be as dense as the quad panels created earlier in this script, but rather more generalized. If desired as well as having the computational processing power, the density parameter, *U* and *V* input on the *Divide Surface* component, can be increased for more detail.

The creation and aggregation of this data is a bit more complex and should follow a set sequence of operations, so the remaining portion of this script in the figure is organized into two groups which will progress from the bottom to the top in the set of operations.

Starting at the bottom of the script, a *Point On Curve* component for the *Flow* component *Curves* is used to extract the end point for all the drainage line curves. All the curves with an end point in close proximity of one another can then be grouped together with the *Point Groups* component to determine the different drainage basins. Depending on the size of the project site, a number slider can be used to gauge how close the points need to be in order to accurately group them together, so for this demonstration a value of *100* worked. The *Indices* (*I*) output from this component can then be used to group the data (end points) by this proximity parameter with a *List Item* component after reducing the tree branching with a *Simplify Tree* component. For the use of this data, only one point out of the group will be needed, so another *List Item* component is used to isolate just the first point from the list. This can be seen in the bottom right image of the terrain model in the figure.

The simplified indices from that previous set of operations will be used again, but this time for the *Flow* component's *Curves* output. The restructured set of outputted curves can then be used to extract a set number of points with the *Divide Curve* component, in this case, *25* points. The number of divided points will dictate the accuracy from which the quad panel center points are measured from. These will be used to simulate how close the drainage curves are to those points, so the denser the points are on the curve, the more accurate the measurements will be.

The points from the divided curves can now be used to find out which ones are closest to all the different quad panel *Centroid* (*C*) points, generated from the *Area* component in Figure 7.4. These will be evaluated as a *flattened Cloud* (*C*) input for the *Closet Point* component to determine the outputted *CP Index* (*i*) which are from the quad panel points. This output value will be a critical piece for the next series of steps as it will first be inputted an *Equality* Component. Since there will be multiple quad panel points closest to a single divide curve point, there will be duplicated indices outputted; however, it will be necessary to only have a single unique index instead. To reduce these duplicates, a *Create Set* component is used where the *Set* (*S*) output contains only the single unique indices. These unique values will next be used to filter out the specific points from drainage lines and their own indices.

The drainage line points from the *Divide Curve* will first be flattened at the input of a *List Item* component for the *Set* (*S*) to be connected as the *Index* (*i*) input. For the next use of this *Set* list, an *Item Index* component is used where the *List* (*L*) input is a *flattened* list of the *Divide Curve* points, but the *Index* (*i*) retains the outputted branching structure. This will ensure the indices for the drainage curves are maintained within their watershed groups. This output will then be *grafted* into the *B* input of an *Equality* component where the *A* input connects the *Set* list. By grafting the *B* input, each point of each curve within the different watershed groups will be evaluated individually with the *Set* indices.

The resulting Boolean pattern from this component is then used for two *Cull Pattern* components that are connected to the *Create Set* component and the *List Item* component containing the *Divide Curve* points. Both are labeled in reference to their data type (indices and points respectively) for future use in the script. The culled indices, however, will also be used immediately for another *Equality* component that is connecting the *Closest Point* component's *CP Index* (*i*) output. This component is also highlighted and labeled *Watershed Pattern* for future reference as well.

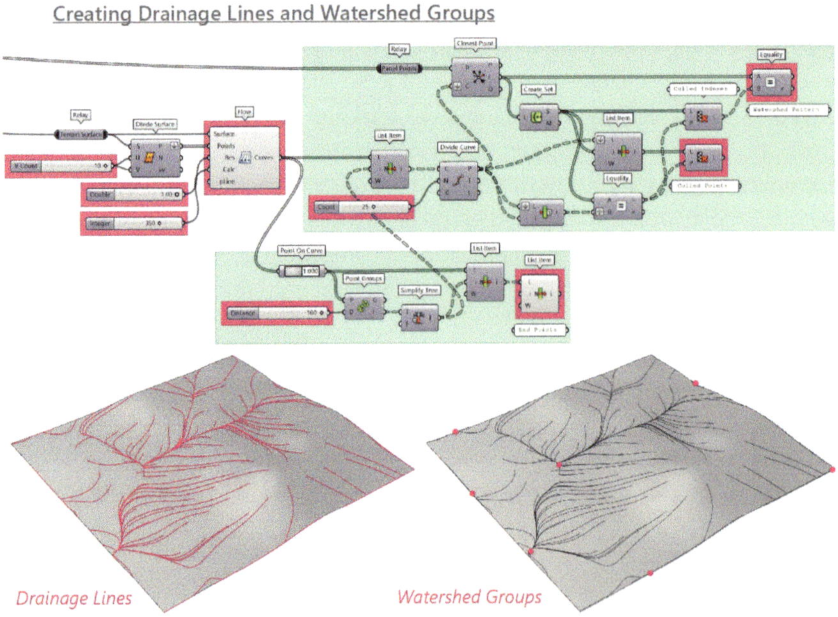

Creating Drainage Lines and Watershed Groups

Drainage Lines

Watershed Groups

Figure 7.13
Generating drainage lines to create watershed groups.

With the previous operation of structuring the quad panel points into their respective watershed regions, the same associated Peak Flow Rate values that were calculated in Figure 7.12 can now be parsed into those same groups and visualized collectively, as shown in Figure 7.14. The first step is to use the *Equality* component labeled *Watershed Pattern* as the *pattern* (*P*) input of the *Cull Pattern* component. The *Peak Flow Rate Relay* will be the *List* (*L*) for culling. Now that the culled values are branched into their watershed groups, the values for each drainage line can be summed up with a *Mass Addition* component. The branch structure is then simplified with a *Clean Tree* component and further reduced with a *Trim Tree* component that removes two lines of branching for each data list from a *Depth* (*D*) input of *2*.

Another *Mass Addition* component is used to now sum up all the individual drainage line values within their respective watersheds together to get a final cumulative value for each basin. These combined values are now scaled with a *Multiplication* component and number slider to scale accordingly, since the output is intended to visualize a comparison of runoff for each watershed. This will serve as the *Radius* (*R*) input for a *Sphere* component that uses the drainage line End Points from Figure 7.13.

In Figure 7.15, the points along the divided curve can be visualized for their contribution to the runoff by creating vertical lines that symbolize their peak flow rates or runoff volume. Since the peak flow rate has been unit of measurement so far, that value will remain for visualization. The *Clean Tree* component's output from Figure 7.14 will be the value for the *Multiplication* component's *A* input and

Modeling Peak Flow Rates by Watershed Groups

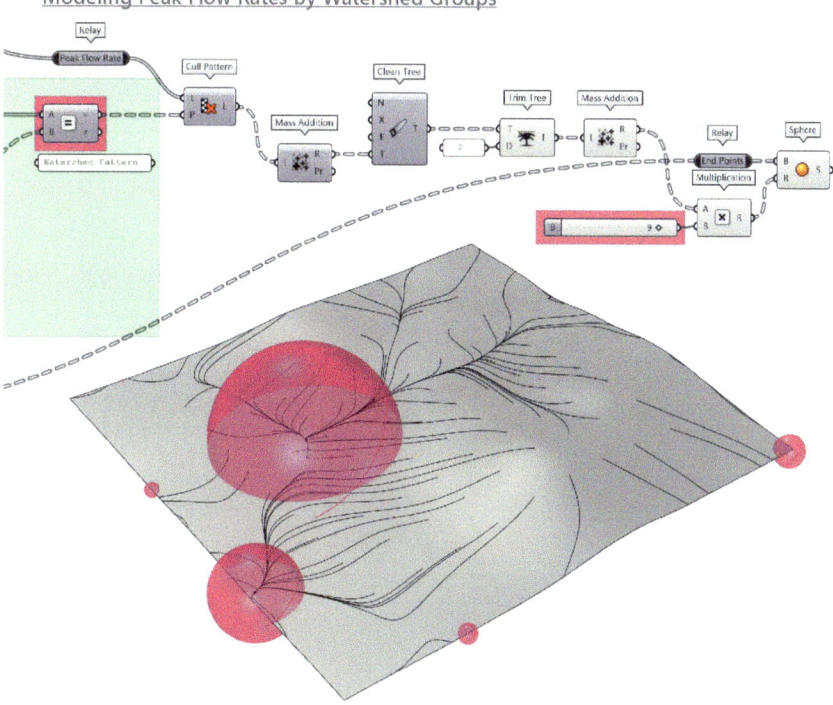

Figure 7.14
Visualizing the rational methods peak flow rate for each watershed region.

a number slider to dictate the size will be the *B* input. This component's output will need to be flattened, since all the remaining operations will analyze all the values together. Like the spheres indicating peak flow rates in the previous figure, the enlargement of this number doesn't matter because the intent is once again to show a comparison. The final image of this performance model will utilize the accurate measured values.

These values can now serve as the vertical (*Z*-unit) *Length* (*L*) input of the *Line SDL* component while a *Clean Tree* component with a *flattened* output for the *Culled Points* from Figure 7.13, is connected to the *Start* (*S*) input. The result of this operation is shown in the bottom left image of the figure where a vertical line is created at every point of the divide drainage curves. The next set of operations will demonstrate how these can be filtered and highlighted for the top contributing runoff locations along these drainage lines.

The first step in doing so is to measure the *Length* of each curve and use it for a *Sort List* component that also sorts the *Line SDL* curves. The outputs from this component then need to be reversed so the largest values are at the top of the list. Finding the top percentage of the contributing factors requires a *Multiplication* of a decimal value, *0.10* or 10% in this case, of the total number of curves generated by a *List Length* component from any of the components from this operation. The sorted curves can now be parsed out with a *Split List* component where the *Index* (*i*) is the numeric value of the top 10% of curves. From this component's *A* output, a *Curve* component is connected to highlight those curves, as shown in the bottom right image of the figure.

These last series of operations will give the option to scroll through the top contributing watersheds to the overall project site while visualizing how much each quad panel contributes to the overall runoff rate. The first step, shown in Figure 7.16, will use the *Equality* component's pattern output from Figure 7.13 for a *Cull Pattern* component with the *Quad Panels Relay* as the *List* (*L*) input. This restructured list

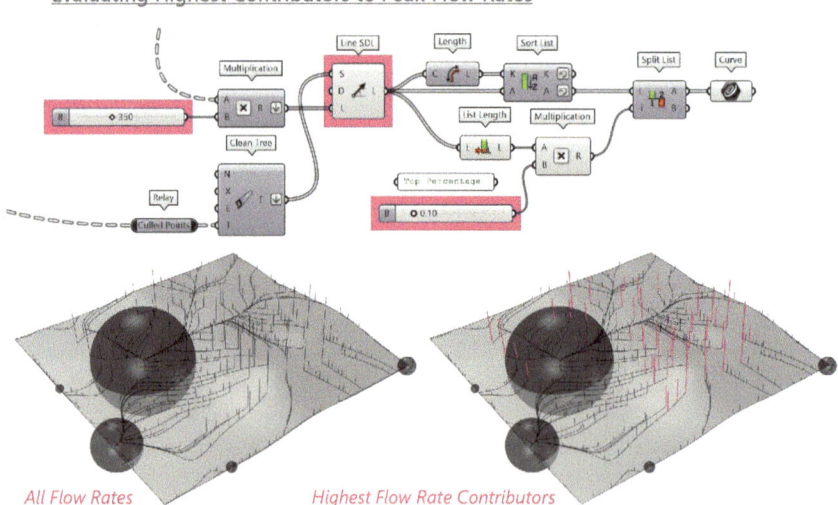

Figure 7.15
Visualizing individual and top peak flow rates contributing to the watershed's total value.

will then be modified with a *Clean Tree* and *Trim Tree* component to simplify the branching structure. The *Depth* (*D*) of the trimming procedure will be to a value of *2*. The process of cleaning and trimming a branching structure will also be performed on the culled list of *Peak Flow Rate* values as well; however, that output will not be used until later in this operation.

The trimmed data will then be *Grouped* for sorting purposes, since the number of lists for sorting needs to be equal to the number of *Key* (*K*) inputs of the *Sort List* component and connected to the *A* input. The *flattened K* input will be from the cumulative value that is outputted from the second *Mass Addition* component from Figure 7.14. The lists that have been grouped also need an assigned index, so an *Item Index*, with both inputs *flattened*, is used for this *Group* component, and connected to the *B* input. All the outputs from the sort lists can be reversed and the *A* output can be connected to a *List Item* component with a number slider as the *Index* (*i*) input. The watershed contributing to the highest rate of peak flow will be the first list item (*0*), so as the slider values increase, the proceeding next highest watershed will be outputted and so forth.

Since these outputs will still read as groups and have no associated data (numeric or geometric data), they need to be *Ungrouped* next. The ungrouped quad panels can now be *Scaled* to represent their own unique contribution to the overall peak flow rates visualized with the spheres in Figure 7.14. The *Center* (*C*) for scaling

Peak Flow Rates as Quad Panels

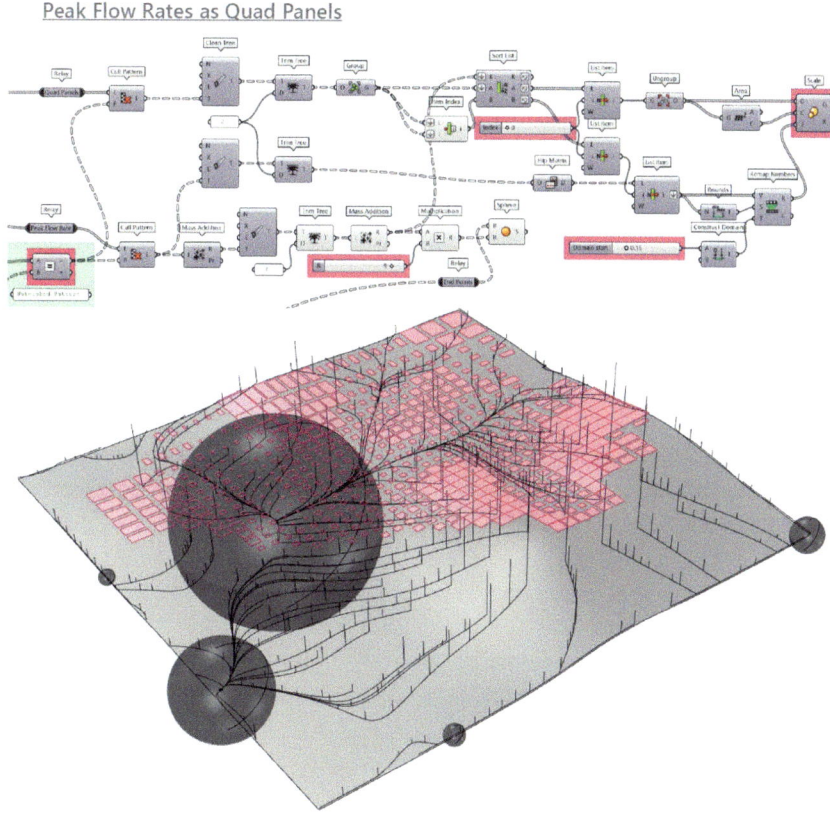

Figure 7.16
Visualizing the different slope contributions to the peak flow rate from a single watershed.

will come from the *Centroid* (*C*) output of an *Area* component from the ungrouped quad panels. Giving them unique sizing, representative of their peak flow rates will be generated from the next set of operations.

The same process of grouping the quad panel geometry cannot be performed the same as grouping and ungrouping data (Peak Flow Rate) because the data will be lost in the process. So that is why the associated watershed groups were assigned an index instead and could be used to sort the 'group' of data. A *List Item* component is first used with the *reversed B* output of the *Sort List* component that contains the same number slider as the other *List Item* component so their values will always match. This will output the watershed's index for the *Index* (*i*) input of another *List Item* component that utilizes the other *Trim Tree* component's output after connecting it to a *Flip Matrix* component as the *List* (*L*) input. After *flattening* the output, it will go through the process of *Remap Numbers*, with its *Bounds* as the *Source* (*S*) input and a *Constructed Domain* ranging between *0*, preferably slightly greater than, and *1* for the *Target* (*T*) input. The output can now be connected to the *Factor* (*F*) input of the *Scale* component. These newly scaled quad panels can be seen in the model image of the figure.

ADDITIONAL CONSIDERATIONS

As these performance models become more complex and output robust sets of information, it is important to consider the legibility and communication of the data through additional colorization and labeling. In Figure 7.17, all of the different outputted data sets are given unique color values and labeling conventions through the various script operations as shown with the different images.

Colorization and Labeling

The script image in the figure shows how the quad panels can both be colorized and labeled based on their respective peak flow rates. The *Location* (*L*) of the text for the *Text Tag 3D* component used the quad panels *Centroid* (*C*) output from an *Area* component, while the *Text* (*T*) is the set of individual peak flow rate values outputted from the shown *List Item* component. Giving the text for each quad panel a unique sized based on the runoff value is another consideration that can be made to give the graphic a more dynamic visualization. This approach uses the same *Remap Numbers* component for the *Scale* component previously, but is also given a *Multiplication* parameter with a number slider to adjust the scale and legibility of the text.

Labeling the watershed spheres is another option that is derived from its own data sets that generated the differing sized spheres. The location for the text will use the same *End Points* for the spheres, but is given an additional *XZ Plane* transformation so that the text is oriented vertically. The cumulative values for each watershed is also given a *Multiplication* parameter with a number slider to next adjust the size of the text.

Figure 7.17
Characterizing
the different
visualization
graphics with colors
and labels.

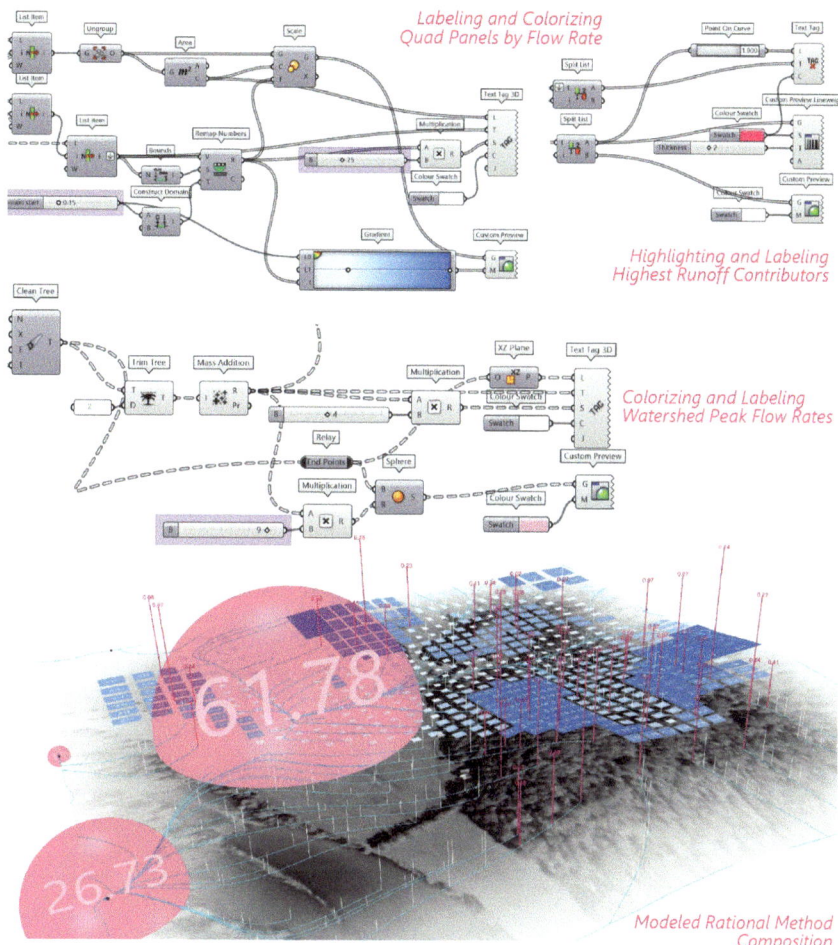

Figure 7.17
Characterizing
the different
visualization
graphics with colors
and labels.

Modeling Additional Annotations and Labels

*Labeling and Colorizing
Quad Panels by Flow Rate*

*Highlighting and Labeling
Highest Runoff Contributors*

*Colorizing and Labeling
Watershed Peak Flow Rates*

*Modeled Rational Method
Composition*

Lastly, the top contributing runoff areas, visualized with the *Line SDL* component, can be highlighted with a *Text Tag* component that is given an accenting color to contrast with the rest of the vertical lines and overall graphic style. This text labels are located at the top of the curves with a *Point On Curve* component.

Although these outputs are valuable in themselves, they become more significant and comprehensive as they get compared and measured with different baseline measurements and regulation standards specific to that context to generate outcomes. The next chapter will include additional visualization techniques of data using charts. Figure 7.18 showcases more detailed views of this comprehensive rational method model.

In Figure 7.19, another option is demonstrated to show all the watersheds, which can then have the previous procedures applied in colorizing and labeling the different geometric outputs.

Figure 7.18
Detailed views
of the graphic
communication of
the rational method
equation.

Applying Peak Flow Rate Visual to All Watershed Panels

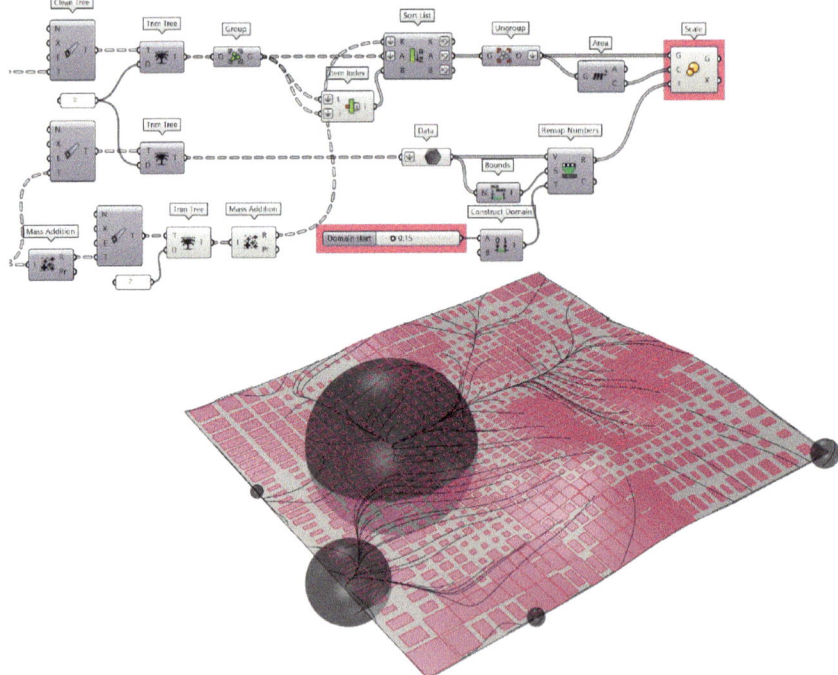

Figure 7.19
Combining all
watershed areas
to visualize peak
flow rates or runoff
volume.

OUTCOMES

With most stormwater analysis, the intent is usually to reduce peak flow rates and runoff volume. This can be achieved in a variety of ways that include adding more pervious or natural land cover and reducing the slope percentages to allow for more infiltration. Based on those metrics, the outputs from this model can begin to inform strategic decision making for reduced peak flow rates and runoff volume.

Currently, this project site is 30% forests, which have a higher coefficient value (runoff potential) than grasslands. If more of the land cover was converted to this lower coefficient value, then that could potentially reduce the site's runoff. Also, 33% of the project site is classified within the extreme slope range of 10%–30%. This also indicates that if those extreme slope conditions in the terrain were regraded or terraced, then that too could potentially reduce runoff.

Lastly, as stated earlier, municipalities often have specific regulations for what the peak flow rate needs to be reduced to from stormwater interventions. The different watershed peak flow rates can be used to compare with those regulations to gauge how much they need to be reduced by if any. The runoff volume that is also generated by the peak flow rate over a specific period of time can be used to evaluate against the earlier suggestions of changing either the land cover or slope conditions. This would inform what the best option could be through design interventions.

FUTURE POTENTIAL

The performance script and model is currently analyzing a very narrow focus of what a site's condition may be – undisturbed natural rolling landscape. However, urban settings and flatter conditions will also need to be considered for different project sites. Another limiting factor in this model is the span of time of the rain event which ultimately dictates the volume of runoff from the peak flow rate. Different hydrographs or designated time intervals will generate different volumes of runoff for a more dynamic comprehension for stormwater management.

Impervious Land Covers

As shown in Figure 7.2, there are also runoff coefficients for the different types of impervious surfaces as well. Although these do not consider soil texture because of their lack of permeability, the range of values provided indicate that the lower values are for flat conditions while the other extreme is for steep conditions. All the ranges are for the most part divisible by three for the three different slope categories, for example, Suburban Residential range from 0.25 to 0.40, suggesting the flat slope coefficient is 0.25–0.30, rolling is 0.30–0.35, and hilly is 0.35–0.40. To make the process easier to transition with the Excel chart, these values can be duplicated for the three different soil textures, since those are irrelevant and won't impair the calculation process.

Hydrographs

Different classifications of hydrographs can help determine runoff by comparing the time of concentration with the duration of the rain event (Strom, 2013). If the time of concentration takes the same amount of time as the rain event, then the time interval, in seconds, is multiplied by the peak flow rate. If the duration of the storm is longer than the time of concentration, then the duration of the storm is multiplied by the peak flow rate. And lastly, if the storm duration is shorter than the peak flow rate, then the duration of time is divided by the time of concentration and multiplied by the peak flow rate.

REFERENCES

Cheng, C, & Trakas, A. (2020). *Arizona State University Orange Mall Green Infrastructure Project*. Landscape Performance Series. Landscape Architecture Foundation. https://doi.org/10.31353/cs1640

Strom, S., et al. (2013). *Site Engineering for Landscape Architects*. John Wiley & Sons, Incorporated. ProQuest Ebook Central, https://ebookcentral.proquest.com/lib/unlv/detail.action?docID=1115640

8   Runoff Volume – NRCS Method

The Natural Resources Conservation Service (NRCS) method is commonly used to determine runoff rates and volumes by government agencies, engineers, and landscape performance advocates. In the previous chapter, runoff rates were determined to calculate runoff volume, but this methodology will focus on runoff volumes from a weighted curve number (CN) value. This calculator and script cannot isolate the source of runoff volume but can output the impact individual surface and coverage areas have on the overall runoff volume. The calculator works by aggregating the individual surface characteristics of coefficients and size to create a weighted or adjusted CN value. Another benefit to this process is the scalability and replicability for a diverse range of project sizes and types.

GOALS

As a result of this performance modeling script, the user will be able to evaluate and assess the following conditions for a strategic and responsive design plan to manage, reduce, and mitigate stormwater runoff. The following are the evaluation and assessment goals from the calculated runoff outputs:

1. Compare the impact different surface types have on stormwater runoff
2. Compare runoff volume from different rain events

NRCS EQUATION

Figure 8.1 shows the commonly used NRCS equation for calculating stormwater runoff by landscape architects, engineers, and other agencies. The solution to the equation is runoff in inches, which will be converted to cubic feet with the model. It considers multiple variables that include inches of rainfall, the potential maximum retention at the start of runoff, and initial abstraction. Initial abstraction refers to all potential constraints to runoff that includes infiltration, evapotranspiration, and interception by vegetation.

NRCS Variables and Equation

Figure 8.1
The NRCS variables
and equation that
will be modeled.

Q = runoff, in.
P = rainfall, in.
S = potential maximum retention
after start of runoff, in.
I_a = initial abstraction, in.

$$Q = \frac{(P - I_a)^2}{(P - I_a) + S}$$

Initial abstraction accounts for all potential water loss before runoff but is usually estimated as 0.2S, so when replaced in the calculation, it becomes the following.

$$Q = \frac{(P - 0.2S)^2}{(P + 0.8S)^2} \qquad S = 1000/CN - 10$$

This variable is often substituted with 0.2S and leads to the modified equation below. The potential maximum retention variable (*S*) converts to 1000/CN – 10 where CN refers to the weighted curve number value that will vary between 0 and 100. This is similar to the coefficient value (*C*) used in the previous 'Rational Method' chapter.

The hydrologic soil group (HSG) categories indicate the potential infiltration or runoff from precipitation in which HSG A usually indicates good drainage and healthy sandy soils, and where HSG C and D refer to disturbed and compacted soils with poor infiltration rates or have a clay-based texture. It is best to refer to a soil survey map or use the Web Soil survey method in Section 2 of this book to find the specific classifications of soils for the project site. It is common practice when doing a project in an urban or developed area to use HSG C, due to the nature of compaction and disturbance to soils from heavy traffic use and construction practices. Figure 8.2 shows a chart of different land cover types with their associated CN values for the different HSG classifications.

Similar to the process stated in the previous rational method modeling chapter, rarely do project sites encompass one consistent surface type or HSG. It is important to classify the Rhino layers by their respective surface type in order to determine the area size of that surface type and respective CN value to generate a weighted CN value relative to all other surface types and size. For example, a project site that is 40% open space in fair condition (79 CN) will have a weighted CN value of 31.6. Once this is applied to all the surface categories, an adjusted CN value will be computed for the NRCS equation.

Model Preparation

Before developing the Grasshopper script, it is important to utilize the same workflow of creating surfaces and assigning them to their respective layers for accurate modeling and calculations, as demonstrated in Chapter 5 of this book. This will ensure complex surfaces that have openings within are calculated correctly for their area size.

Although trees do play a role in stormwater runoff reduction and interception, for the purposes of this model and calculator it is not necessary to include them as part of computing process. In Figure 8.3, the Rhino model will contain the different surface types organized for the scripted operations. It is important to use a simple naming convention for these layers since the referencing of the geometry from these layers will require a plugin that is case sensitive to Rhino layers.

Cover Description	Curve Numbers for HSG—Hydrologic Soil Group			
Cover Type and Hydrologic Condition	A	B	C	D
Open space (lawns, parks, golf courses, cemeteries, etc.)				
Poor condition (grass cover <50%)	68	79	86	89
Fair condition (grass cover 50 to 75%)	49	69	79	84
Good condition (grass cover >75%)	39	61	74	80
Impervious areas				
Paved parking lots, roofs, driveways, etc.	98	98	98	98
Streets and roads				
Paved; curbs and storm sewers	98	98	98	98
Paved; open ditches	83	89	92	93
Gravel	76	85	89	91
Dirt	72	82	87	89
Western desert urban areas				
Natural desert landscaping	63	77	85	88
Artificial desert landscaping	96	96	96	96
Urban District				
Commercial and business (average impervious cover 85%)	89	92	94	95
Industrial (average impervious cover 72%)	81	88	91	93
Residential districts by average lot size				
1/8 acre or less (average impervious cover 65%)	77	85	90	92
1/4 acre (average impervious cover 38%)	61	75	83	87
1/3 acre (average impervious cover 30%)	57	72	81	86
1/2 acre (average impervious cover 25%)	54	70	80	85
1 acre (average impervious cover 20%)	51	68	79	84
2 acres (average impervious cover 12%)	46	65	77	82
Developing Urban Areas				
Newly graded areas (pervious areas only, no vegetation)	77	86	91	94

Figure 8.2
A matrix of the different NRCS CN values for different land cover types.

Modeled and Organized Landcovers

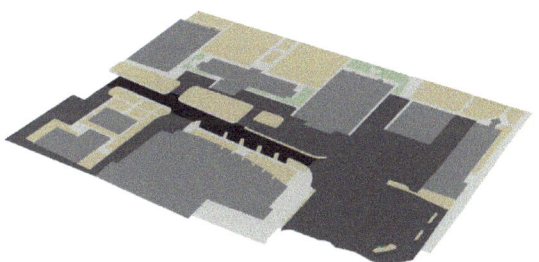

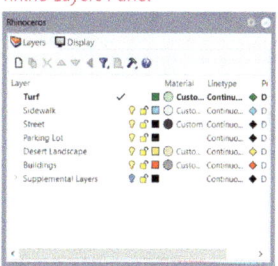

Rhino Layers Panel

Figure 8.3
Rhino model of the different surface cover types organized by layer.

Modeling Methods

The first part of this script will calculate runoff volume for a given site from multiple rain events to show the significance of how differing rain intensities will impact surface runoff. The second part will cover visualization and communication techniques to make the data tangible and relative for the user and intended audience.

After preparing the model by organizing the different surface typologies onto their respective layers, the process will begin with a panel and *Elefront* plugin's *Reference by Type* component, as shown in Figure 8.4. The panel needs the text *Surface* inputted into this component. Remember, these are case sensitive and will get an orange cable if the geometry can't be referenced or has invalid geometry. Next, *right-click* on the icon of the component to disable the auto update function, so that if surfaces are added, moved, scaled, or deleted, the script will update based on those changes.

The *Filter By Layer* component is now used to begin the process of referencing the different layers starting with one of the categories and repeating the next few steps for all the other remaining layers. The *Filter By Layer* and *Mass Addition* component have been highlighted (*Ctrl+G*) because their values will be needed in later steps and will make it easier to find with the repetition of operations for other surface typologies. As shown in the image, all these components can have the *Preview* disabled to reduce visual clutter in the Rhino model.

Now in Figure 8.5, a series of multiplication and division components are used to calculate for the weighted CN value along with the individual impact of each surface within that typology. The highlighted *Mass Addition* component will be inputted into another *Mass Addition* component that will eventually connect all the area values for each typology, as shown in the next step. It will also be connected to the *A* input of one *Division* component and the *B* input of another as well. For the one that it is connected to the *A* input, it will be divided by the *Mass Addition* component that has been labeled *Total Surface Area of Site*. This will give the percentage coverage of the overall site so that it can be weighted with a CN value. It is also highlighted as a visual reference because it will be connecting several other values to the input in the next steps.

The output from this first *Division* component will now be connected to a *Multiplication* component with its respective CN value from the *Cover Description* table in Figure 8.2 for this operation. Choosing the correct CN value will often require some ground truthing with the actual project site but since this demonstration is hypothetical, the turf condition was classified as *Open Space in Good Condition* with a CN

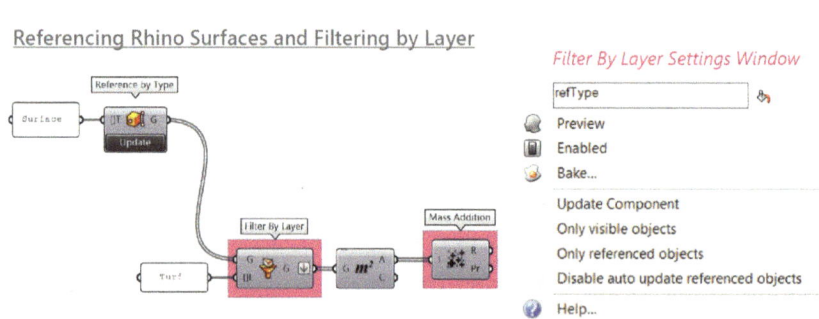

Figure 8.4
Referencing Rhino surface cover by type with the *Elefront* plugin.

value of 74. This outputted value can now be connected to another *Multiplication* component that connects the *Division* of each individual surface size from the previous *Area* component with the *Mass Addition* of all those same surface types. The operation of determining the *Weighted CN Value* and *Surface Impact* values for the turf surface is grouped for organization and copying purposes as well.

The highlighted group of operations is now copied for the next surface typology in Figure 8.6, where the initial surface layer panel is changed to another layer

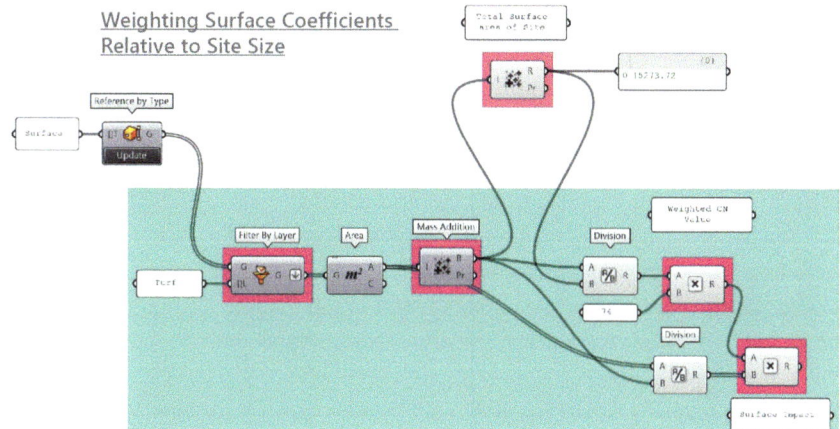

Figure 8.5
Separating the referenced surfaces by the first land cover layer – Turf.

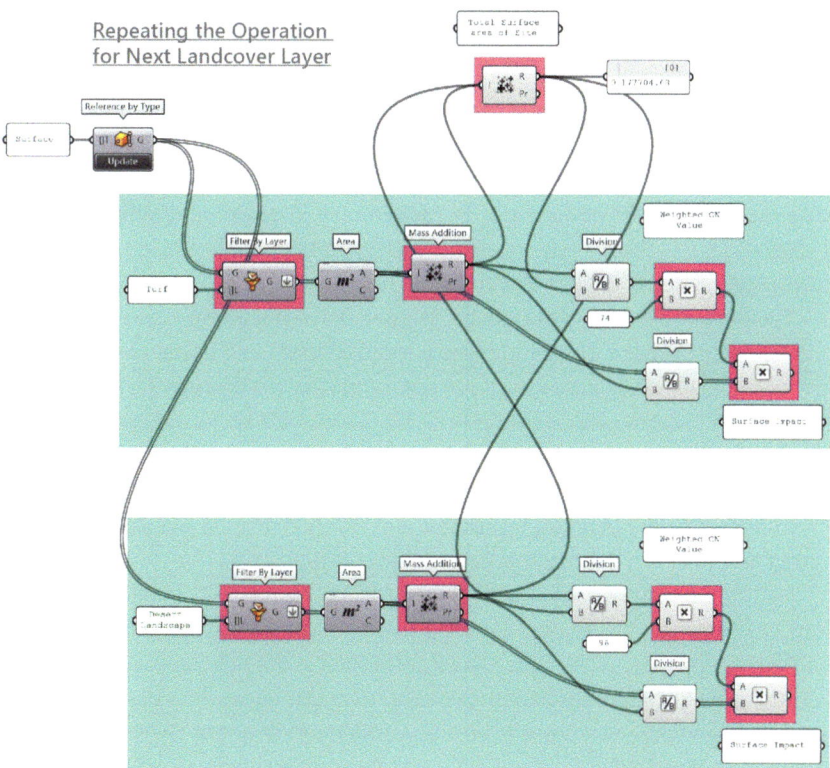

Figure 8.6
Repeating the same script operations for the next Rhino layer.

name, in this case, *Desert Landscape*. This is where the highlighted *Mass Addition* component from that group is connected to the *Total Surface Area of Site Mass Addition* component, as shown in the script. The CN value will also need to be changed to the correct value; in this case, desert landscapes have a value of *96* so that will replace the previous panel value.

This operation will be repeated for all the remaining respective Rhino layers: Sidewalk, Street, Parking Lot, and Buildings. Although those surface types will have a similar if not the same CN value, it is still important to separate them in order to evaluate how each surface type impacts runoff on the site. These findings can then influence responsive design strategies that require unique approaches. For example, even though sidewalks and parking lots both have a 98 CN coefficient, they will most likely require different design interventions to address stormwater runoff and other landscape performance metrics.

Figure 8.7 provides a preview of the collection of surface calculations that have all been connected to that *Mass Addition* component.

The benefit to copying this group of operations is because the output from the collective *Mass Addition* component will automatically be added to the required *Division* component within that surface type's group. This eliminates the challenging process of finding the correct one and maintaining all the correct connections within the script.

The next step in Figure 8.8 will begin by adding all the labeled *Weighted CN Value*, *Filter by Layer*, and *Surface Impact* component values from each surface typology into the *D1* input of their three respective *Merge* components. In that same order, include panel labels titled *Combined Weighted CN Values*, *Combined Surface Geometry*, and C*ombined Surface Impact Values*.

For organizational purposes, it will be best to begin with the first surface type, Turf, and sequentially move to the next surface type for the *D2* inputs of these merge component and so forth.

Figure 8.9 shows how this process is repeated for all the surface type values while maintaining the correct order. In this example, it is demonstrated how all the *Turf* values go into the *D1* input of all three *Merge* components, while all the *Desert Landscape* values go into the second input (*D2*), etc.

These merge components can be dragged down the Grasshopper workspace to the different surface groups to make it more efficient and easier to connect the correct inputs. Once all the values have been imputed, the preview can be disabled for the three merge components even if the *Combined Surface Geometry* is the only one that will have geometry to display. As stated earlier in Chapter 3, as these scripts flow from left to right and become more complex, turning off the preview of components will help with the clutter and recognition that those components will not be necessary for the final performance model.

Now is the first step of calculating these values for the NRCS method equation from the beginning of the chapter using the different variables with exception of the *P* variable (precipitation). This will be performed in the next step of the script to include multiple values instead of the common singular rain event value. Figure 8.10 demonstrates how the output from the *Combined Weighted CN Values* is connected to a *Mass Addition* component to determine the overall weighted CN values

Figure 8.7
Demonstrating all
the site's different
referenced land
cover layers.

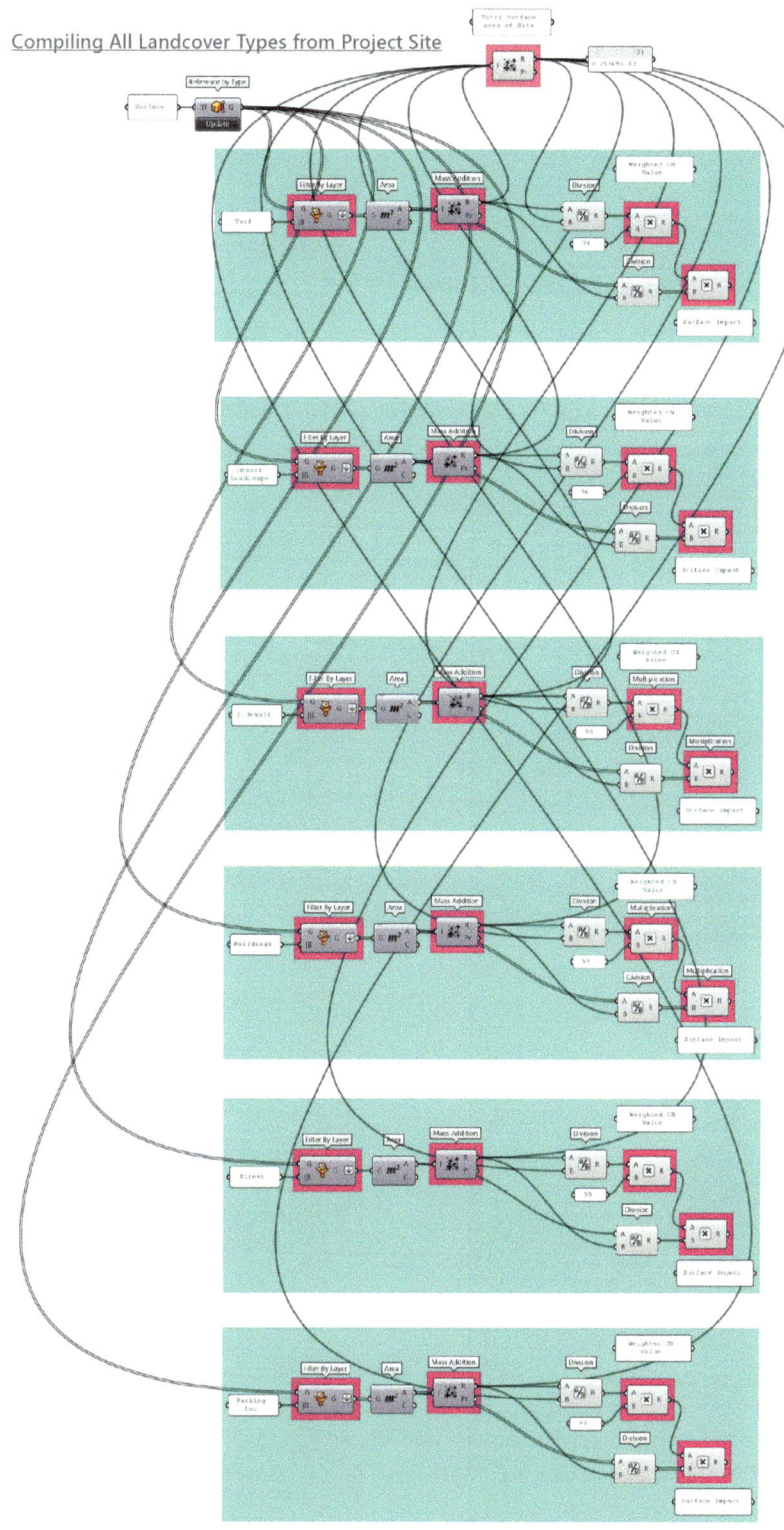

Compiling All Landcover Types from Project Site

Beginning to Merge All Landcover Outputs

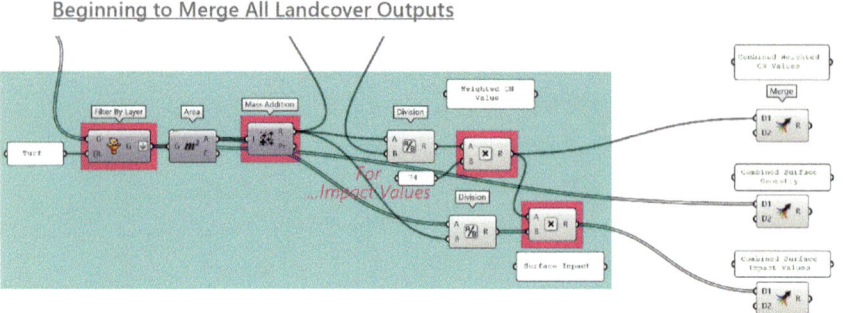

Figure 8.8
Beginning the process of merging all the pertinent layer information for the model equation.

Repeating the Process for Remaining Landcover Outputs

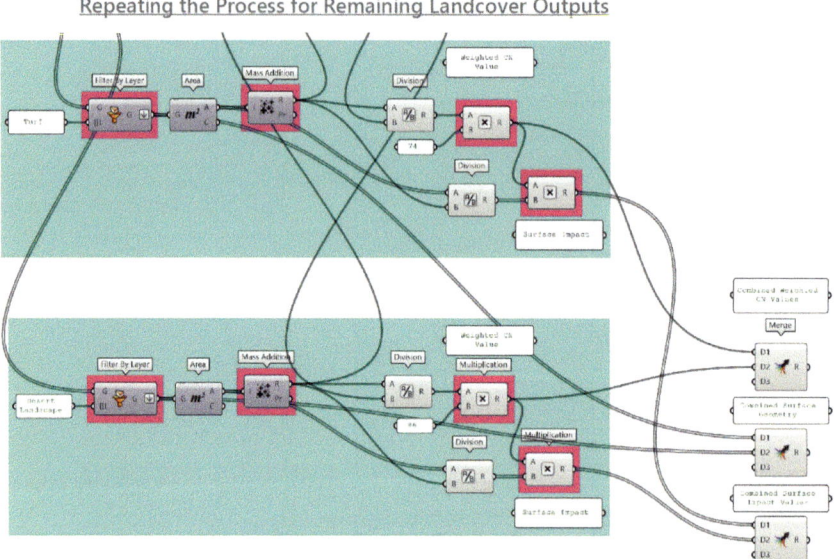

Figure 8.9
Repeating the previous procedures with the next surface layer and so on.

Applying the NRCS Equation to the Script

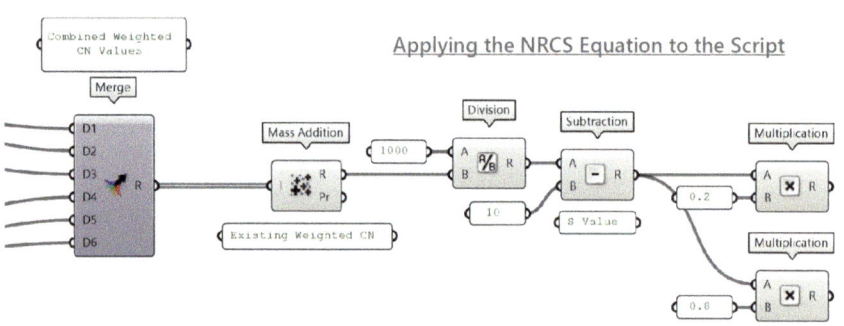

Figure 8.10
Calculating all the merged weighted CN values for the first steps of the NRCS equation.

that considers all the different surface coefficients and their prevalence within the project site. This will then serve as the *B* input to a *Division* component where the *A* input is the given value of *1,000*.

The divided value can then be subtracted with the *Subtraction* component using another given value of *10* to compute the *S* Value of the equation. This

Figure 8.11
Incorporating
multiple rain event
values to analyze
their relationship
and impact on
runoff volume.

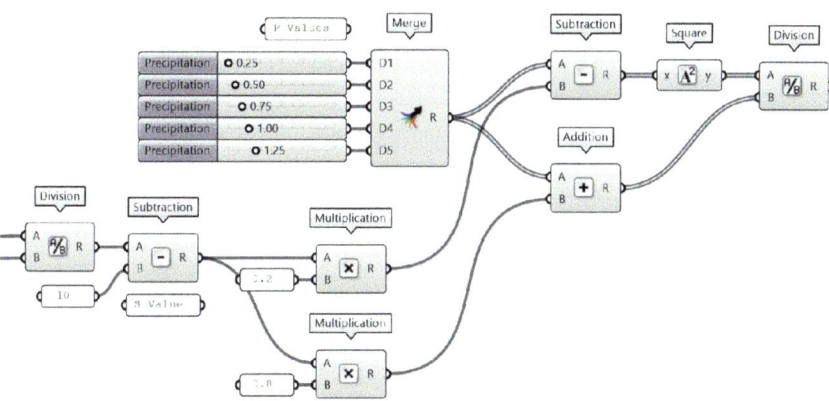

Applying Different Rain Event Values (P) to the Equation

Figure 8.12
Incorporating the
total area size of
the project site to
the equation.

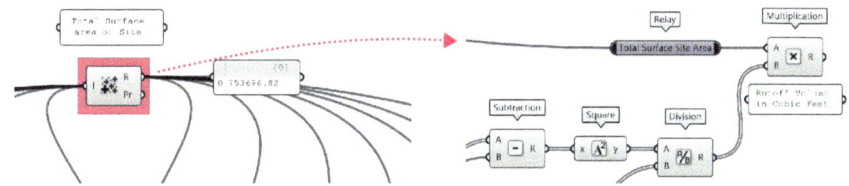

Including Total Surface Area of the Site to Equation

calculated value will now be connected to individual *Multiplication* components with a value of *0.2* and *0.8* respectively.

A series of precipitation values can now be inserted into the equation to get multiple runoff calculations. Even though the equation generally only accounts for one precipitation value, the inclusion of multiple rain events will not distort the final calculations. This is first done in Figure 8.11 with a *Merge* component to add multiple number sliders representing different rain event values; these values can be to discrepancy of the user and audience. The merged precipitation values will be first subtracted with the *Multiplication* component that contains the value 0.2. The resulting calculation will then be connected to a *Square* component to complete the top portion of the NRCS equation.

Finally, the precipitation values will also be added to the *Multiplication* component that contains the value of *0.8*. The squared values can then be divided by this last calculation using a *Division* component to complete the runoff volume in inches.

Figure 8.12 shows how a *Multiplication* component is now used to convert the values from the scripted NRCS equation's *Division* component and the labeled *Total Surface Area of Site Mass Addition* component. This will output the respective surface runoff volumes from the different rain events in cubic feet.

Now that all the data has been computed, the next steps will demonstrate the visualization of the data values to communicate the significance of the runoff volumes. The method in which this portion of the script has been developed allows the user to continually change the site's surface conditions and simulated rain event values for a dynamic and responsive calculation of stormwater runoff volumes.

There are a variety of ways to graphically communicate runoff volume, some of which will be demonstrated specifically in this chapter but could also be included from other chapters. The emphasis of these graphic communication pieces will be the impact of size and surface type have on runoff volume. This is obviously influenced by the ratio or percentage of hardscape materials to softscape, so it is important to visually communicate that with supplemental graphics like charts. It is also valuable to show the nonlinear relationship between runoff volume and rain events, as larger precipitation values will demonstrate a more compounding contribution to runoff volume rather than a more direct distribution. Many of the chart settings for this phase of the model will be contingent on the Rhino workspace display and user preferences on size, location, and color.

The first graphic visualization will be to chart the impact that the different surface types have on stormwater runoff volume, signifying a hierarchy of surface categories, as demonstrated in the initial steps in Figure 8.13. This will first involve dividing the individual CN values from the *Combined Weighted CN Values Merge* component by the *Existing Weighted CN Mass Addition*. Once these values have been inputted into the *Division* component, it is best to move it into a large open area within the Grasshopper workspace to accommodate for the larger charting components from the *Conduit* plugin.

As stated in Chapter 4 on different plugins, the *Conduit* plugin is a great option for creating user interface charts. A variety of chart options can be explored; however, in this example the *Draw Pie Chart* component will be used. The output value from the *Division* component in the previous step will be used for the *Data* input and a panel listing all the surface types in the same order as the script will be used for the *Labels* input. By default, the panel won't structure text as a list, so it is important to *right-click* the panel and deselect the *Multiline Data* option, so the different surface types coincide with a list value. This pie chart component will register orange with a warning note until a *Bounds* value is inputted but that will be resolved in the next step.

The next series of *Conduit* components will set up an operation to locate and display the different charts, in this case, at the bottom of the Rhino window using the *Horizontal Tiles* component, as shown in Figure 8.14. This will involve a variety

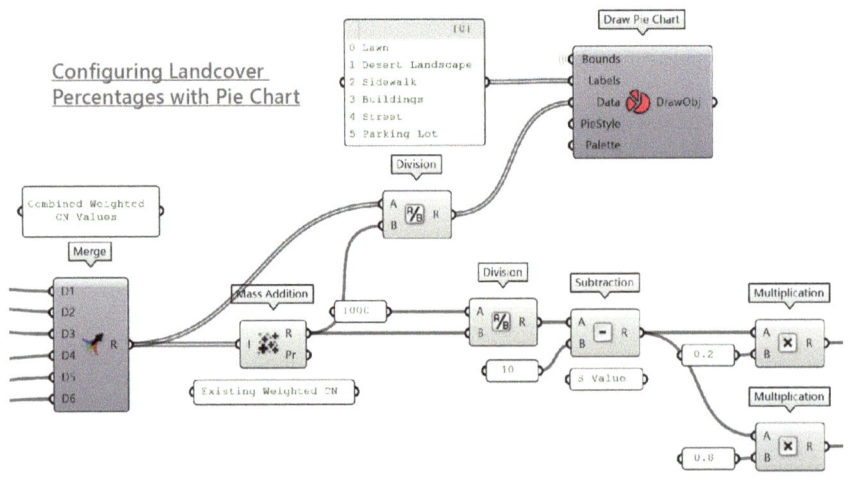

Configuring Landcover
Percentages with Pie Chart

Figure 8.13
Creating a pie chart from the *Conduit* plugin to visualize the percentage of the different land cover's weighted CN values.

of dynamic settings from different sliders to dictate the size, height, and padding of tiles for the charts to be nested in. The process begins with a *Get viewport boundary dimensions* component to size out a workspace from the Rhino viewport. The component's *Bounds* output is then inputted into a *Horizontal Tiles* component to generate a set number of tiles to configure specific sizing parameters.

For this demonstration, an effective parameter for the *RelSize* input uses seven equally proportionate sizes by using the *Duplicate Data* component with a *Data* (*D*) input of *1* and *Number* (*N*) of *7* to repeat that proportion that many times. The *Height* input will be a percentage of the tile from the bottom of the viewport where a number slider value of *0.37* makes their size almost a third of the window. The *HeightPad* input uses another number slide value of *0.09* to ensure the charts don't crowd the edges of the viewport. The outputted *Tiles* can then be connected to a *List Item* component to select a specific tile to embed the pie chart in. An *Index* (*i*) value of *1* is used to give it an appropriate placement relative to the Rhino model of the project site. Many of these parameter decisions won't be finalized until the end when all the chart data is placed within the viewport and can be sized accordingly as supplementary communication of stormwater runoff.

The *List Item* output can now be connected to the *Bounds* input of the *Draw Pie Chart* component to complete that part of the chart operation. The pie chart

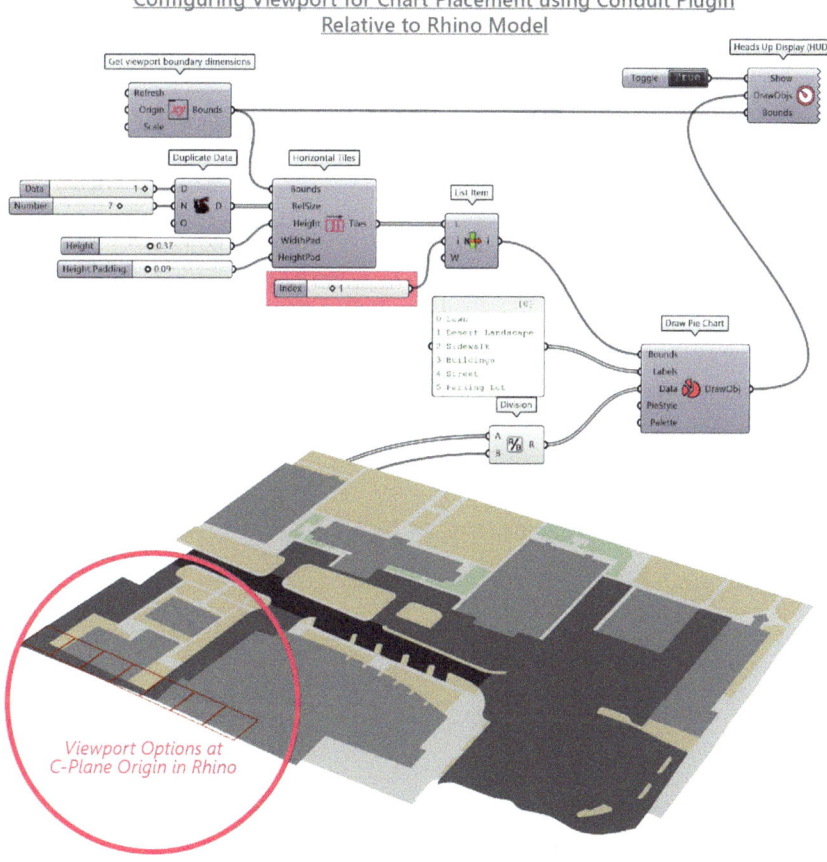

Figure 8.14
Configuring the layout of the charts to the Rhino viewport.

component's output can next be connected to the *DrawObjs* input of the *Heads-Up Display* (HUD) component. The initial viewport boundary *Conduit* component can then be connected to the HUD's *Bounds* input as well and a *Boolean Toggle* set to *True* can finally *Show* the chart.

Having both the *Horizontal Tiles* and *List Item* component preview on will display the array of tiles and the isolated tile for the chart at the 0,0,0 axis point in the Rhino workspace as shown in the model image of the figure. These components can have the preview disabled once the appropriate tile is determined.

The next step in Figure 8.15 will create a style and color palette for the pie chart, which is an elaboration of the previously demonstrated operations for this plugin in Chapter 4. The *Conduit Pie Style* component will most likely need adjustment with number sliders for at least the InnerRad, LabRad, and LabHeight inputs to output either a traditional pie chart or a doughnut style and resize the labels. These again will be contingent on the previous chart size settings and location within the viewport.

The colorization of the chart can be done with a *Gradient* component that uses the same *Division* component for the *Parameters* (*t*) input. Its lowest and highest values for the *Lower* (*L0*) and *Upper* (*L1*) input using an operation with a *Bounds* and *Deconstruct Domain* component. For most effective communication and consistency purposes, the gradient's color style will be copied for the other demonstrated graphic chart visualizations.

The *Conduit Chart* component in Figure 8.16 will be used to visualize the different runoff values from different rain events. Similar to the pie chart in the previous steps, a *List Item* component connected to the *Horizontal Tiles* component is used for the *Bounds* input to place the chart by using a number slider for the *Index* (*i*) input. A number slider input makes it easy to scroll through the different tile locations for best placement of the chart; however, it won't be noticeable until it is connected to a *HUD* component for display.

The *Values* input for this chart component will use the *Runoff Volume* output from the highlighted *Multiplication* component at the end of the scripted NRCS calculator. The *Categories* input will use the *Merge* component that contains all the rain event number sliders. The titles *Rain Event in Inches* and *Runoff Volume in Cubic Feet* are used respectively for the *CatTitle* and *ValTitle* inputs. It is also recommended that once all these scripted operations have been completed, this

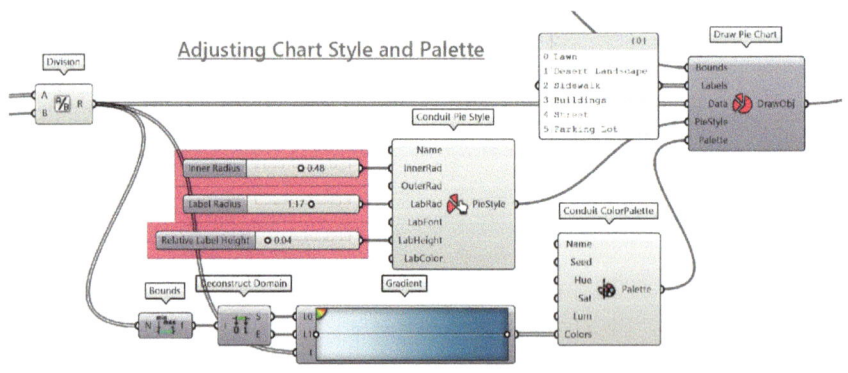

Figure 8.15
Configuring the different settings for the pie chart style and palette.

Figure 8.16
Applying the runoff
volume values to
the *Conduit* pie
chart option.

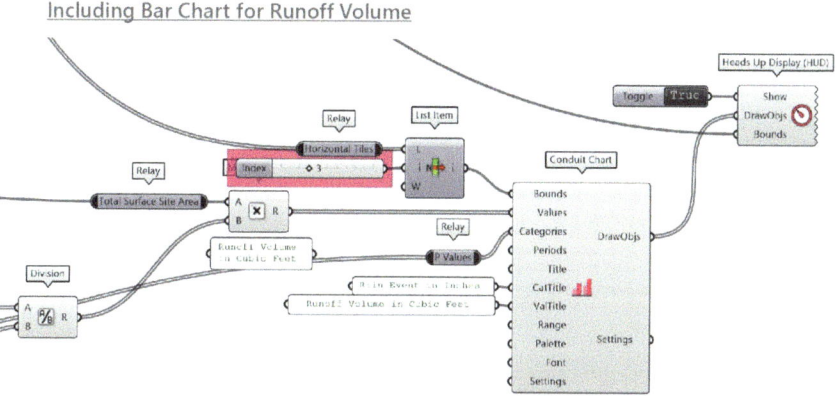

Including Bar Chart for Runoff Volume

group of components is moved to the right to give space for the chart style shown in the next step.

Stylizing the chart in Figure 8.17 uses the same blue tone *Gradient* component for the bar chart *Colors* input where the same *Runoff Volume* output is connected to the *Parameters* (*t*) input. For the *Range* input, a *Construct Domain* component is used with the default *0* as the lower domain value and the upper domain value being the highest runoff value. This input can be retrieved from the *Deconstructed Domain* component that is used as part of the *Conduit ColorPalette* operation.

Lastly, a *Conduit Font* component is used to determine the font settings for the chart. Number sliders are once again used to gauge proper sizing for the *Height* and *Align* inputs of the component; however, their values will be contingent on legibility within the Rhino viewport. A *Colour Swatch* component is also used to apply a specific *Color* as well. Below are the two charts within the Rhino model workspace. The two chart types communicate the significance of surface type impact and runoff volume from different rain events; however, it is also important to communicate that spatially within the project site in the next steps.

The final step of the NRCS script is visualizing the impact that each individual model surface has on the total surface runoff volume. This will create a robust and comprehensive graphic that visually communicates the data through size, color, and numeric display. This will give a relative relationship between the different individual surfaces but not necessarily be representative of the runoff volume. Once again, this is important to understand since the NRCS calculator does not indicate point sources of stormwater runoff but, rather the overall impact of a site's surface composition on runoff volume.

The other two remaining *Merge* components in Figure 8.18, labeled *Combined Surface Geometry* and *Combined Surface Impact Values*, will now be used to complete the visual communication of this performance modeling script. First, a *Multiplication* component is used with a number slider to adjust the relative heights of the surfaces that will be extruded. Again, this will be contingent on the size and scale of the project site and is intended to show a proportionate relationship between the surfaces. This scaling factor will be used for *Unit Z* component value that can be inserted into *Direction* (*D*) input of an *Extrusion* component. The *Base* (*B*) geometry of this component will be the output from the *Combined Surface Geometry Merge* component. The *Gradient* component is slightly edited by *right-clicking* the two color values (white circles on the gradient component) to adjust the alpha channel for different transparency levels, maintaining the dark blue (right-hand side) as more opaque and the light blue (left-hand side) as more transparent. The list and range of these values will follow the same process as the previous gradient operations with a *Bounds* and *Deconstruct Domain* component for the *Combined Surface Impact Values Merge* component.

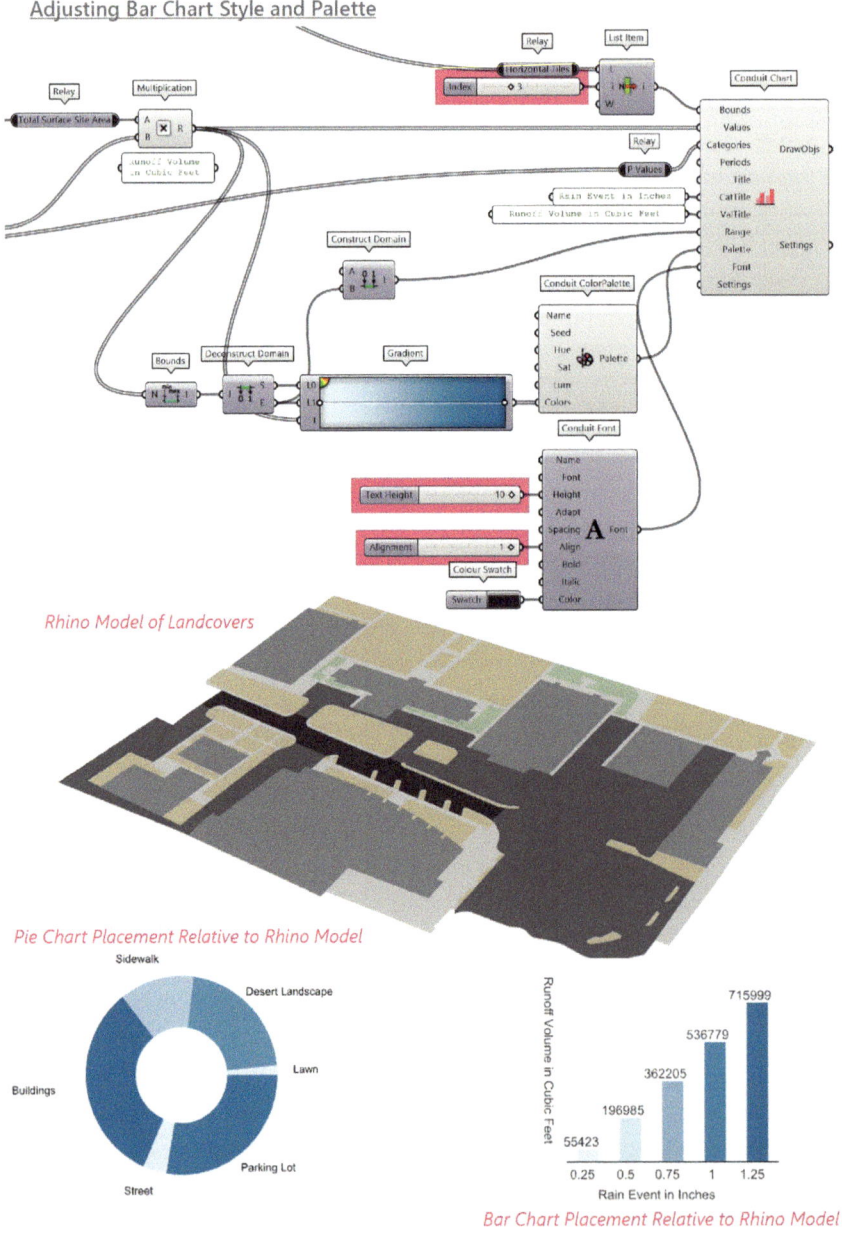

Figure 8.17 Composing the collection of charts and Rhino model for graphic communication purposes.

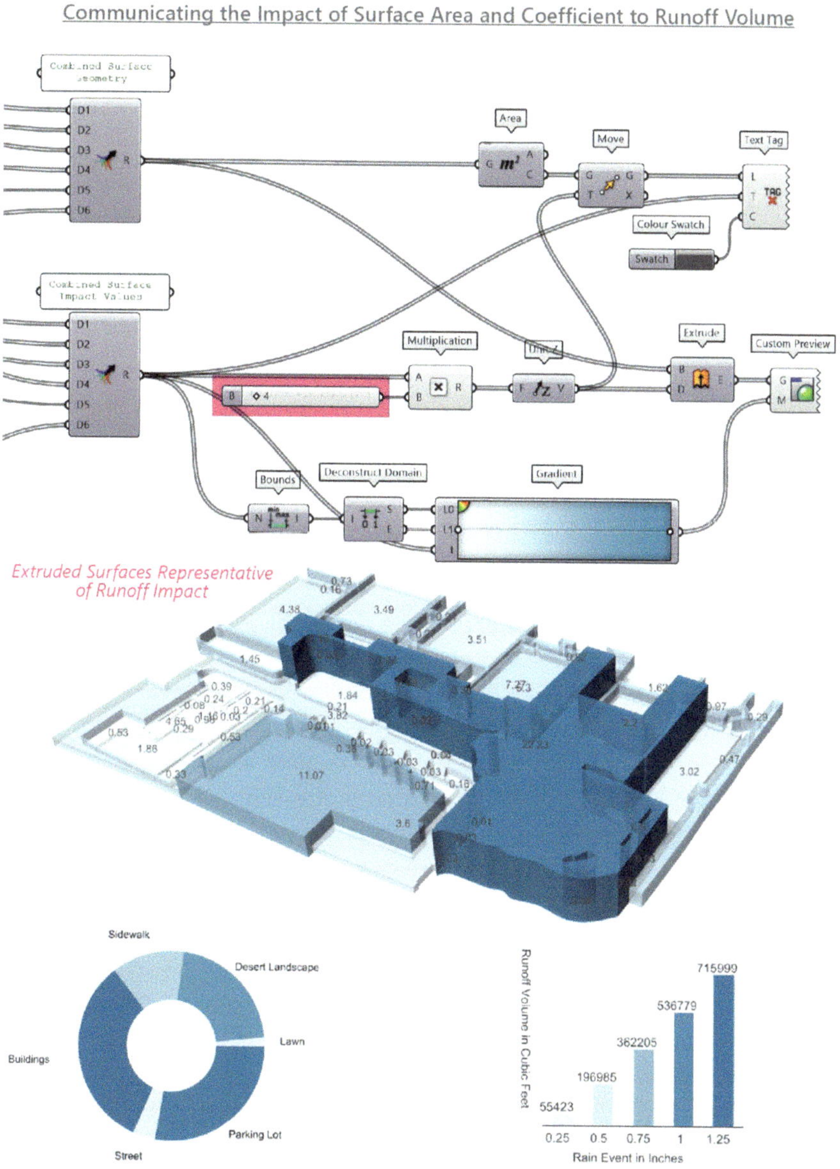

Figure 8.18
Modeling the
impact of different
surface's CN value
on runoff volume
with extruded
geometry.

Communicating the Impact of Surface Area and Coefficient to Runoff Volume

To label the extrusions, a *Text Tag* component is used to display the weighted individual CN values for the surfaces. The *Merge* component labeled *Impact Values* will be the *Text* (*T*) to display, while the *Centroid* (*C*) from an *Area* component for the surface geometries is moved on the z-axis using the same previous value for the *Extrude* component. This can, however, use a different z-value, depending on the scale of the project and display clarity.

As a result from this final operation, each surface from all the layers referenced is extruded, colored, and annotated as its individual impact to surface runoff volume. Graphically, the extruded geometry's color gradient will visualize the relationship of size and color between the different individual surfaces.

Comparing the Relative Impact of Size and Coefficient
between Turf and Impervious Surfaces

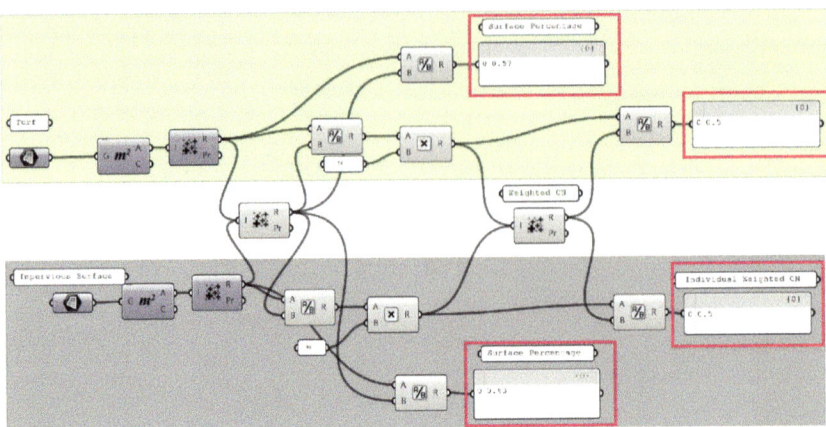

Figure 8.19
Comparing the
impact of different
surface coefficients
and size to runoff
volume.

OUTCOMES

By visualizing the impact of different surface typologies along with individual surface impact, strategies can be deployed to prioritize green infrastructure and best
management practices to reduce surface runoff at specific locations indicated in
this model. As a result of this modeling method, this stormwater calculator can
serve as both an operational equation, which is abstract, and a spatial tool to integrate the values into tangible and visual results. This stormwater surface impact
model will create a datascape of individual contributions, with the higher the value
meaning, the higher the singular impact relative to one another.

As mentioned throughout, the relative surface impact (combination of surface
area and weighted CN value) is another significant outcome from this model. For
example, in Figure 8.19, regardless of surface area, when an impervious surface
with a CN value of 98 is compared with a grass surface with a low CN value of 74,
the impervious surface needs to be 14% smaller to have the same runoff impact
as the pervious grass surface, as shown below.

This demonstrates that it is not simply about determining the composition of
different surface types and their respective surface area, but also how the ratio
or relative relationship between surface types and size can impact runoff volume
from the weighted CN value. This complex function can be instantly computed
and updated as surface conditions change within a project site leading to a robust
comprehension of data and runoff processes.

FUTURE POTENTIAL

This script serves as an effective introduction to modeling and visualizing stormwater runoff volume using the NRCS calculator. However, as a main point to landscape
performance, the script simply generates outputs instead of outcomes. Although
there is a series of different runoff volumes from different rain events, this information is irrelevant and relative until compared with other surface conditions to
formulate a baseline for comparison.

In order to provide performative outcomes, it is important to compare the existing runoff volume that is currently being calculated from this script with both a predevelopment or natural condition along with a green development model. By doing so, the user can understand the significance of the current runoff volume in relation to predevelopment conditions and how that has been impacted by the addition of impervious surfaces. Including green development strategies can begin to formulate and recalculate the runoff volume reduction from alternative surface types that include permeable pavers, rain gardens, and other green infrastructure applications. Modifications to this script would be minor in order to accommodate these additional outputs by adjusting the weighted CN value to represent both the predevelopment and retrofitting existing surface typologies with appropriate green infrastructure CN values. These adjustments will simultaneously be re-represented in the different model graphics and charts.

One of the primary principles behind this book is to graphically represent and communicate the metadata of a computational model. With the data of this script focused on surface area, this data can be cross-referenced with a material cost sheet by square footage that includes construction, maintenance, and lifecycle costs. As green improvements are implemented in this model, a tabulated chart can record and compare the cost viability for green infrastructure to incorporate the economic benefits of stormwater reduction.

9 Erodibility

Erosion of unprotected or disturbed land from surface runoff can have a detrimental impact to the environment beyond just soil loss. Eroded sediment in the runoff is one of the top contaminants to water quality due to its high concentration and its role in acting as a vehicle for other contaminants to latch on to and travel further downstream to other receiving water bodies (Environmental Protection Agency, 2021). There are also valuable nutrients in this soil that helps sustain the local flora and fauna of an area so that element of the ecology becomes impacted as well. This becomes even more evident during the construction phase of a project when vegetation is removed and regrading practices are performed in preparation for the site's development.

This chapter will demonstrate how to model and evaluate critical areas where there may be highly erodible land (HEL). Identifying these areas can suggest mitigation strategies to prevent or reduce erosion potential through different design tactics. The chapter will not demonstrate the modeling of these tactics; however, the preliminary process of assessing a project site for HEL becomes valuable for responsive landscape performance interventions.

GOALS

The purpose of this performance model is to evaluate potential erosion or soil loss to a project site due to regrading, construction, or vegetation loss as it leaves bare and unprotected soil for a duration of time. By implementing the erodibility index (EI) calculator to this project site, the following can be achieved to strategize erosion control interventions such as retaining walls, regrading, or plant communities.

1. Aggregate the appropriate terrain attributes for calculating the EI of the site
2. Assess the impact differing cell or plot densities have on generalized or detail erosion potential
3. Determine which areas within the project site have HEL
4. Evaluate how rainfall-runoff throughout different time intervals impact the EI

PERFORMANCE METRICS

Defining erodibility for this model uses the National Food Security Act manual to calculate an EI for HEL. HELs are classified as either agricultural lands or construction

DOI: 10.4324/9781003208020-12

sites with an index of eight or higher and can be calculated by dividing the potential erodibility for soil by the loss tolerance (T) using the following equation:

$$EI = (R \times K \times LS)/T$$
(Natural Resources Conservation Service, n.d.)

Erosivity Factor (R)

This R Factor is essentially the rainfall-runoff impact on the project site. The force and amount of rainfall is what contributes to different levels and has a direct relationship of higher rainfall with equally higher R Factor. These values can be found from either weather records, maps, or even Environmental Protection Agency's (EPA's) Rainfall Erosivity Factor Calculator for Small Construction Sites (Environmental Protection Agency, n.d.).

Soil Erodibility Factor (K)

This factor is a measurement of how susceptible soils are to erosion to the amount and rate of runoff. Both fine clay and coarse sandy texture soils have low K values that range from 0.02 to 0.15 and from 0.05 to 0.2, respectively. Silty loam soils have a moderate K value that range from 0.25 to 0.40 because they can detach and contribute to moderate runoff. Soils that are high in silt content often have the highest K Factor, greater than 0.4, because they easily detach and contribute to high rates of runoff.

Slope Length of the Gradient (LS)

Although they are separate variables, they are measured as a combined factor, as shown in the table of Figure 9.1, which is pulled from the Agriculture Handbook 703 for construction sites. L factors the slope length and S factors the slope gradient. Slope length is the horizontal distance runoff travels before there is significant change to the gradient, as shown in the table. Slope gradient is the change in elevation over a specific distance, representing it as a percentage.

LS Factor Chart

Figure 9.1
Matrix of different LS Factors for the modeled equation.

Soil Loss Tolerance (T)

The T factor of this equation determines the maximum rate of annual soil erosion that will permit either crop productivity or healthy vegetation growth. These factors range from 1 to 5 in units of tons per acre per year. Since the rest of the equation is divisible by this value, a 1 value represents shallow and fragile soils that are highly erodible versus a value of 5 that classifies deep soils not susceptible to erosion.

MODELING METHOD

Like Chapter 7, modeling the potential erodibility of the landscape first requires the aggregation of multiple factors starting with the terrain itself which, begins in the first portion of the script in Figure 9.2. The process is no different than the previous demonstration of utilizing topography lines to create a mesh, so in this portion of the chapter instructions will be more generalized but can be referred back to Chapter 5 for more thorough explanations.

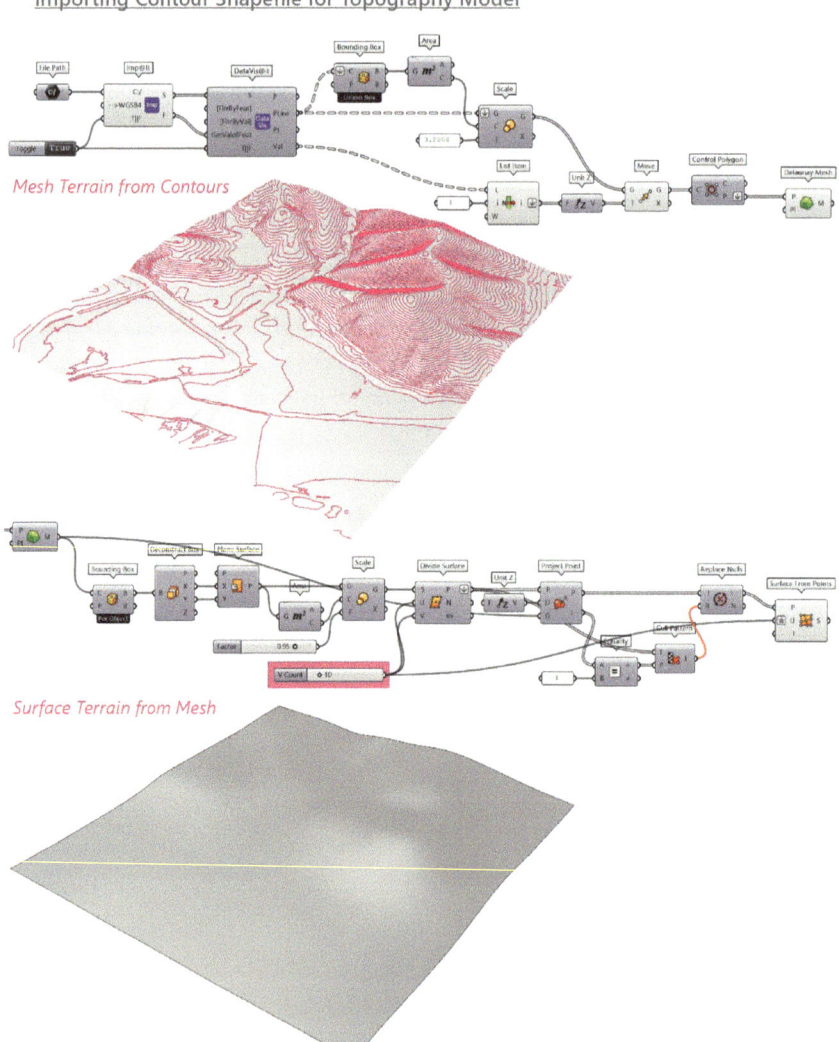

Figure 9.2
Generating a surface terrain model from imported contour shapefile.

The contour elevation metadata brought in from the shapefile using the *DataVis@it* component is used for the *Unit Z* input to the *Move* component after the *PLine* output has been *flattened* and resized correctly with the *Scale* component to a factor of *3.2808*. Once those transformations have been done to the contour lines, the *Control Polygon* component can be used to extract all the pline points and *flatten* them for the *Delaunay Mesh* component. Since the purpose is to visualize the comprehensive impact different erosion factors have on a landscape, the most effective way is to next convert the mesh into a surface which is more conducive for that type of representation. This conversion operation is also a repeated process from previous chapters, as shown in the second script image of this figure.

The Delaunay Mesh component provides parameters to inform the creation of a surface by starting with a *Plane Surface* component. It is recommended that the planar surface is scaled down enough to ensure that points can be projected onto clipped corners or other obscurities in the mesh terrain. In this example, a scale factor of *0.95*, a 5% reduction, was adequate enough to all points that could be projected in the later steps. This scale factor can be adjusted throughout this process with a number slider in order to retain as much of the existing mesh terrain as possible.

The scaled planar surface can now be converted into a grid of points with the *Divide Surface* component where a set density of points can be determined with a number slider. A value of *10* will generate less detail in the terrain but ensure that the later drainage component performs accurately. It is not essential to get exact drainage lines with a high detail terrain model, as this is near impossible to replicate, so it is preferred to get the general flow of drainage on the terrain. This number slider can obviously be adjusted throughout the duration of the script development to find the best fidelity of this model.

From the *Divide Surface* component, the points can be *flattened* and projected in a positive *Unit Z* direction onto the Delaunay mesh. A quick operation then tests if any points missed the projection for non-orthogonal site boundaries or for other conditions and removes them for the *Surface From Points* component. The expression for the *U* input of this component is *x+1* since this component is asking for the number of points instead of where the previous component is asking for the number of segments while using the same number slider. Now that the mesh has been converted to a surface, the next step integrates the soil profile extracted from an external source such as Web Soil Survey (WSS).

The script in Figure 9.3 shows that the same process occurs when bringing in soil characteristics, in this case, the K Factor for erosion, as was also demonstrated in Chapter 5. This next operation has two main benefits. First is the removal of duplicate curves and the second is for a more effective evaluation of points in these soil profiles. When a specific soil profile has two curves – an outer and an inner – the latter can be removed with a *List Item* component since the inner curve is the second item which gets omitted from in process. From that component, the flattened curves can be extruded and capped for the *Solid Difference* component to remove any duplicate geometries. This operation is also beneficial for the next step for 'projecting' these now surface volumes onto a surface terrain.

Importing and Morphing Soil Profile to Surface Terrain

Figure 9.3
Morphed soil profile
curves to surface
terrain model.

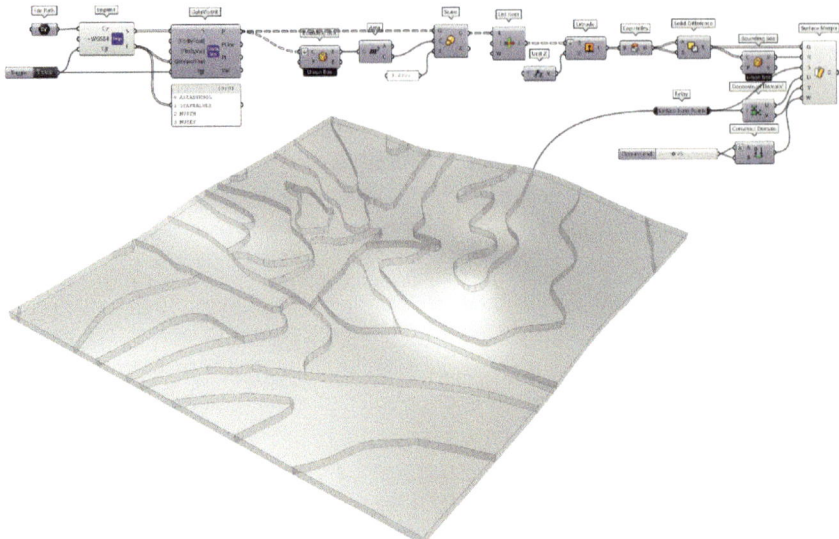

The process of projecting these geometries is performed with the *Surface Morph* component. A *Bounding Box* component is used as a *Reference* (*R*) while the previously created terrain surface geometry, illustrated in the image with the labeled *Relay*, is connected to the *Surface* (*S*) input. The main benefit of performing this operation is to ensure all points are accounted for when determining which are located in a specific soil profile curve in the later steps. By giving a *W* domain value in both the negative and positive direction from a *Construct Domain* component, the morphed surface will extrude both below and above the surface terrain. The extent of this domain with a number slider is subjective to the project site, as long as all three-dimensional points fall within the closed volumes. The *expression* for the *A* input of this component is −*x* so that it extrudes in both directions. This will be utilized and elaborated further in the future *Point in Breps* component that outputs more concise data than the evaluation of points in a curve.

The next step of aggregating all the erosion factors is cross-referencing the soil profile geometry with its associated metadata. Both sets of data will usually come from the same source, such as WSS, so it is fairly simple in connecting the two. Figure 9.4 shows how a *List Item* component is used from the *Val* output of the *DataViz@it* component to isolate the common variable (*mukey*) between the two data sets. It is best to use a panel from the *F* output of the *Imp@it* component to ensure the correct index is used for the list item, which is *3* in this demonstration.

First, the data from an external spreadsheet downloaded from WSS needs to be filtered after referencing it with the *ExcelDynamicRead* component and plugin. An operation is first done to filter out the *mukey* column values, where the *List Item* component has an index of 0 representing the first column. The same set of components is repeated for both the *K Factor* and *T Factor* column values where the *List Item* component has an index of *3* and *4*, respectively, since that data is contained in the fourth and fifth columns of the spreadsheet. These indexes may vary from the download, so it is best to first check and confirm which list column contains the numeric K and T Factor values so that the correct index value is assigned.

A panel from the *B* output for each *Split List* component is used to visualize the filtered datasets. A *Match Text* component is now used to cross-reference and structure the data into their respective soil profile branches. The *mukeys* from the soil profile geometry is first connected to the *Text* (*T*) input as separate branches to maintain a consistent structure throughout the rest of the script. The split list containing the *mukeys* is then connected to the *RegEx* (*R*) input of that component to evaluate which values match by outputting a Boolean pattern for two *Cull Pattern* components, labeled *K Factor* and *T Factor*. These components won't be utilized until later in Figure 9.8, so the labels will help in referencing them further down the script.

Going back to the *Surface From Points* component, Figure 9.5 illustrates how the terrain surface can be divided into equal-sized cells with the *Lunchbox Quad Panels* component. The sizing of these quad panels is highly contingent on the LS factor of the performance calculation, which is measured in specific intervals shown in the table of Figure 9.1. Equally dividing this surface first requires the *Dimensions* component to output a width (*U*) and length (*V*) value that can be *divided* by a specific interval. This sizing interval will be refined later, but for now a value of *100* from a number slider is used to dictate 100-foot dimensioned cells as inputted into the *Data* component labeled *Length*.

The quad panels can now be used to extract the *Centroid* (*C*) point with the *Area* component for input into the *Point In Breps* component. The grafted *Brep* (*B*) input for this component will be from the previous *Surface Morph* component, as illustrated with the *Relay*. The reason for grafting the input is so that the points can

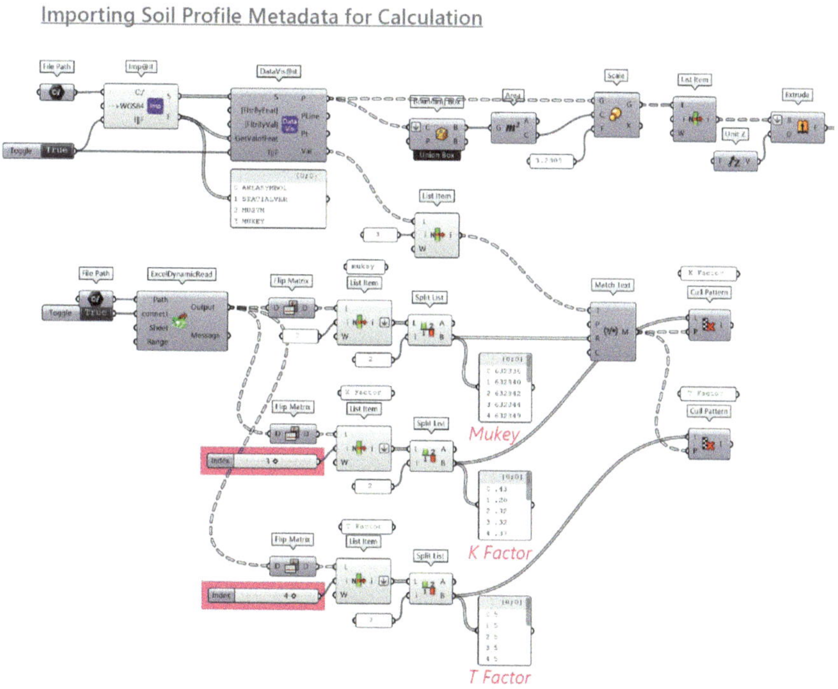

Figure 9.4
Cross-referencing soil survey metadata with the soil profile model.

be evaluated for inclusion for each soil profile Brep. This will assist in restructuring the data to coincide with the soil metadata from the previous script figure.

The outputted evaluation of points generates a Boolean pattern that can be used to restructure the flattened list of points and quad panels into their respective grafted soil profile groups, as shown in the model images of the figure, with a *Cull Pattern* component. Both components should be labeled with their respective geometry type (points or surfaces), as they will be referenced later. The right-side model image is not a specific outcome, but is intended to illustrate how all the data is parsed out into separate groups with one of those groups being highlighted.

Since there may be some Breps that contain no points, it is important to remove any branches with zero data with the *Clean Tree* component, with the N, X, and E input settings set to *True*. Although this will physically remove those empty branches, it is best practice for this demonstration to also renumber the branches to account for the gaps with the *Path Mapper* component. The default setting for this component is not the correct one, so the component needs to be *right-clicked* and have the *Create Renumber Mapping* selected. If this component is copied and pasted in the script, it still needs to have that setting reapplied since it won't follow suit. These two *Path Mapper* components are also labeled as they will be referenced in a later step.

From the newly filtered and renumbered branches of data, the quad panels can now be analyzed for their slope conditions to determine their LS Factor for erodibility, as shown in Figure 9.6. This is also a familiar operation process of evaluating each quad surface from their points to generate a vector or Normal (N) direction. This output can then be measured against a *Unit Z* value with the *Angle* component. Since this angle is in radians but the erodibility calculator use percentage, the output needs to be converted using the *Tangent, Multiplication*, and *Minimum* components to prepare the script for evaluation.

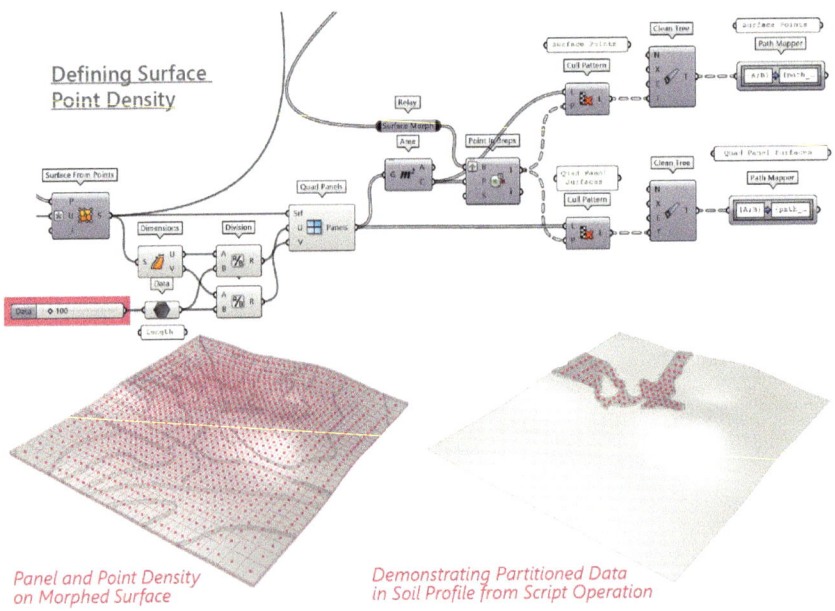

Figure 9.5
Creating and partitioning point grid with the morphed soil profile geometry.

The *Multiplication* component needs a *B* input of *100* to convert the decimal values into percentage and the *Minimum* component needs a *B* input of *60* since the LS Factor table has a maximum slope consideration of 60.

The LS Factor table found in Figure 9.1 can be applied to an Excel table as is and referenced with the *ExcelDynamicRead* component as shown in Figure 9.7. Similar to the soil profile metadata, the output from this component is restructured with the *Flip Matrix* component and first inputted into a *List Item* component with *0* as the index to isolate the table to the first column of slope percentages. The output is *flattened* and then removes the header row of the table with the *Split List* component connecting a panel value of *1* for the *Index (i)* input. The B output now shows just the slope percentages, as visualized with the panel while also being structured as *Consecutive Domains*. The *Additive (A)* input of this component needs to be set to *False* so that they remain consecutive instead of cumulative as visualized by its connected panel. Finally, the consecutive domains can be connected to the *Domains (D)* input of the *Find Domain* component with the *Minimum* component connected to the *Number (N)* input.

The next step is finding the LS Factor that relates to all the unique slope conditions of the terrain surface analyzed previously. A *List Item* component is connected from the *ExcelDynamicRead* component with a *flattened* output. A panel is connected to showcase which index values is associated with the different length values. As shown in the script image of the figure, a length of *100* is index item *9*,

Figure 9.6
Using restructured dataset of points and panel surface to evaluate their slope percentages.

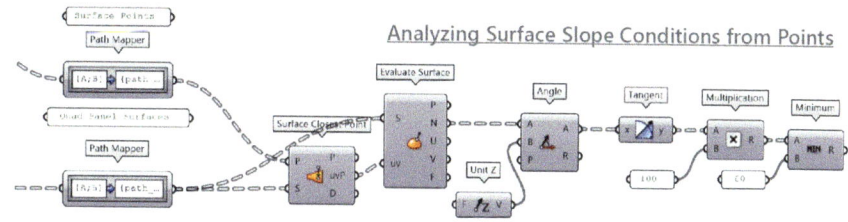

Figure 9.7
Cross-referencing an imported Excel file of LS Factors with the model's slope conditions.

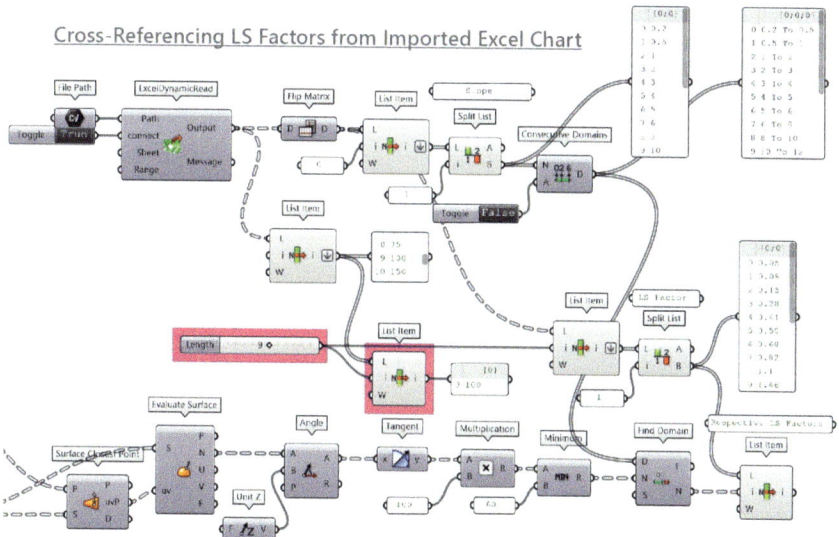

which correlates with the previous dimensioned quad panels into 100-foot cells. This next set of operations makes the process more dynamic with this parameter by adjusting which length value to apply to the project site and have the corresponding LS Factor reflect that change.

Another *List Item* component is connected from the previous one with a number slider associated with the length index items, which can be seen in the script that a value of *9* outputs the *100*-foot length dimension. The next step is key in making this dynamic by connecting the highlighted *List Item* output to the *Data* component labeled *Length* in Figure 9.5. The same number slider is then connected to another *List Item* component that is connected to the *Flip Matrix* component to isolate the table to that respective LS Factor for that specific length value. That can be better understood once *flattened* and connected to a *Split List* component with the *Index* (*i*) set to *1*. The panel connected to the *B* output now shows all the different LS Factors for different slopes at a length of 100 feet. The *B* output is connected to one final *List Item* component of this operation that uses the *Find Domain* output *Neighbor* (*N*) as the component *Index* (*i*). This component is also given a label of *Respective LS Factors* for reference in the next step of the script.

In this portion of the script in Figure 9.8, the different factors are calculated through the erodibility index equation. This will include both the K and T Factor values from the soil survey data along with the LS Factor from the previous step. These values, as illustrated in the figure, are labeled in the *Relay* components as they are brought from the different parts of the script.

Since both the K and T Factors are only single instances from the map, they need to be duplicated to match the length of the number of points identified within each soil profile. To perform this action, the points coming out of the previous *Cull Pattern* component labeled *Surface Points* in Figure 9.5 are also *relayed, Points Cull Pattern*, to serve as the branching structuring for duplication. That relay is connected to a *List Length* component that is inputted into a separate *Repeat Data* component for both the K and T Factor *Relays*. Like what had been done previously with data branching that had empty or *null* values, an operation with a *Clean Tree* and *Path Mapper* component set to *Create Renumber Mapping* is used for each factor. All the settings for the *Clean Tree* component should also be set to *True*.

Per the erodibility index equation, a *Multiplication* component is used to first compute the R, K, and LS Factors together. The R Factor can be attained from either a rainfall factor (R Factor) map or by following the instructions from EPA's Rainfall

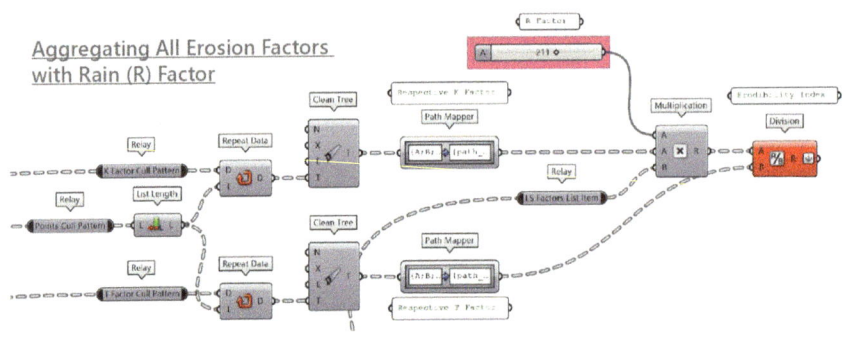

Figure 9.8
Aggregating all the erosion factors for the equation with the rainfall (R) factor.

Erosivity Factor Calculator for Small Construction Sites website to find the value for a specific date and location. The static maps will give an annual average value, whereas the online map can output a value specific to a single or range of dates.

This specific location during the month of March had an R Factor of *211* according to this website, so a number slider set to that value is used as the first input to the *Multiplication* component. The second input is the cleaned and renumbered K Factor from the *Path Mapper* component, and the last input is the LS Factor from the *List Item* component labeled *Respective LS Factors* in Figure 9.7, illustrated as a connection from a *Relay*. The values outputted from this component can finally be divided by the T Factor values with the *Division* component with the output *flattened*.

In this example, the project site contains water with a 0 T Factor, so when divided by that value, the erodibility index will also be 0. This will result in a red *Division* component, noting an error; however it works to its advantage in making the geometric cells for water disappear in the later steps, since this equation only considers erosion from rainfall drainage lines.

Now that the calculations have been made for all the points within the site's designated soil profiles, they can be visualized through appropriate color, sizing, and annotation of their respective erodibility index. For colorization and scaling, as shown in Figure 9.9, the values are first remapped to correspond better with the scaling transformation since that parameter prefers to be within a range of *0–1*, whereas the erosion values can be as high as 300 or greater.

The *Remap Numbers* component inputs the *Erodibility Index* values coming out of the *Division* component and then uses the *Bounds* component to calculate the full range of values. For legible visualization of erosion occurring at each surface cell, a new range of values from *0.25* to *0.95* are used for the *Construct Domain* component as well as for the lower (*L0*) and upper (*L1*) limits of the *Gradient* component. The reason for these range of values is so that none of the cells gets scaled down to 0 since that can create errors and with a slightly lower value than 1 the cells will mostly retain their original size. These values can also be set to number sliders, as demonstrated, to adjust accordingly to an appropriate range and contrast of erosion values.

The new remapped numbers are then connected to the *Parameter* (*t*) input of the *Gradient* component where the colors are adjusted for the most effective visual communication of erosion, as starting to be shown in the model image of Figure 9.9. This component will serve as the *Material* (*M*) input of the *Custom Preview* component.

The scaling portion of this script is not going to create a geometry size representative of erosion, but rather a subtraction or perforation of geometry representative of that soil loss. In other words, the larger the hole in the geometry, the higher the soil loss. Representing this first requires the original *Surface Points* (*Relay*) and *Panel Surfaces* (*Relay*) retrieved from their respective *Path Mapper* components in Figure 9.6. These data sets will both be flattened at the input of their geometry components (*Point* and *Surface*). The surface geometry will be converted to a *Curve* and connected to the *Geometry* (*G*) input of the *Scale* component. The points will serve as the *Center* (*C*) of scaling and lastly the remapped

Visualizing Erodibility Index with Colorized and Scaled Panels

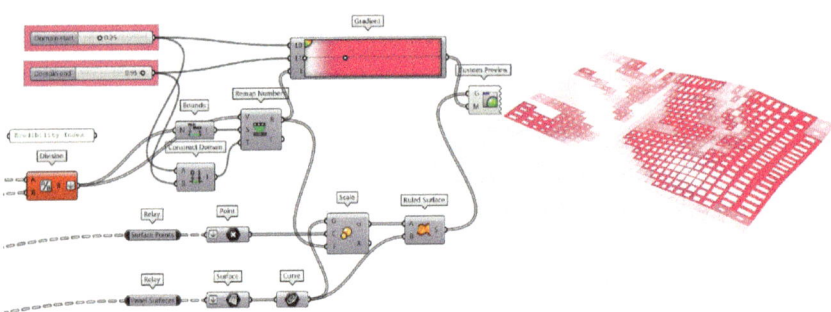

Figure 9.9
Using the calculated
erosion factors to
parameterize the
surface quad panels.

values will be the *Factor* (*F*) input. Creating this soil loss effect will use the *Ruled Surface* component to loft a surface between the original *Curve* component and the scaled curve. This surface is now connected to the *Geometry* (*G*) input of the *Custom Preview* component.

Visually, in the model, there is a clear distinction between which areas have a higher erodibility index than others. However, these values are still relative until they are measured against a baseline comparison index value of 8, which is considered highly erodible land (HEL) (Natural Resources Conservation Service, n.d.). This baseline value can be used to evaluate which areas of the project site are susceptible to erosion and require control devices or preservation of vegetation to prevent potential soil loss. This remaining portion of the script will be able to identify and visualize those areas for future design interventions.

To begin this evaluation process in Figure 9.10, the *Erodibility Index* values are inputted into a *Larger Than* component with the *B* input being *8* to represent the HEL baseline value. This will create a Boolean pattern that can be used to isolate the data equal to or greater than this input. Multiple *Cull Pattern* components use this pattern for the original *Surface* and *Point* components along with the erodibility values.

Next, text labels will be created where their size corresponds with the amount of soil loss, so that higher soil loss have larger labels to emphasize these critical areas further visually. The culled points will be the *Location* (*L*) of the *Text Tag 3D* component and the culled erodibility indexes will be the *Text* (*T*) to display. Another remapping operation is performed to give size variants to the displayed text. The erodibility values are first inputted into the *Remap Numbers* component and their *Bounds* used as the *Source* (*S*) domain. For the *Target* (*T*) domain input, a *Construct Domain* is used with number sliders to gauge the smallest and largest text sizes. Color and justification can also be assigned to this component.

The final step is to highlight the HEL areas with a gradient boundary, as shown in the model image of the figure. Creating this boundary first requires the merging and deconstructing of the culled surfaces. The *Brep Join* component is used to merge all the surface cells and then the *Brep Edges* component extracts the different types of surface curves, such as exterior (naked) and interior. The *Naked* (*En*) exterior curves are then inputted into the *Join Curves* component. The joined curves can now be *extruded* to an appropriate and legible height, in this case 100 feet.

Figure 9.10
Highlighting the
designated HEL
within the site
model.

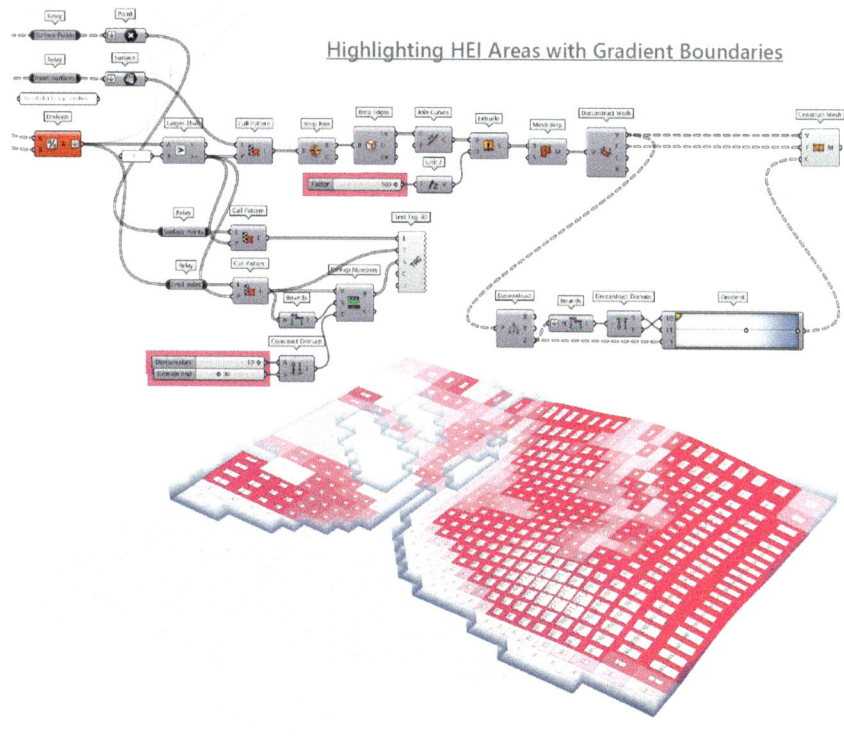

Surfaces cannot be visualized as a gradient; however, the conversion to mesh and utilization of its characteristics can be used for this effect. The extruded surface is first converted to a mesh with the *Mesh Brep* component for extracting the points with the *Deconstruct Mesh* component. The points or *Vertices* (*V*) are also *deconstructed* to extract their *Z* coordinate value.

These values are first inserted into a *Gradient* component's *Parameter* (*t*) input as grafted branches. This is important so that the various strands of vertical points as part of the mesh are retained in respect to each other. However, when inputting these same grafted values into a *Bounds* component, they need to be *flattened* to determine the overall range of vertical heights versus each individual branch's range. The range of values are then parsed with the *Deconstruct Domain* component and inputted inversely of each other. The purpose for this is to colorize the lower vertices with a highlighting color, whereas the higher values become transparent. The gradient itself can be adjusted to visualize this appropriately once connected to the *Colours* (*C*) input of the *Construct Mesh* component along with the other respective *Vertices* (*V*) and *Faces* (*F*) values from the *Deconstruct Mesh* component.

Graphically, as part of the complete visualization, an aerial image can be used as a material application to the terrain surface, as shown in Figure 9.11. As demonstrated in Chapter 7, an aerial image is used for tracing out the different land

Graphic Compositions of Erodibility Model

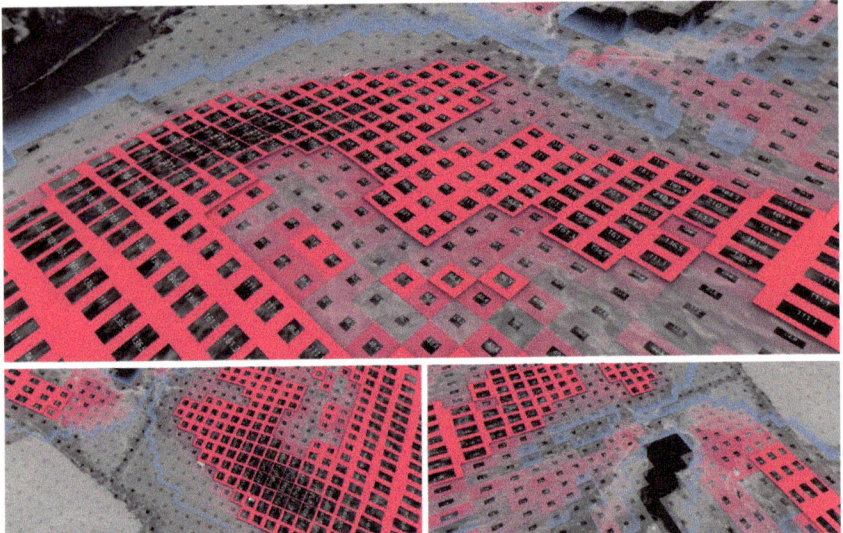

Figure 9.11
Composition of
detailed views of
the model with
labeled and scaled
quad panels on a
rendered surface
terrain.

covers. Since both the terrain and aerial share the same boundary, the aerial texture can also be applied without issue of scale and size. The best approach to this is to *Bake* the *Surface From Points* component, from Figure 9.2, to a separate layer. The aerial texture can then be found in the *Material Editor* option under the *Render* tab in Rhino. For this example, when applying this material, the texture is also set the *Grayscale* option.

This is just one way of further expressing the topic of erodibility within a project site; however, it provides the foundation for additional exploration with three-dimensional geometry and annotated information.

OUTCOMES

Like all these performance models, it is important to evaluate different outcomes generated from the performance calculations. The script in Figure 9.12 shows how different outputs can be used to inform these analytical outcomes. In this example, the erodibility index is evaluated; however, this same approach can be applied to the model's slope, K, T, and LS Factor as well. For the erodibility index, values equal to or greater than 8 are significant to HELs, so the same operation as shown before is used to filter down the data to only those values. From the culled output, a *List Length* component can be used to calculate the number of data items there are within that threshold as well as the number of items for the entire project site. Those values can then be *divided* to determine the percentage of HELs, which resulted in *0.6* or 60% of this project site.

The lowest and highest HEL indexes may also be significant to the erosion outcomes. This can be found with the *Bounds* and *Deconstruct Domain* components which range from *8.6* to *271.2*. Although these values provide a wide range in erodibility indexes, this data alone does not provide a thorough characterization of the

Figure 9.12
Using a
grasshopper bar
graph component
to chart the range
of erodibility index
values.

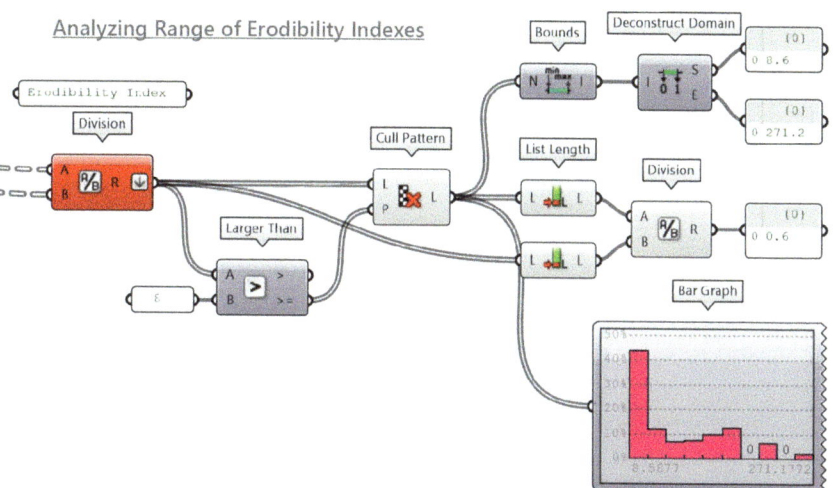

Figure 9.13
Comparing the
different erosion
effects of varying R
Factors.

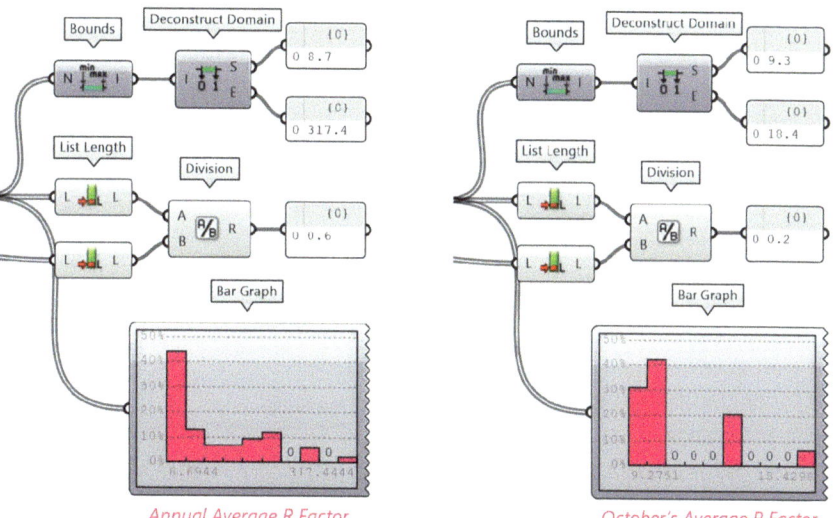

HEL indexes. So, by also connecting a *Bar Graph* component to the culled value, a breakdown of the different indexes can be better understood where the majority, over 40%, of the HELs are within the 8.6 range, whereas the other values are within 10%. Also, the highest value of 271.2 is minimal throughout the project site.

Another consideration that can be made through this process is how the R Factor of rainfall contributes to these erodibility indexes. Figure 9.13 shows how these differing R Factor values impact the erosion of the site. In this demonstration, a specific date was used to determine an R Factor of 211; however, when using the same online resource of this value for the annual average or specific month, significantly different values are computed.

For the entire 2021 calendar year, an overall R Factor of 247 is generated but during the driest month for this location, October, a value of 14.34 is calculated. The left-side image of the figure shows the overall annual erodibility index values and the right-side image shows the month of October. Between these low and high R Factors, the percentage of erosion occurring on the site significantly differs between 20% and 60%.

SUPPLEMENTAL AND FUTURE POTENTIAL

This performance model can be further expanded by identifying the key factor contributing to HELs and utilize that output to inform responsive erosion control strategies. This addition can create a strong and informed decision-making tool for a direct link between performative analysis and design. If key factors can be identified for contributing to HELs, then appropriate design interventions can be deployed to reduce or prevent substantial soil loss.

Contributing Factors

When it comes to assessing proper design interventions, it is important to first determine what may be contributing to high soil loss. In Figure 9.14, a *Sort List* component is used with all the outputs *reversed* to simply reorder all the unique variables by the highest erodibility index. For this example, the top ten highly erodible area have the same values, so it is safe to assume those areas can be resolved with the same or similar design interventions.

But if there are discrepancies in these values where the LS factor may be low but the K factor is high, this may suggest that the design intervention utilizes soil amendments to change the texture composition to something with less silt or more coarse.

Erosion Control

As suggested with the previous step, different design interventions or strategies can be applied to prevent soil loss. If the LS factor is the main contributor to erosion, those areas may require regrading to a gentler slope condition or more

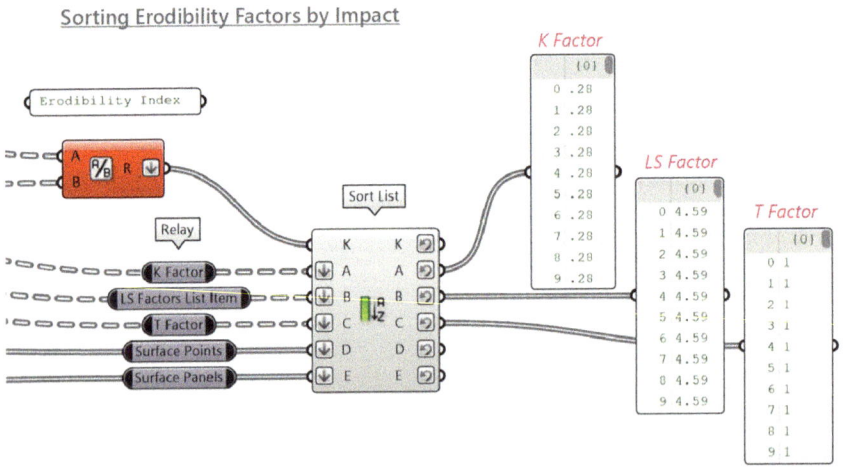

Figure 9.14
Sorting the different erosion factors to determine their impact on the project site.

aggressive techniques with retaining walls if there is not adequate space to regrade. Additionally, if all the factors are high, it may be more conducive to preserve the vegetation or minimize removal of established plantings.

From these assessment and strategic tools within this performative model, a project site can be analyzed for erosion potential and suggest strategies to prevent soil loss. This serves as a very robust and comprehensive model that establishes a strong connection between these steps in the design development process.

REFERENCES

Environmental Protection Agency. (2021, May 26). *Climate Adaptation and Erosion & Sedimentation*. EPA. Retrieved from https://www.epa.gov/arc-x/climate-adaptation-and-erosion-sedimentation

Environmental Protection Agency. (n.d.). Low Erosivity Waiver (LEW). Rainfall Erosivity Factor Calculator for Small Construction Sites. Retrieved from https://lew.epa.gov/

Natural Resources Conservation Service. (n.d.). Highly Erodible Land Definitions. Retrieved from https://www.nrcs.usda.gov/wps/portal/nrcs/detailfull/pr/soils/?cid=nrcs141p2_037282

10 Tree Benefits

Whether it's the US Forest Service or the landscape architecture profession, as a society we inherently understand trees help the outdoors with wildlife habitat, aesthetics, and shade. Although most of these benefits may be anecdotal, measuring quantifiable evidence of tree's ecosystem services is becoming more attainable and publicly acknowledged. Trees and green infrastructure are also being relied on more heavily for climate change mitigation, as they have been identified to provide 15 direct and indirect benefits of the 17 internationally supported UN Sustainable Development Goals for cities and countries (Turner-Skoff & Cavender, 2019).

This chapter demonstrates how quantified tree benefits can be calculated and utilized to assess the performance of outdoor spaces. This information becomes invaluable to developing planting and preservation strategies for urban forests and tree inventories to ensure project sites are not only benefiting from trees but are needs for that specific context. A tree's ability to effectively manage stormwater runoff does not necessarily equate to it being able to sequester a healthy amount of carbon from the air. Since tree benefits are unique to its species and size, this modeling method can begin to output robust sets of information for individual trees and the surrounding context.

GOALS

Tree benefits can be calculated and measured in a variety of ways to evaluate for project goals. These benefits can range from individual trees to an overall impact on outdoor and indoor spaces. The following will be the sequence of goals achieved with this performative model:

1. Measure different tree benefit categories (stormwater, carbon dioxide sequestration, energy savings, and property value increase)
2. Show outputted benefits individually for each tree
3. Demonstrate the cumulative impact trees have on the surrounding buildings
4. Calculate the overall performance for each category

PERFORMANCE METRICS

Through the mapping and inventorying of urban forests, the Forest Service, state agencies, and even local municipalities are quantifying tree benefits as they relate to stormwater management, air pollution, energy savings, and even increased

Figure 10.1
Image of tree
canopy that was
part of the project's
evaluation process
on carbon dioxide
capture.
Source: Özer et al.
2014.

property value (https://www.fs.usda.gov/learn/trees). The LAF performance guidebook and website showcase projects and methods that have been measured for their tree benefits specific to air quality and carbon sequestration (LAF guidebook reference).

The 1100 Block of Lincoln Road Mall from the Landscape Performance Series website is a great example where the i-TreeStreets software program was used to inventory and evaluated the project's tree for their ability to capture carbon dioxide (Özer et al., 2014). Figure 10.1 shows some of the large mature trees that were part of the evaluation process while also providing comfortable outdoor conditions and increased the adjacent assets value.

One useful resource for acquiring unique tree benefits, specifically energy (kWh) savings for buildings, is i-Tree Design (https://design.itreetools.org/), shown in Figure 10.1. After inputting the location of the project site, the first step will ask for the building footprint to be drawn to measure energy savings. The next step will ask for a specific tree species to use along with its diameter at breast height (DBH). Once these values are set, the instructions will ask to have a tree placed on the map using the indicated icon. As soon as the tree placement icon hovers over the map, a radiant graphic will appear, as shown in the left-side image of Figure 10.2. The radiant is divided into the eight cardinal directions with a subset of distance variables in increments of 20 feet.

At the bottom of the map, highlighted with the pink box, savings will be shown in both dollars and kWh. Depending on the need for the project, these values can be documented in an Excel file. This demonstration will focus on kWh.

To match the order in which Grasshopper reads data structures, row 1 should start with the west side of the building starting from the closest distance, 0–20 feet, and radiating out to 21–40 feet, then 41–60 feet for columns A to C, with column D having a 0 for no energy savings. Moving down the rows, the directions should be inputted clockwise so that row 2 is northwest, row 3 north, and so on for all eight cardinal directions, shown in the right-side image of Figure 10.2. This Excel file will serve as the referenced file later in the modeling script.

i-Tree Design Online Calculator

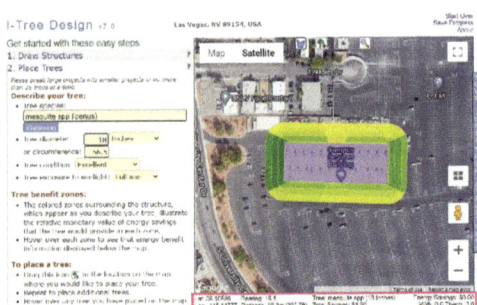

Translating Data to Excel

	A	B	C	D
1	613	625	558	0
2	401	286	264	0
3	259	253	251	0
4	368	275	253	0
5	527	502	441	0
6	407	318	266	0
7	392	340	266	0
8	290	344	428	0

Extracting Cost Savings Data

Figure 10.2
Extracting tree benefit values on energy savings for buildings from i-Tree Design.

MODELING METHODS

To prepare this model for measuring tree benefits, there should be both tree geometry and building footprints in Rhino. The trees can be modeled in a variety of ways that best suits their visual representation which may include points, closed curves, surfaces, closed polysurfaces, or meshes. Regardless of which geometry type is used, an extracted point will be type of data used from the script. The building footprint can also be any of those geometry types, except for a point, but it is preferable to simply use a closed curve or a closed polysurface that has been exploded. The reason for that is because at the end of the scripting process, the top surface of the buildings will be colorized based on the degree of benefit the surrounding trees provide.

As shown in Figure 10.3, the modeled mesh trees have been referenced and have a *Centroid* (*C*) point extracted and projected to the c-plane. If only points are being used to reference the center of a tree, then the extraction and projecting operation is not necessary. The key thing to keep in mind is that the new point outputted from the *Project* component will be the tree measurement for the remainder of the script.

The referenced building footprint curves need to be *grafted* because a tree can provide benefits to multiple buildings simultaneously, such as energy savings and property value. If these curves are not grafted, then the function of this operation with a *flattened* branch of data will only measure the tree benefits for the closest building, even though there may be multiple buildings within a required distance.

The center points of each curve can now be measured for their distance to multiple buildings using the *Brep Closest Point* component. This component and operation can be a little misleading, since curves are being used for the building; however, it is still able to recognize that geometry type and perform the necessary procedure. This makes the versatility of different referenced building geometry more seamless with the workflow of the script. With the *Vector 2Pt* component, the tree point from the *Project* component can be connected to the *Brep Closest Point* (*P*) output to find the directional relationship between the two.

Similar to the explanation in the 'Elementary Operations' section of Chapter 3, the normal values need to be measured against a *Unit X* with the *Angle* component. These angles are then converted with the *Degrees* component and translated into index values using the *Remap Numbers* component. The reason for this is because the values collected from the i-Tree Design model, Figure 10.1, are imported as index columns. By converting the angle of degree to index values, they can all be cross-referenced in the next step to determine the tree's specific benefits from the external table. An Integer component is used last to round the values to a whole number, since the previous step will create decimal values that won't match the external data in the next step.

Figure 10.4 shows how the *ExcelDynamicRead* component is used to import the external Excel data sheet containing different energy savings from i-Tree Design resource in Figure 10.1. This data is then parsed with a *List Item* component that uses the previously created integer values as the *Index* (*i*) input after being reformatted with the *Flip Matrix* component. This operation is used to first determine the directional relationship between the trees and buildings as a filtered row based on that vector value. The next series of steps will then filter down the row based on the distance the trees are from their respective buildings. This is achieved by using a panel list of given distances (*0, 20, 40, 60, 1,000*) to create a range of values with the *Consecutive Domains* component. The *1,000* value can be any significantly high value to ensure trees farther than 60 feet will generate zero energy savings. Each tree's distance to its respective building is matched with the correct domain to output the appropriate index value.

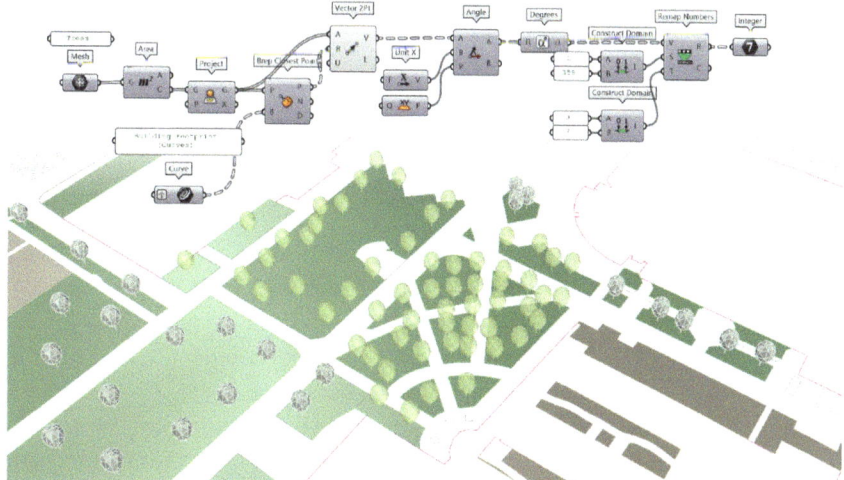

Figure 10.3
Referencing mesh tree geometry for measuring and assessing their proximity and relationship to surrounding buildings.

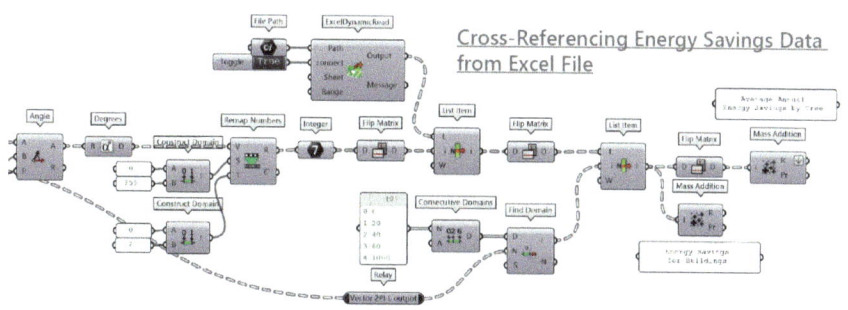

Figure 10.4
Cross-referencing energy savings Excel file with tree geometry.

These index values will be connected to another *List Item* component that is using the previous list item's *flipped* values. These values will be *flipped* one last time to find the cumulative benefit for each tree using the *Mass Addition* component with a *flattened Result* (R) output. Using another *Mass Addition* component without *flipping* the values will result in the cumulative energy savings for each respective building.

After this operation and the similar ones to follow, it is important to create Panel labels for the concluding components to help identify which will be connected to the next series of operations. For energy savings and property value, a panel is used for both the individual tree values and the building benefits.

When measuring the property value trees can add to a building, the direction is not a factor, but rather just the distance. The value also doesn't vary with different ranges of distance like energy savings as long as it is within 60 feet of the building. To evaluate this tree benefit for a project site, a *Construct Domain* component is used for both a range between *0* and *60* and *60* and *1,000*, as shown in Figure 10.5. The *60–1,000* range acts the same as the one for energy savings to ensure any trees further than 60 feet will not contribute any property value increase. Once these range values are *merged*, this can follow the same logical operation as before with the *Find Domain* component to determine which range the *Distance* (D) from the *Brep Closest Point* component is in.

The resulting index values from this component can then be flipped, like before, to reference the property value increase. For the panel being used to reference a monetary savings, the 50.16 is the average annual property value increase from a Honey Mesquite *(Prosopis glandulosa)*. This value can be attained from either i-Tree Design or other tree benefit calculator. Finding the cumulative tree benefit and building value increase also follows the same set of operations as before using the *Mass Addition* components, with the one for buildings be *flipped* prior.

The last two tree benefits being measured is stormwater runoff reduction and carbon dioxide (CO_2) sequestration. Since these values are not contingent on distance or location within a project site, the quantity of trees is the only value needed to measure both individual benefits and overall site benefits. As shown in Figure 10.6, this can be achieved using the *List Length* component to count the number of trees, from either the initial referenced tree geometry or superseding geometry extracted such as points. The numeric value outputted from this component can then be used to duplicate the given data with the *Repeat Data* component for stormwater reduction and CO_2 sequestration for each number of trees.

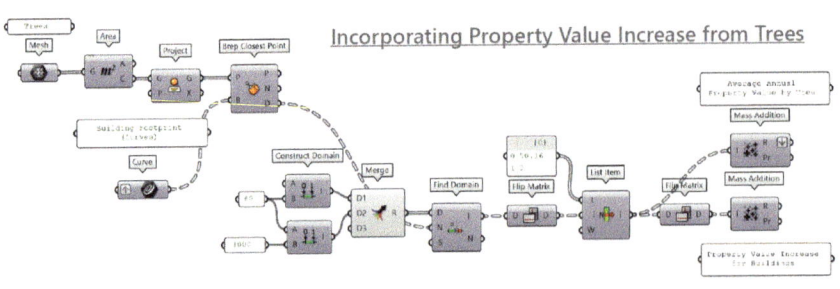

Figure 10.5
Creating a script operation to compute the property value increase from trees.

Figure 10.6 Integrating individual tree benefits on intercepting rainfall and sequestering carbon dioxide.

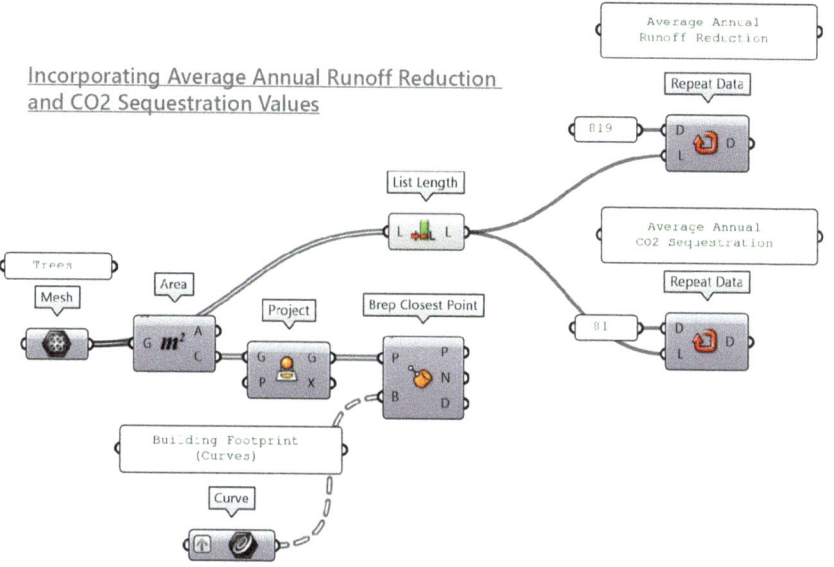

These respective given values are also attained from i-Tree Design for a Honey Mesquite's average annual benefit. The *819* value is the number of gallons storm-water is reduced by and *81* is the number of pounds of CO_2 captured by trees.

With all the different benefits calculated for each tree, the data can now be combined for a variety of legible displays that include both individual and cumulative performance to the surrounding context as well as overall performance, as shown in Figure 10.7. These three data scales of tree performance can be helpful in demonstrating which trees are having the highest positive impact on the site, which buildings in the surrounding context are benefiting the most from trees, and how the overall performance compares to goals through the evaluation of these tree outputs.

The figure shows each benefit organized in the sequence of stormwater runoff: CO_2 sequestration, energy savings, and property value. It is important to maintain the order of merging the data within the different evaluation measurements to ensure accurate results.

The first data scale of tree benefits will be individual performance, as shown in Figure 10.8. The first step is to graphically represent each benefit differently through colorized spheres. This helps identify each benefit with a unique attribute for legible communication. To do so, each tree is given the same random point by connecting the initial referenced tree geometry to a *Populate Geometry* component. It is important to note that this component has by default *100* points assigned, so before connecting it with the tree geometry, a value of *1* is inputted for the *Number (N)* to avoid overpopulating the model with data which can potentially crash the file. Another parameter to modify is the *Seed (S)* to ensure the sphere is placed in an appropriate location, preferably at the top of the tree crown versus near the trunk. This component can now be connected to the *Mesh Sphere* component where a proper *Radius (R)* is assigned as well. Both the seed and radius input should be a number slider to adjust accordingly once this operation has been completed.

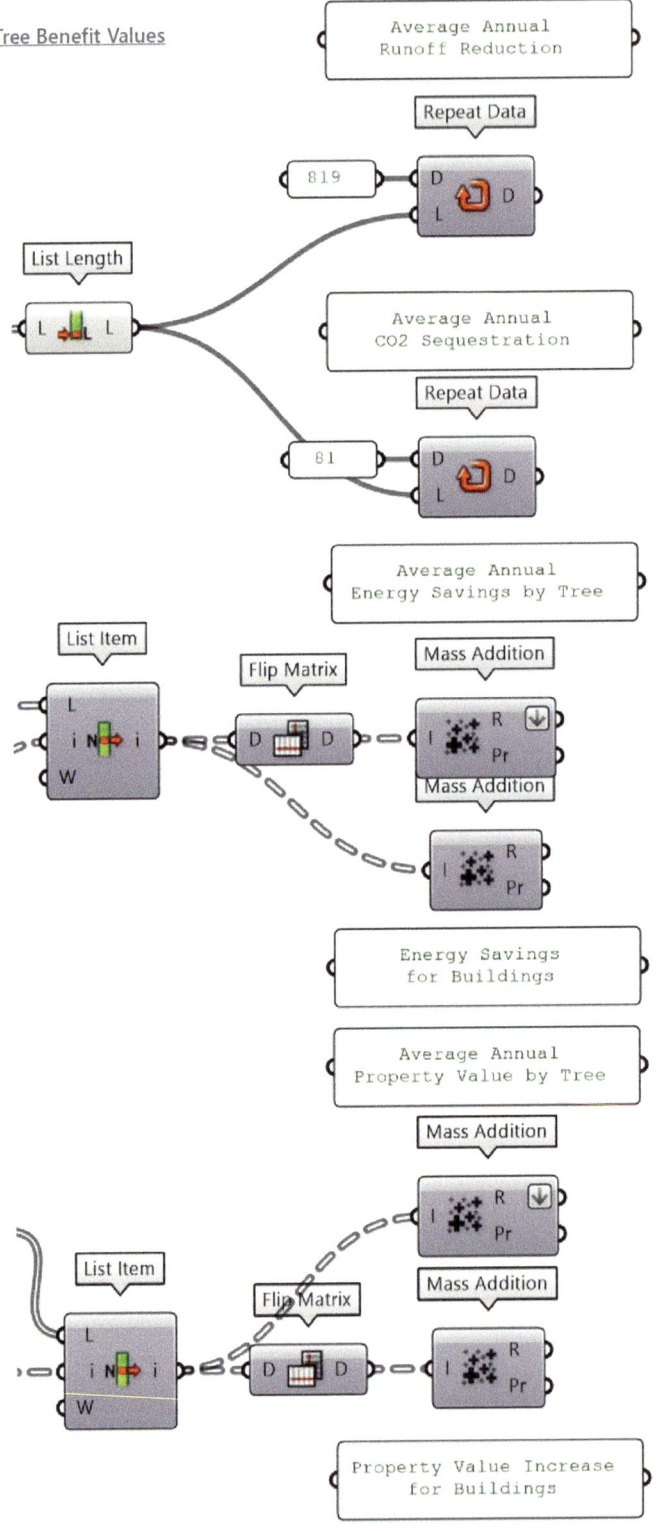

Organizing All Tree Benefit Values

Figure 10.7
Organizing all the
individual tree
benefits and impact
on surrounding
buildings for
visualization.

Next, different color values are assigned for each benefit category. One option is to first create a *Colour Swatch* that is white and fully transparent to give a blank display of benefits when it is unnecessary to show numeric data as part of the model display. After that, different color preferences can be made for the four corresponding tree benefits, as shown in the below figure. All the swatches are then *merged* and inputted into a *List Item* component. This will give the ability to cycle through which benefit to display by using a *Value List* component.

For this component to function correctly, it is important to use recognizable text labeling with a corresponding index value. The labels do not have to match the ones for this demonstration, since it will not affect the selection of data as long as some similar nomenclature is used with the corresponding tree benefit. By *right-clicking* on the component and choosing *Edit* from the pop-up window, the following can be written:

No Values = 0
Runoff Reduction = 1
CO_2 Sequestration = 2
Energy Savings = 3
Property Value = 4

This sequence follows the order of connected data in the final step of this operation while beginning with the option to show no data with the *No Values = 0* line. This first *List Item* component will dictate the coloring of the mesh spheres when connected to the *Custom Preview* component.

The final step is matching the corresponding data with the colorized spheres. An *Entwine* component is used to function like the *Merge* component but the output will maintain the benefits within their respective branch, whereas merging them will combine all the values from each benefit category into one branch list. The first input of this component is left empty, so that when the *Value List* component is set to *No Value*, there will be no display of data. The connection will then proceed as stated earlier and as shown in Figure 10.8. For the *Energy Savings* and *Property Value* categories, the *Mass Addition* components that were labeled *by Tree* is used, not *for Buildings*. The output from this component is then restructured with the *Flip Matrix* component and inputted into another *List Item* component with the *Value List* also connected to the *Index (i)* input. The values can now be *flattened* and connected to a *Text Tag* component to match the data branching of the tree geometry and mesh spheres. This type of tag will always align the text with the Rhino Viewport opposed to being orientated on a specific plane. That is also an option, however, with the *Text Tag 3D* component as demonstrated in previous chapters.

To evaluate how the trees provide benefits to the surrounding context, energy savings and increased property value are used since the correlation between runoff reduction and CO_2 sequestration does not have a direct link to buildings like the aforementioned benefits. In Figure 10.9, the tops of the buildings are colorized based on the cumulative benefits they are receiving from the surrounding trees. One option to visualize this is by taking the initial building footprint curves and raising them to their respective height with the *Move* component and *Unit Z* value. These curves are then converted to a *Mesh* for colorizing.

The coloring of the roofs uses a separate *Gradient* component with a range of lower and upper limits for both benefits. The total cumulative potential is not a given, so a number slider is used to adjust accordingly to the highest potential. For example, once a building is wrapped with the maximum number of trees, there may only be a total property increase of $1,500. If the upper limit is set to a

Communicating Different Tree Benefits Through Color, Size, and Labels of Spheres

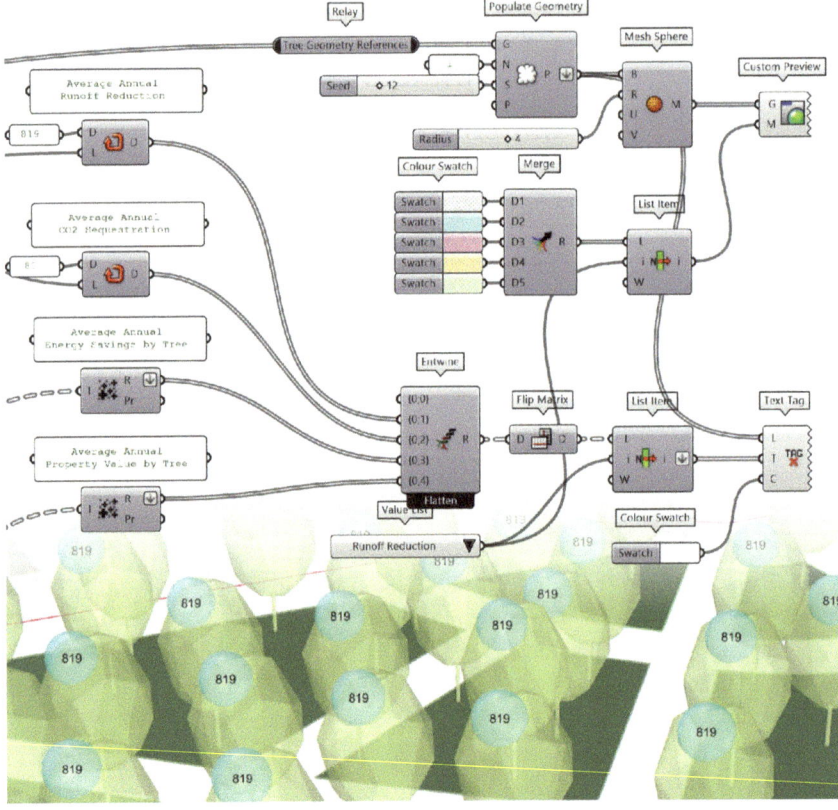

Figure 10.8 Visualizing individual tree benefits with colored and labeled spheres representative of their impact.

significantly higher value like *10,000* or greater, it can be very difficult to discern a color contrast between all measured buildings, as shown in the bottom image of the figure. Similarly, a color gradient with significant contrast is preferred to show a range of value.

The output from these gradients can then be connected to an Entwine component like previously done with the individual tree benefits. Since there is no data for the first three categories, those inputs are left empty. The addition of a gradient that displays no color requires more effort than is needed for this demonstration, so that is why they are left disconnected. The *entwined* data is then *flipped* once again and connected with a *List Item* component with the same *Value List* inputted for the *Index* (*i*). These color gradient values are then rendered out with the *Custom Preview* component once the *Material* (*M*) is *flattened* to match the data structure of the building rooftops.

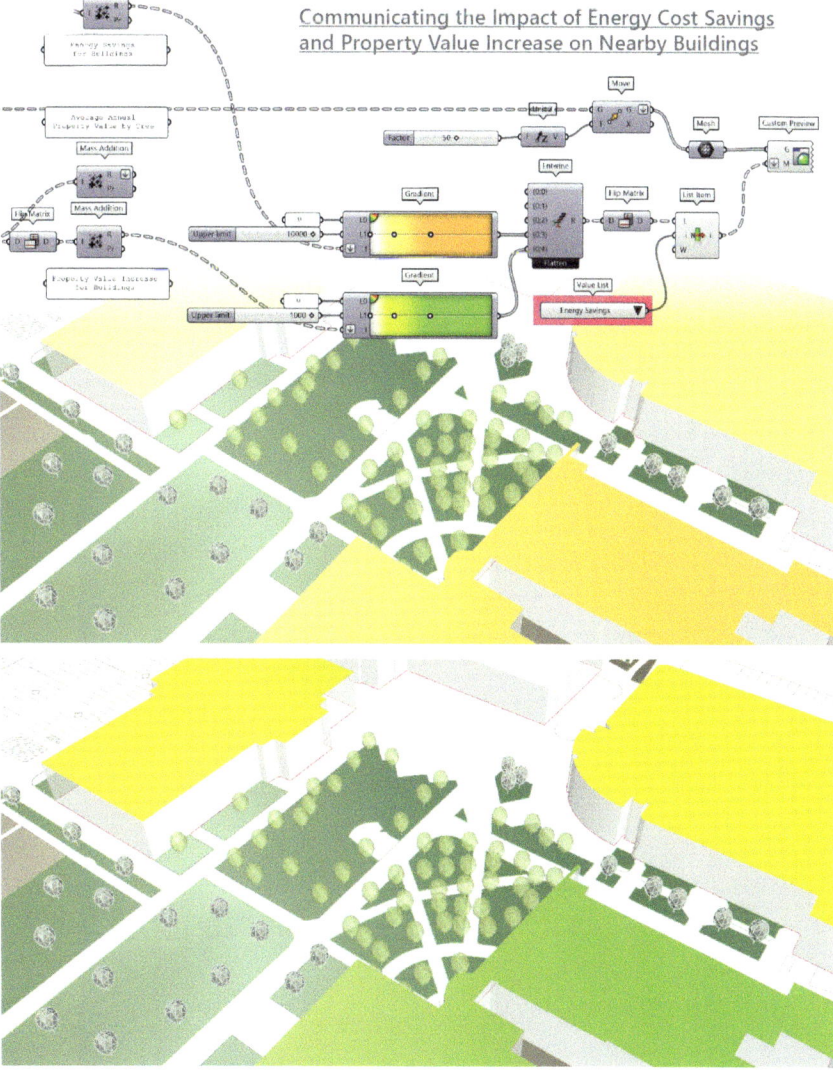

Figure 10.9 Colorizing building rooftops to demonstrate the differing trees benefits.

The same *Mass Addition* components, shown in the script of Figure 10.10, are used to display the cumulative values for each building as well. Before these values are entwined, labels for the unit of measurement are added using an operation that duplicates the unit to match the number of buildings using the *Cross Reference* component with the unit (*kWh* and *$*). When merging the units with the values, it is important to keep in mind the order in which they are connected to the same input of the *Text Join* component. For energy savings, the unit kWh comes after the values, so the output order from the *Cross Reference* component first goes *A* then *B*, whereas for property value, the *$* unit will come before the value, so the order is *B* then *A*.

The values are then connected in the same order as before with an *Entwine* component, *flipped*, and connected to a *List Item* component. The same operation used before to raise the rooftops to their respective height with the *Move*

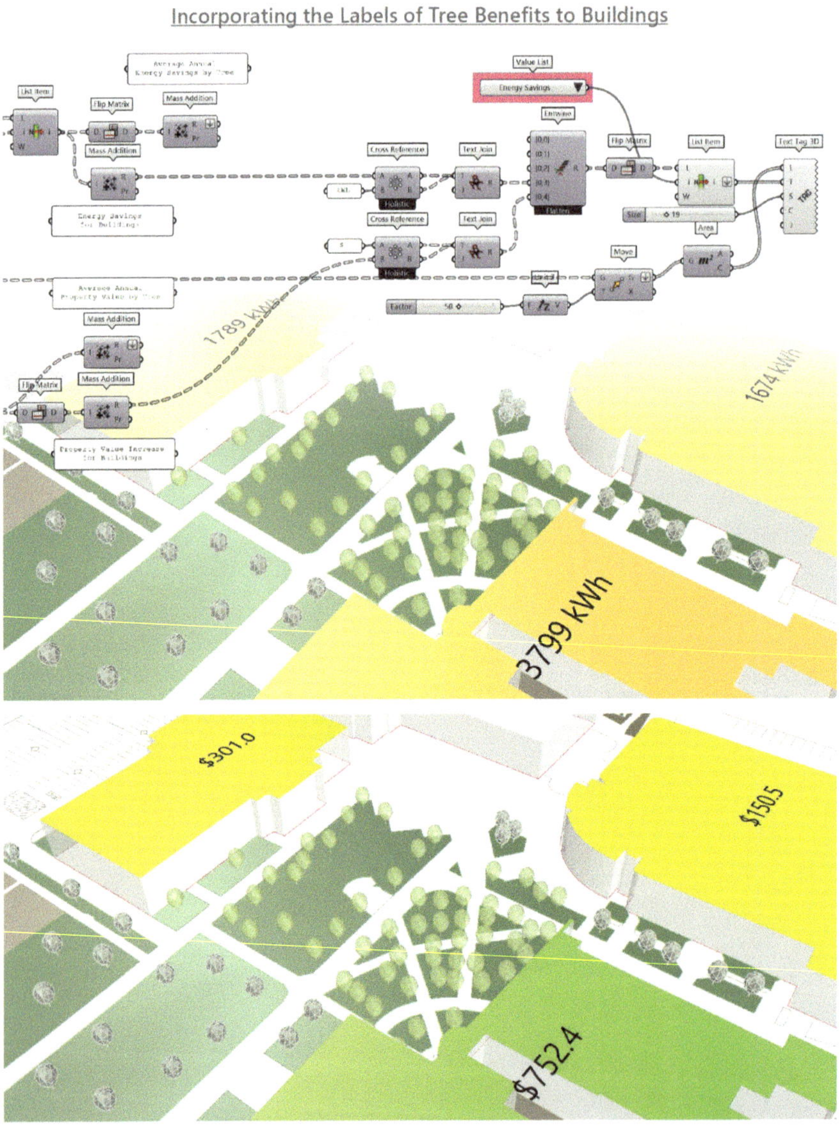

Incorporating the Labels of Tree Benefits to Buildings

Figure 10.10
Including labels to the different building benefits.

component is used to extract a *Centroid* (*C*) with the *Area* component for the location of the values to be displayed. In this instance, a *Text Tag 3D* component is used to place the text on the default *xy plane* where the size is adjusted with a number slider for an accurate reading. If the *xy plane* is not a conducive way for displaying the text, then other vector planes can be used in lieu of the default and rotated as well if needed.

The image displayed in Figure 10.11 shows the comprehensive benefits of trees to the project site by demonstrating from the script individual performance, benefits to buildings, and the overall output to assess the impact of trees. To display the overall output of tree benefits, the *Mass Addition* components used for the individual trees is connected to another series of *Mass Addition* components to find the total output. To ensure all data is structured the same, each component's output is *flattened* and *merged*. The same ordered sequence is then done for the unit of measurement labels. Both *Merge* components contain an empty *D1* input to maintain the first list item as having no values.

Even with that input left empty, after merging the first value will be the runoff reduction data instead of a *null*. So, when these components are both connected to a separate *List Item* component, an expression of *x–1* is given so that the *No Values* option from the *Value List* component does not display the incorrect data. The images in Figure 10.12 show more detailed views of the tree benefits on the buildings.

Next, a panel containing *Overall:* is connected to the *Text Join* component followed by the benefit values and unit labels. A *Text Tag 3D* component is then used with a referenced point given a *xz plane* orientation with an adequately sized font.

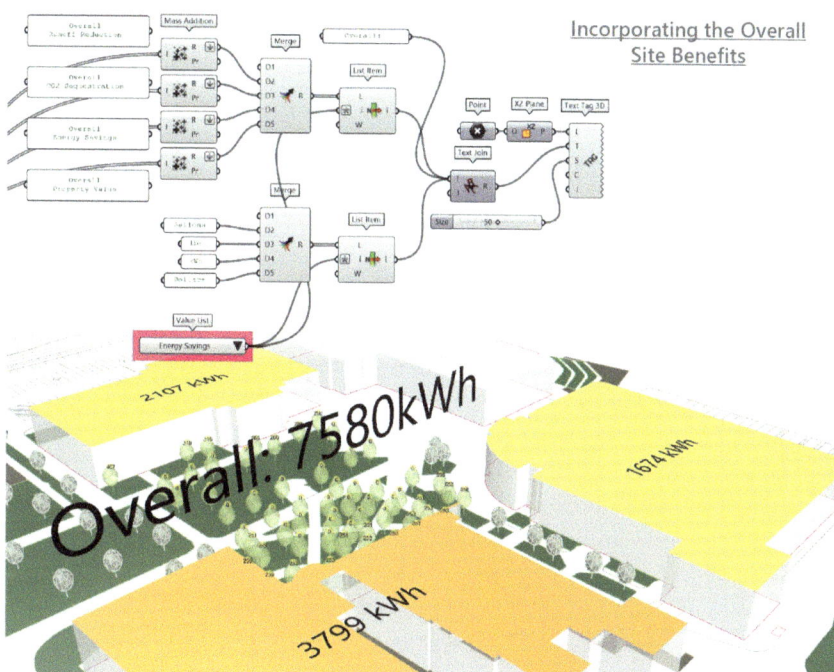

Figure 10.11 Including the overall site benefits from the different categories.

Figure 10.12
Detailed views of
tree benefits on the
project site.

OUTCOMES

The model creates a variation of performative outputs that contribute to the environmental, social, and economic impact of outdoor and indoor spaces. Their value can be understood both individually and collectively. As individuals, the model can calculate which trees in specific locations provide the highest level of benefits to the space and vice versa. The accumulation of multiple trees can then be used to gauge which and how surrounding buildings may benefit from nearby trees. Lastly, the overall performance can be used as metric to evaluate against project improvement goals.

Like most of the performance outputs from these models, the tree benefits do provide substantial data for a variety of ecosystem services; however, they are all still relative until evaluated against a baseline measurement. This will require outside resources, field surveys, and further analysis to assess the significance of the tree benefits. One example is to calculate the runoff volume from the same project site and determine what the overall runoff reduction from trees equates to. This can be substantiated with project goals aimed at reduction percentages or fulfilling municipal regulations related to stormwater management. The example in Figure 10.11 is showing the kWh savings per building; however, how much energy are the buildings consuming? Once that value is calculated, the savings from trees becomes more relevant.

ADDITIONAL POTENTIAL

Along with showing the percentage of change in different savings and reduction benefits stated in the previous section, other procedures can be derived from this to make the modeling outcome more robust for design and planning strategies. The individual trees themselves can be subdivided into groupings or clusters based on their performance. This model is also only demonstrating one type of tree with an average annual performance; however, i-Tree Design has the ability to meticulously pull annual values through a tree's lifespan as it grows larger within its large database of different tree species. Trees can also be evaluated for their performance from all the categories to determine which are most valuable in their collective benefit to the space.

Percent Change

Incorporating percent change for the different tree benefits, with the first example in Figure 10.13 showcasing energy savings, only requires a small mathematical operation and additional labeling for clarity. In this demonstration, the *Mass Addition* component *...for Buildings* is *relayed* to a *Division* component and the first input of the *Entwine* component. According to the US Energy Information Administration (https://www.eia.gov/tools/faqs/faq.php?id=97&t=3#:~:text=How%20much%20 electricity%20does%20an, about%20893%20kWh%20per%20month), the average household consumes 10,715 kWh annually. Although the buildings in this example are not residential units, the same principles and function of the script still apply. After dividing each building's savings by that average consumption, they can be multiplied by *100* to convert the output from decimal to percentage values. A *Cross Reference* component is then used for each of the following labels: *kWh* (%) and connected to their respective inputs: *{0;1}*, *{0;2}*, and *{0;4}*. For the *{0;3}* input, the percentage value is connected to the nest that is between the parentheses. After following the same steps as previously demonstrated labeling operations, the text display for the buildings will now include both the kilowatt hours and annual percentage saved.

High Performance Clusters

Showing groupings or clusters of trees can also be valuable when it comes to strategic planning for inventories by allocating efforts to maintain and preserve high performing areas or improving upon low performing areas as shown in Figure 10.14. The model image in Figure 10.14 demonstrates how these clusters can be visualized

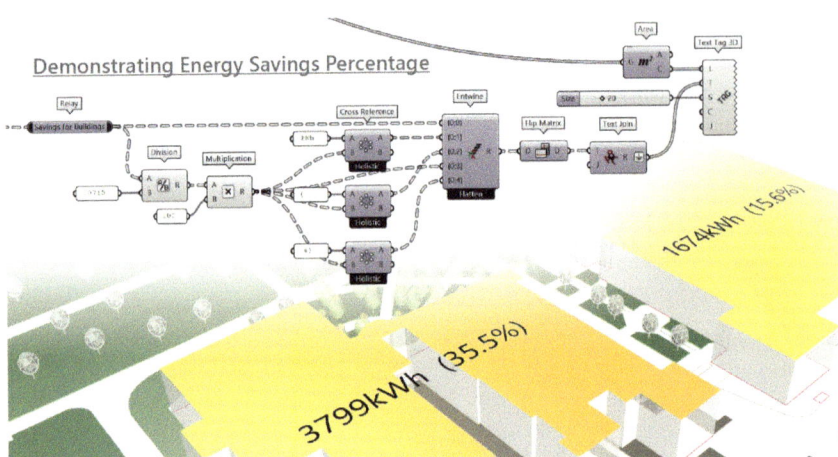

Figure 10.13
Including percent savings to the buildings from tree benefits.

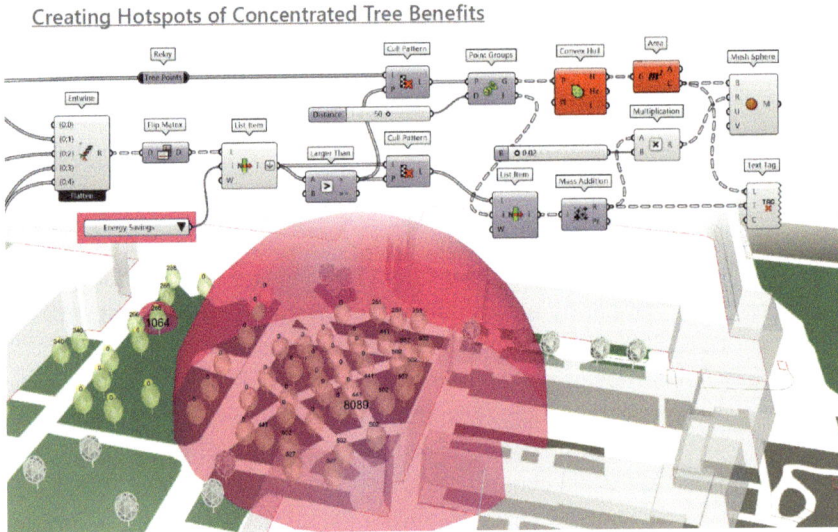

Figure 10.14
Highlighting tree clusters of concentrated benefits with modeled geometry.

by performance values. The *Entwine* component is merging all the values from the respective components *...by Trees*. With the individual tree values by category, they can be filtered using a *Larger Than* component to only include trees with a performance higher than zero. The Boolean pattern created from this can then be used to *cull* both the tree points and performance values.

From the culled tree points, a *Point Groups* component can be used to begin parsing points by a set distance for the clustering effect. The grouped points can then be connected to a *Convex Hull* and *Area* component to create an enclosed curved for each point group and create a center point. This operation will only function on groups with three or more points since one will just be a point and two will only create a curve. This inadvertently filters the data even further to only consider groups with more than two trees.

The culled tree benefits can then be grouped the same way from the *Point Groups* component's *Indices* (*I*) output for a *List Item* component. The grouped

194 □

values can then be summed up with a *Mass Addition* component. This will serve as both the text label and size of the sphere's radius. Because these values will most likely be substantially larger than the scale of the project site, a *Multiplication* component is used with a number slider to scale down the size of the mesh spheres to read legibly in the model.

Tree Variety, Size, and Age

With i-Tree Design and other tree benefit calculators, it is possible to retrieve annual tree benefits as trees begin to mature. This combined with showing a variety of different tree species in the model can create a more comprehensive model that substantiates the dynamics of tree benefits as they change over time rather than seen as a static snapshot of performance. For example, some trees begin to capture stormwater runoff at an early age, whereas some are never able to make a significant runoff reduction. Other trees may reach the national CO_2 sequestration average of 48 pounds after a few years but then digress and capture less than that at a mature age. These variations in tree benefits further reinforce the need to model these significant benefits and utilize them in time-based scenarios.

Comprehensive Tree Benefits

Another option in the communication of tree benefits is to show all the performative categories simultaneously to deliver a more comprehensive depiction of trees. This can be done for each tree individually or for all the trees being measured; however, this output can become very busy and inconclusive visually. This process can also be further refined by measuring how each tree compares to the collective performance. In other words, how does each tree rank amongst all the trees when it comes to energy savings. The significance of this is because depending on the location, distance, and relation to multiple buildings, trees can have significantly different benefits. This coupled with the species variety and size can further compound the robustness of this technique.

REFERENCES

Özer, E., Alvarez, V., & Gonzalez, G. (2014). *1100 Block of Lincoln Road Mall.* Landscape Performance Series. Landscape Architecture Foundation. https://doi.org/10.31353/cs0810
Turner-Skoff, J. B., & Cavender, N. (2019). The benefits of trees for livable and sustainable communities. *Plants, People, Planet* 1(4): 323–335. https://doi.org/10.1002/ppp3.39

11 Forest Succession

Modeling forest growth is key to evaluating healthy wildlife habitats, ecosystem services, and plant diversity. This can be implemented after a forest fire, restoring a natural ecology, or post-construction of a project site. By simulating forest succession through tree growth patterns, these different metrics can be measured and assessed for their outputs toward project goals and objectives. This is one of the few demonstrations in the book that relies more on future projections as well as modeling hypotheticals instead of existing conditions. This same process and methodology can, however, be replicated once a tree inventory of a project site is created with an accurate characterization of tree species, size, and condition.

As it was alluded to already, this is a model simulation of tree growth and succession within a project site. This can be modified to simulate conditions along a water body, as demonstrated in this chapter, or as more of an overall growth succession. The end of the chapter also covers the future potential of integrating a Floristic Quality Assessment (FQA) component to the model to align more seamlessly with LAF's performance guidebook.

GOALS

The purpose of this script is to simulate different forest or planting growth scenarios for a diverse number of tree species. The growth rate and placement will be contingent on tree characteristics that include size and canopy shape. The following are the evaluation and assessment goals from the simulated forest succession model:

1. Create a diverse and proportionate tree canopy
2. Simulate tree growth on a project site
3. Establish a healthy urban forest for additional ecosystem services

TREE PARAMETERS

The project location for this demonstration will reside in the southeast region of the United States where plant communities and forests of Pin Oak, Red Maple, and Dogwood trees could potentially coexist. As stated earlier though, a proper and accurate tree inventory for an existing project site can be used or external resources may be required to determine appropriate tree selection for simulation. Below is a list of these trees with their defining characteristics for modeling parameters in this chapter.

DOI: 10.4324/9781003208020-14

Pin Oak
Canopy Diameter/Radius: 40/20 feet
Height: 70 feet (Minckler, 1965)

Red Maple
Canopy Diameter/Radius: 30/15 feet
Height: 70 feet (Hutnick & Yawney, 1961)

Dogwood
Canopy Diameter/Radius: 20/10 feet
Height: 30 feet (Chapman & Bessette, 1990)

These characteristics will be parameterized throughout the chapter in relation to their respective characteristics.

MODELING METHODS

This chapter will require a plugin not covered in Chapter 4. The Kangaroo 2 is a physics simulator plugin that has an extensive range and use of modeling different scenarios related to geometric interactions. Because of its elaborate collection of unique components, it would not have been conducive to go into detail of this plugin since this chapter's script will only use a small sampling from it. One of the advantages of this plugin versus similar operations of Grasshopper's default components is that it can use tree size parameters to simulate proper location so as not to intersect or overlap unnecessarily like real forests. Some of the default component can randomly place points to represent tree locations, but is challenged with the ease of implementing multiple size parameters for trees or growth succession from an origin point.

For the purposes of this model, forest succession will originate from a curve drawn in Rhino. This curve can be specific in representing a water source or considered random. As shown in Figure 11.1, this blue curve will simulate the origin point as it bisects the project boundary site. The placement, size, and number of origin curves can be done in a variety of ways to best suit the intended outcome of simulating forest succession.

The origin curve is then populated with a random number of points. The first part of this operation will be for one specific tree but then demonstrate the addition of other tree species in the next figure.

As part of this forest succession simulation, it is important to first determine the number and proportion of tree species. Using number sliders and the *Merge* component, a ratio of species can be made for this modeling of Pin Oak, Red Maple, and Dogwood. This starting ratio can then be increased to simulate the population growth of these trees by connecting to a *Multiplication* component with another slider serving as a growth parameter. This slider will need to begin at *1* at the start of this simulation, but can have a wide range depending on the tree types and size of the project site. This number slider is given a panel labeled *Growth Simulation* for organizational purposes and reference later in this script.

The next part of this operation is the placement of these points on the curve using the *Random* component. Both the *R* and *S* input of this component can be a random number slider value. The *R* input will determine the range of random values and the *S* input will determine the random seed. It is important to make sure that the *N* input is representative of the number of tree species to be simulated. Since the previous step showed three different trees, this number slider is set to *3*.

A *List Item* component will be used for both the proportion and random operations and connected to their respective inputs on the *Populate Geometry* component. The utility of this component will be expanded on in the next figure as well. The *Populate Geometry* component will perform for the first tree species, even though the initial steps are for the overall proportion of tree species.

When zooming into the *List Item* component, there is an option to add additional outputs for each index item (trees) by clicking on the *+ icon*. For both *List Item* components, there should be an equal number of outputs to the number of tree species being used, in this case, three as shown in Figure 11.2. Two additional *Populate Geometry* components are used for the other remaining two tree species where the newly added outputs from the *List Item* components can be connected respectively. Essentially, each *Populate Geometry* component will serve the initial random placement of points for each tree species.

Additionally, a *Repeat Data* component will be used to input the tree's mature canopy radius for each tree species. This component will then match the number

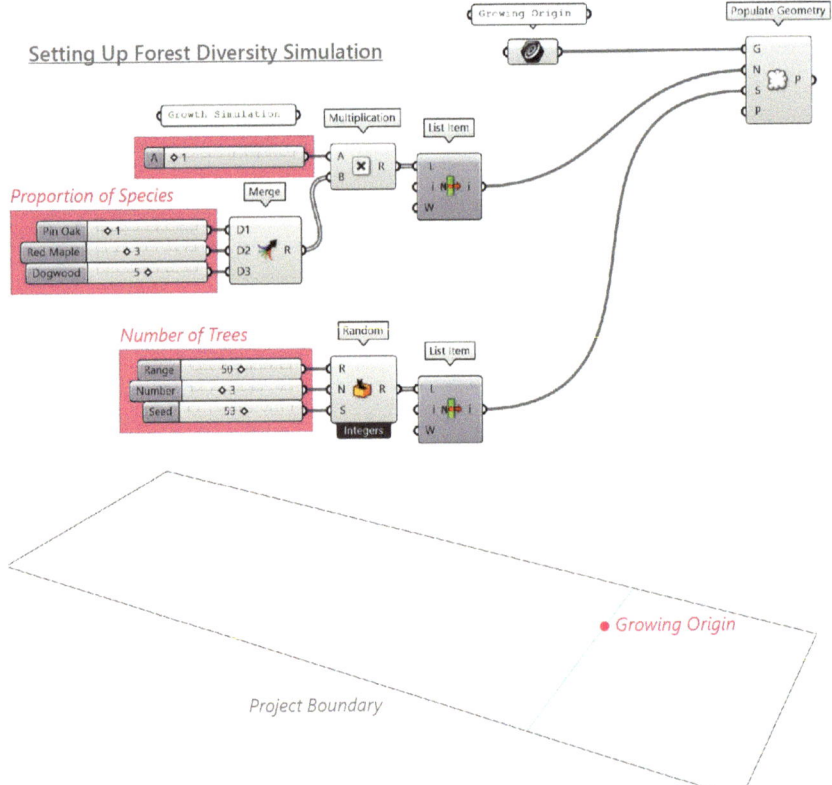

Figure 11.1
Setting up simulation settings with tree diversity ration, growing origin, and project boundary.

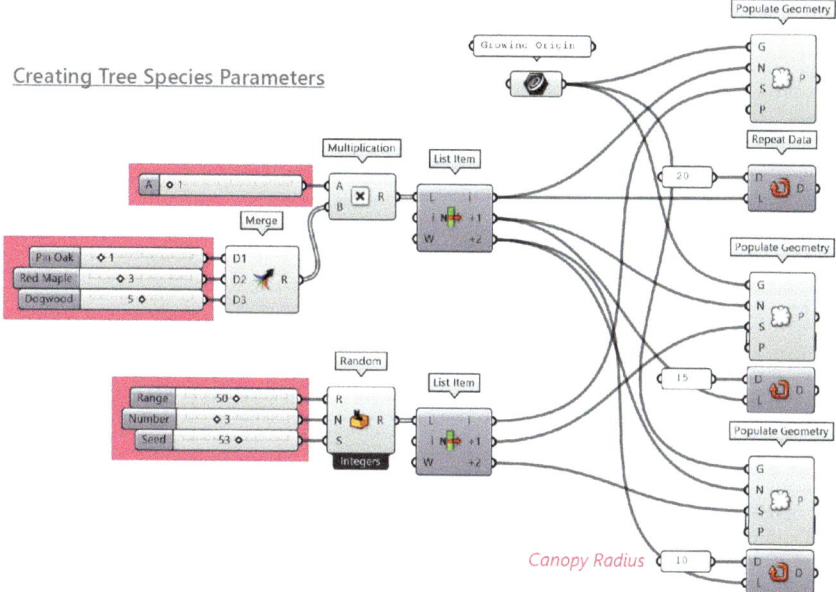

Figure 11.2
Creating individual tree canopy sizes for their respective species.

of radii as the tree simulation grows to ensure each tree has an associated radius. The outputs from the *List Item* component can be used for the *Length* (*L*) input of this component. The select values are based on the average canopy diameter of a Pin Oak, Red Maple, and Dogwood being 40, 30, and 20 respectively or half that to represent the radius.

The next part of the script will introduce component from the Kangaroo 2 plugin to use physics parameters to space points for each tree based on their canopy size to ensure trees do not overlap, but rather space accordingly. The first step for this physics simulation is to connect each *Populate Geometry* output to a separate *Show* component in Figure 11.3. This component allows the viewing of the points and their associated geometry after the simulation has run. These same outputs will also be connected to a *Merge* component in the same order. This is important for all the steps because if components are connected out of order then the parameters will be mixed and not provide accurate results. Another *Merge* component will be used to connect all the *Repeat Data* components as well.

The two *Merge* components will be used for the *Collide* component to develop the physics engine of assigning collision distances for each point using their respective diameters. The rest of the inputs for this component can remain at their default values.

For the next component in this operation shown in the figure, an *OnMesh* component is used to contain all the tree points within a project boundary. After the points from the *Merge* component are connected, a closed curve for the site boundary modeled in Rhino can be referenced and converted into a mesh and connected to this component.

From the referenced curve, a centroid can be created with the *Area* component to maintain that all the points stay within a plane as well by using the *Anchor* component. Although it is not required, the addition of the *Grab* component can

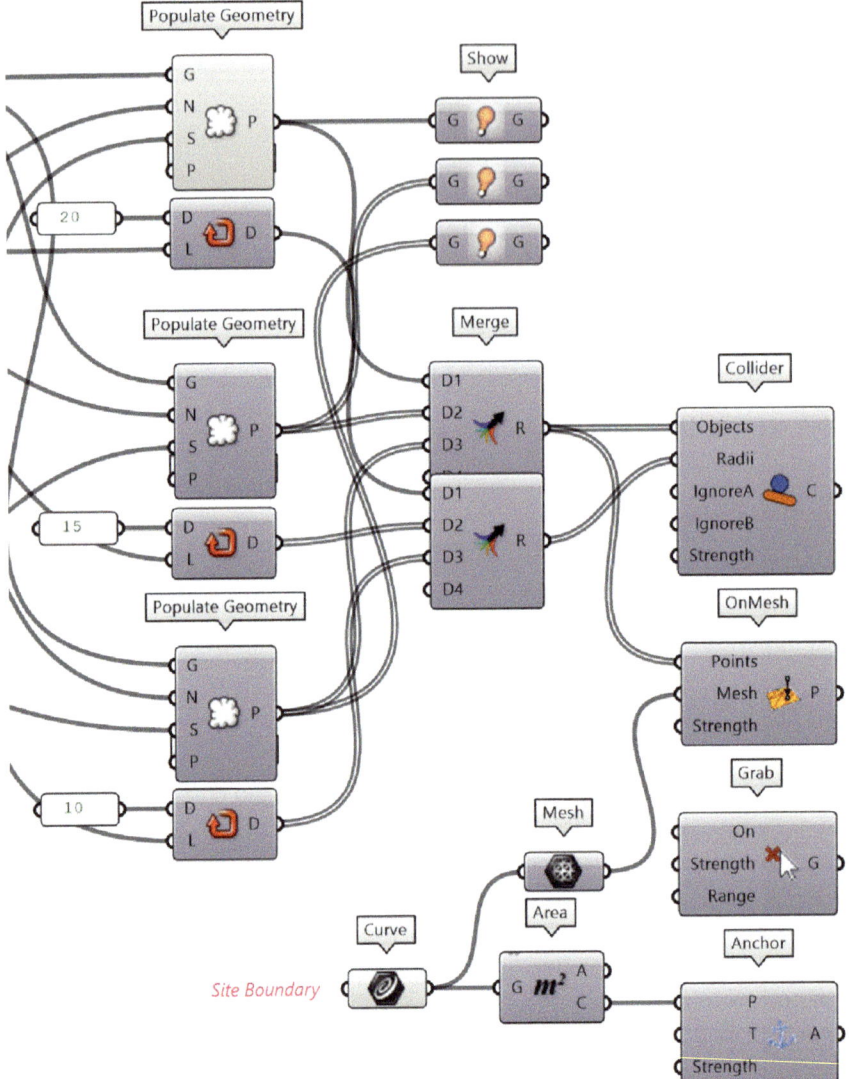

Introducing Kangaroo (Physics Simulator) Plugin

Figure 11.3
Setting up physics simulation generator with the Kangaroo plugin.

allow for the replacement and movement of trees after the simulation by using the mouse pointer to click and dragged to a newly desired location, if the simulated outcome is not preferred.

Now that all the physics engine parameters are established, they can all be joined using the *Entwine* component so that each tree species maintains its own separate data branch after the simulation, as shown in Figure 11.4. This is important since additional parameters to represent the other unique tree characters will be modeled and need to be done separated from each other. Each *Show* component will have its own separate input in this component, whereas all the other remaining components (*Collider, OnMesh, Grab*, and *Anchor*) will be merged into the last input.

The *Entwine* component is then connected to the Kangaroo 2 *Solver* component to run a simulation of forest succession within the project site boundary. The other parameter inputs for this component will determine the run time and resetting of the simulation. A *Button* component is used to reset the simulation after it has completed its operation and another outcome is desired. Another step will also need to be done prior to this, but will be elaborated on as shown in the bottom image of this figure. A low value number slider is then used for the *Threshold* input to determine when the simulation should stop based on the frequency of movement from the geometry. For this model, a value of 0.004 works well as a run time stopper. Last, a *Boolean Toggle* set to *True* is used to run the simulation. Once the simulation has stopped, it can be turned to *False*.

To simulate the growth of forest suggestion, the initial number slider at the beginning of the script, labeled *Growth Simulation*, is used to show the progression of point populations as the slider value increases. This progression can be seen in the side image of Figure 11.4. Once the simulation reaches the optimal number of

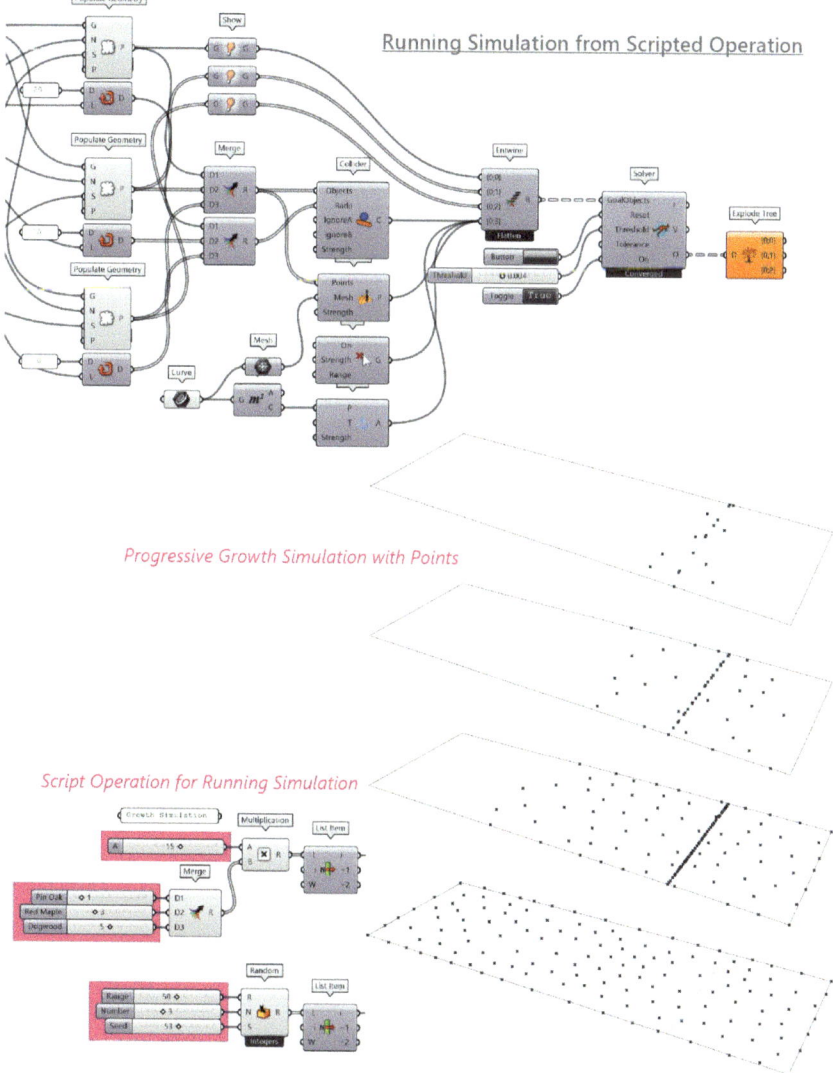

Figure 11.4
Visualizing the growth simulation as snapshots using points to represent tree locations.

points, no additional points will be added into the project boundary. This simulation completed when the number slider reached *15*.

To reset this simulation to its origin, as shown in the beginning, the *Growth Simulation* slider first needs to be returned to the *1* value. Next, the button can be pressed to reset it. If the *Boolean Toggle* has been set to *False*, it will also need to be turned back to *True*.

Lastly, as stated earlier, the *Entwine* component keeps the data branched according to the species type and number of simulated points as a merged set of data. This data now need to be separated back to those respective species using the *Explode Tree* component. Just like in the previous steps, this component will need to have the same number of outputs as there are tree species. As seen in the above figure, this component may appear orange since there will likely be *null* points serving as points attempted to be added to the project boundary, but the tree sizing parameters prevented that to occur. This will not create any issues in the next steps of this script.

Just like previously demonstrated, the next step in this script will go into detail on adding the other unique tree characteristics for the first tree species, Pin Oak, and then that same operation can be copied for the other species with the adjusted values. The values that will require adjustment for each species is highlighted in Figure 11.5.

The simulation will output only points, so this data will serve as the starting point for the tree height and size. The first step is to model the height of the trees using the *Line SDL* component with a *Length* (*L*) input of a mature tree. For a Pin Oak, it is 70 feet. The next step is more subjective in modeling the canopy extent and density. Using the *Evaluate Length* component, a range of values between *0* and *1* is used, since this component uses percentage and not a specific distance to generate points along the curve. With the *Construct Domain* component, the starting and end point can first be determined. Because most tree canopies will end at the top of the tree, the default value of *1* can be used; however, a number slider for the starting point can be used to determine where the bottom of the tree canopy should begin. With a value of *0.10* inputted, the canopy will begin 10% up the curve.

The domain component can then be inputted into a *Range* component with a number slider for the *Number* (*N*) input to determine the density of the tree canopy. Once these parameters are set, the component can be inputted and *grafted* at the *Length* (*L*) input of the *Evaluate Length* component. The grafting parameter is necessary to ensure that each curve gets evaluated at those respective points instead of trying to connect mismatching list lengths, as discussed in the 'Parametric Modeling' section of this book.

A *Range* component will be used again to connect with a *Graph Mapper* component to model the form of the tree canopy. When first using this component, it will appear blank, so it will need to be *right-clicked* to open the pop-up menu where *Graph Types* can be highlighted to select a variety of graphing options. For this first tree canopy demonstration, the *Bezier* option is selected. The graphing component can be seen as a sideways profile of the tree canopy where the leftmost value (0) will serve as the starting point or bottom of the tree canopy and the rightmost

Figure 11.5
Modeling a tree's
parameters to
include canopy
profile, height, and
width.

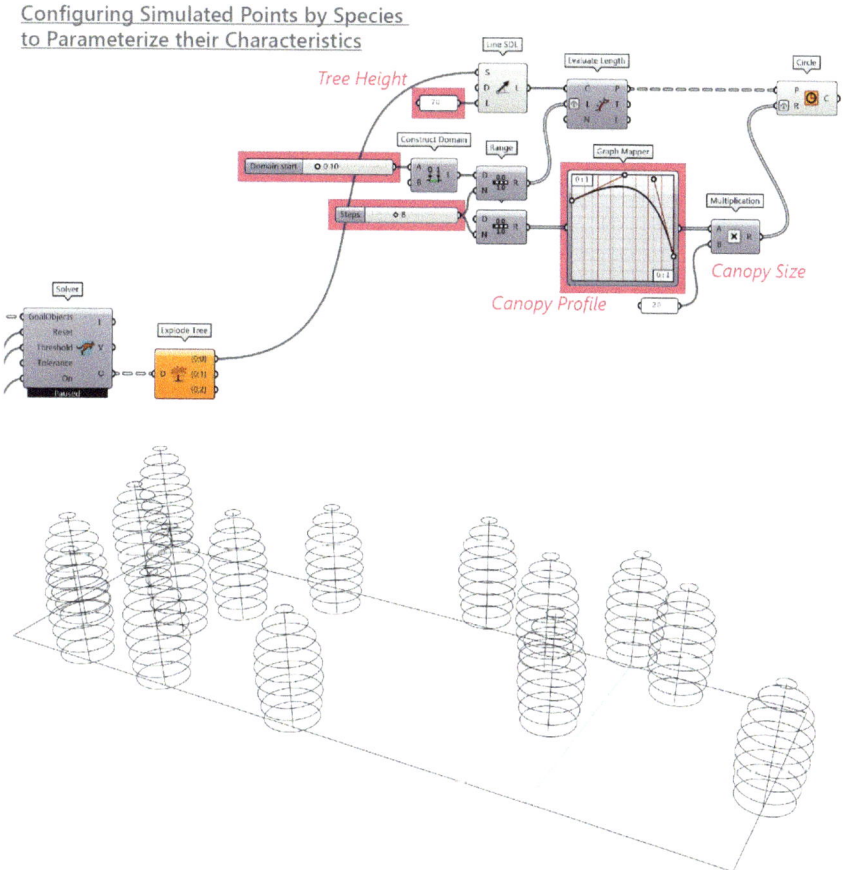

Configuring Simulated Points by Species
to Parameterize their Characteristics

value (1) will be the end or top of the tree canopy. When looking at the graph, the tree canopy will start relatively wide and increase slightly and then taper off to a narrower crown. These outputted values are only proportional, so a *Multiplication* component with the tree radius value inputted will model the proper width of the tree canopy.

Now, the *Evaluate Length* and *Multiplication* components can be connected to a *Circle* component. The *Radius* (*R*) input of this component also needs to be *grafted* to match the branching of the points input. With this component now modeling the general form of these trees, the highlighted parameters as part of this operation can be adjusted accordingly to fine-tune toward a desired tree figure.

Additional graphic elements can be incorporated into the simulation to accurately and legibly communicate the diversity and density of the forest succession process, as shown in Figure 11.6. The circle curves for the tree canopy are first converted into surfaces and assigned a color with the *Custom Preview* component. The color for the *Swatch* component should have an *Alpha* (transparency) value in the middle range (~135) to show density and layer of the tree canopy. This color and transparency settings can be adjusted by *right-clicking* on that component.

Colorizing Simulated Trees and Quantity with Labels

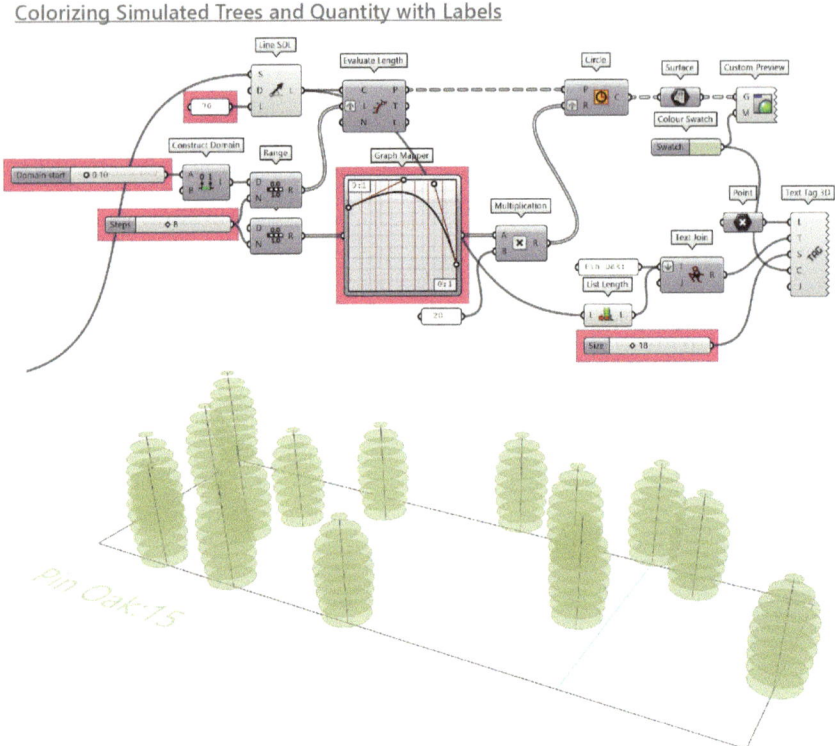

Figure 11.6
Colorizing and
labeling the
quantity of first tree
species.

Labels and numeric values should also be included to give more comprehension to the entirety of the growth simulation. From the *Line SDL* component, a *List Length* component can be used to count the number of trees for this select species. This value can then be merged *after* inputting the name of the tree with a colon, for this example *Pin Oak:* is used. A location for this label will also need to be determined, but instead of placing a point in Rhino and referencing it, a *Point* component can be placed, then *right-clicked*, and select the *Set one Point* option. This will give the option to place a Grasshopper point inside Rhino instead of vice versa and allow for easy adjustment as other labels are included to the process. In this example, the text is also colored the same as the trees for consistency and visual connection. Lastly, a number slider is used to determine the size of the text according to the size of the project model.

As stated earlier, this series of operations can now be copied for the other remaining tree species, as shown in Figure 11.7. Different parameters are highlighted to indicate what should be changed due to its unique characteristic in comparison to other tree species such as height, width, density, and form. Different Grasshopper points are used for the other labels so that there is no overlapping text as well. It is important to consider where the bottom of a tree canopy begins using the *Construct Domain* operation as well as the form with the *Graph Mapper* component.

Figure 11.7
Repeating the
process of scripted
operations for the
remaining tree
species.

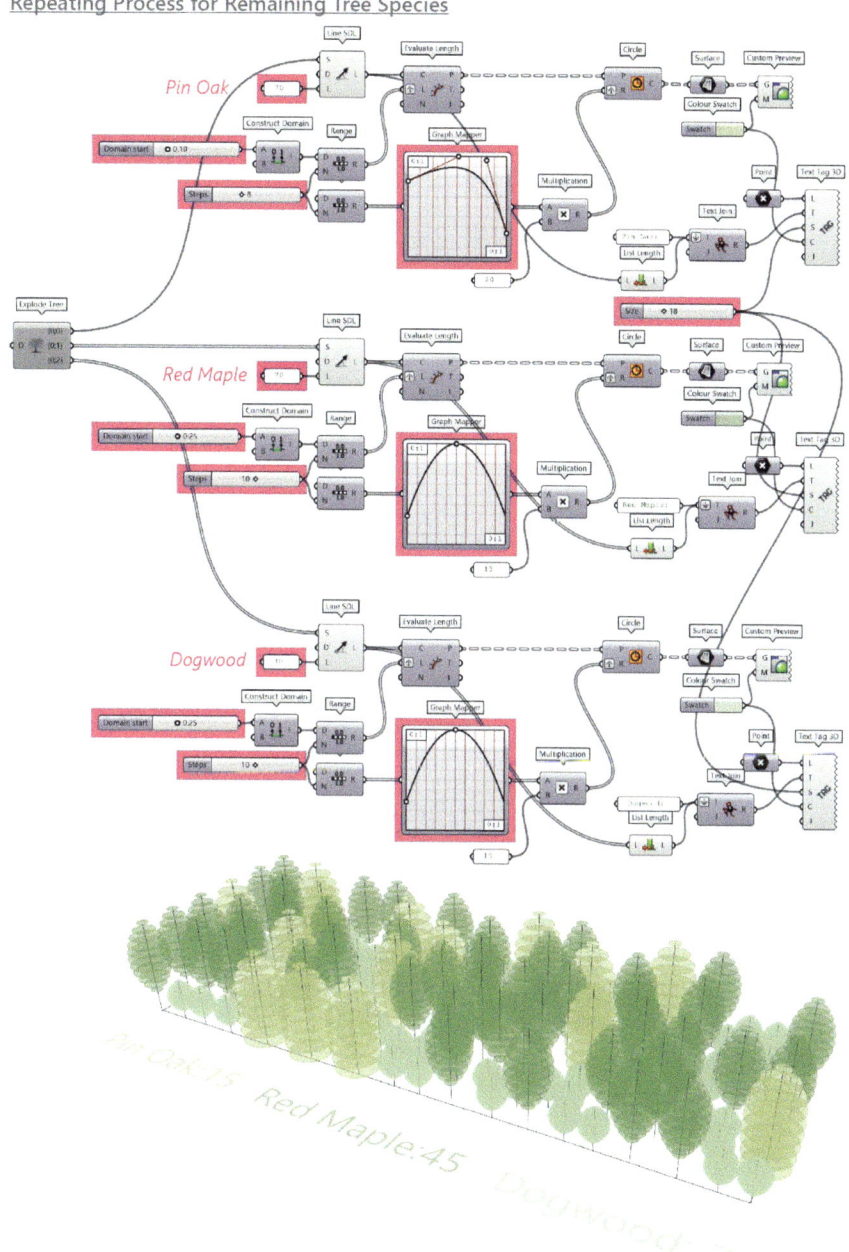

OUTCOMES

From this modeling simulation of tree growth and forest succession, one can begin to plan and strategize around natural processes of ecosystem services from trees. When this growth simulation originates from a water source, it can suggest

Reapplying Simulation with Tree Characteristics

Figure 11.8
Re-simulating
the tree growth
succession from
origin point with all
the tree parameters
visualized.

stormwater management and water filtration opportunities. It is intended to be applied to a vacant lot in an urban setting. It can suggest infill growth and mitigation efforts to repair soils and clean the air. These are just some of the outcomes that can model through this scripting process.

As shown in Figure 11.8, a variety of different growing patterns can be tested as a forest begins to succeed along an empty space. It can also showcase planting diversity in size, color, and density for spatial performance and aesthetics. Forests and nature are dynamic living systems that are constantly in flux, and although this does not simulate every variable or process, it can serve as a foundation for other potential processes that may impact an outdoor space. These potential opportunities are not always benefits to the environment but could also include deficiencies such as tree mortality, outdoor discomfort, or even pest inhabitation.

When considering parametric modeling of landscape performance, it is important to account for a time or other changing variable to output a variety of scenarios to plan for either holistic and comprehensive design strategies or focused efforts that revolve around a specific project goal. Additionally, the simulated growth pattern can also be run in tangent with other ecosystem services, as demonstrated in the concluding *Future Potential* of this chapter.

ADDITIONAL CONSIDERATIONS

This modeling process serves as a foundation for other ecosystem services and other landscape performance measurements through its simulation of dynamic forest growth and succession of empty space. As different growth patterns are simulated, other benefits and references can be cross-referenced with the tree population to assess additional benefits to a project site. The following figures show how other opportunities are modeled to demonstrate that comprehensive outcome as well as the fluid workflow of integrating other parametric operations from other chapters into another script.

Although there are a variety of different incorporations of these performance metrics, this example will demonstrate reduced planting densities, wildlife habitat, and tree benefits. The example will show how the previously constructed script can be revised and added to for a more robust model.

Reduced Density

The *reduced density* operation is a very simple and straightforward procedure that can be integrated at the beginning of the script as shown in Figure 11.9. The initial use of the radius values was intended to serve as the size of the tree canopy; however, this value can be either increased or decreased with a *Multiplication* component and number slider to change the density of the forest succession. Each tree species can have a different density value for more space or even have a value less than *1* to allow for overlap with an understory canopy layer.

Wildlife Habitat

In Figure 11.10, a bird flight path is simulated toward the end of the script when the tree points are assigned height values with the *Line SDL* component. Different point elevations are determined to simulate potential nesting or perching locations for a specific bird species. This is done using a range of random values between a lower and upper value along the tree height. Each tree can be given a unique random value if the *Length* component is used, or they can all be the same elevation if that component is omitted. The curve is then assigned these point values with the *Evaluate Length* component and connected for a flight path using the *Interpolate* component.

The Cornell Lab of Ornithology (https://www.birds.cornell.edu/home/) is a great resource to get specific bird habit metrics for more accurate depiction of a bird habitat. Although this is specific to bird habitats, the same principles can be applied for other wildlife species.

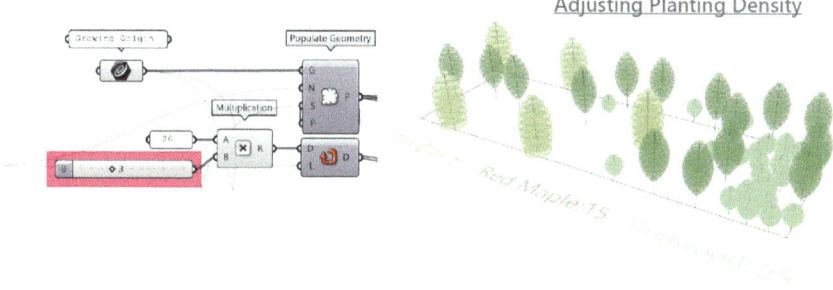

Figure 11.9
Including additional parameter to adjust planting density.

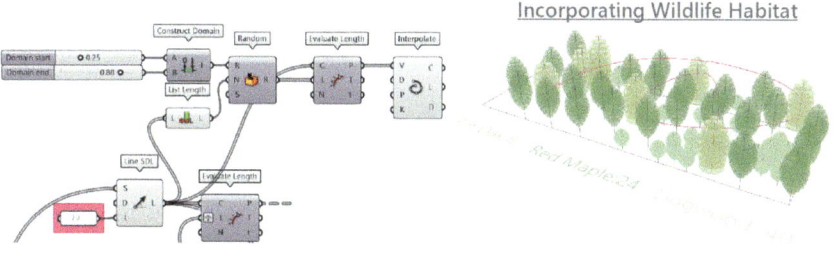

Figure 11.10
Modeling a potential bird nesting and feeding habitat based on parameters from external resources.

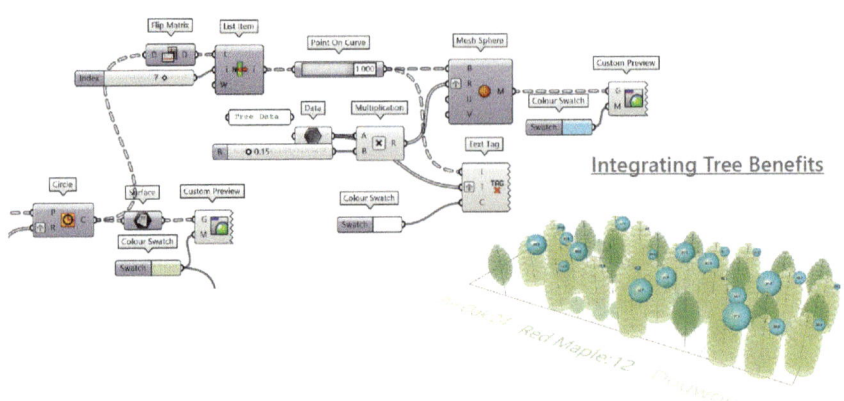

Figure 11.11
Integration of tree
benefits to tree
growth succession.

Integrating Tree Benefits

Tree Benefits

Different project site may require different tree benefits in addition to simulated growth patterns. The previous chapter on tree benefits is a great starting point when considering integrating other performative values to this script. This script operation will not repeat the process of incorporating tree benefit data, as already elaborated on, but will take that previously collected data and internalize it within a *Data* component labeled *Tree Data* in Figure 11.11.

Before that data is modeled, the circle curves used to model the tree forms are flipped and inputted into a *List Item* component where a number slider can be used to determine which elevated circle will be used to place the tree benefits. A *Point on Curve* component is then used to determine a legible location on the curve for the geometric and numerical representation process. The radius of *Mesh Spheres* is determined by scaling the tree benefit data for appropriate sizing and used for the text to be labeled. Because each tree has a unique benefit, both the mesh sphere *Radius (R)* and *Text (T)* inputs need to be *grafted*.

FUTURE POTENTIAL

Other future considerations not demonstrated in this model could also include growth rates and FQA, a metric used in several LPS case studies and a part of LAF's guidebook.

Growth Rate

Another element that can be considered for future model simulations is the incorporation of tree species growth factors. Growth factors are essentially used to determine the age of a tree based on its diameter at breast height (DBH) and its unique growth factor. The equation for this is *DBH × Growth Factor = Age*. For the purposes of modeling the growth and aging of a forest or outdoor space, a number slider can be used to simulate years of growth starting at 1 and progress for a set number of years. These values can then be divided by the growth factor value to determine what size (DBH) the tree should be at a certain age. So, the equation would now be *Age (Year)/Growth Factor = DBH*.

Figure 11.12
Landscape
Architecture
Foundation
Courtesy of Landscape
Architecture
Foundation: Christina
Sanders, 2019.

The application of this equation can also be applied in a variety of ways. The random set of points illustrated earlier that represent tree locations can all begin growing at the same time or they can occur in a sequential pattern. In other words, one tree begins its growth simulation and then after two or three years the next tree of that same species can begin to grow to have an offsetting and seemingly more realistic simulation.

Floristic Quality Assessment

This method uses a calculator to measure the condition of either a natural habitat or plant community within a designated project site (Freyman et al., 2016). They are commonly used for land management strategies by government agencies or conservation groups to evaluate the overall health of an area's ecology. The tree species and quantity can be measured to determine the conservatism-based metrics for estimating native plant probability or habitat quality. These measurements output a Floristic Quality Index (FQI) that uses a Total Mean C factor and multiplying by the *square root* of the total number of species. The resulting values can then be used to evaluate against baseline measurements of natural areas, in which the higher the FQI value, the higher the habitat quality, as assessed in the Glenstone project in Figure 11.12.

This measurement can also be adapted to consider both human disturbance and non-native species, which can be found at both the Universal FQA website or specifically within the Glenstone case study methods (Mendel & Nagraj, 2019) on the LPS website.

These integrations amongst many other metrics, measurements, and spatial conditions can easily develop significant and robust models that simulate all the performative benefits of trees and planting communities in natural and developed environments.

REFERENCES

Chapman, W. K., & Bessette, A. E. (1990). *Trees and Shrubs of the Adirondacks*. Utica, NY: North Country Books, Inc. 131 p. [12766]. Retrieved from https://www.fs.fed.us/database/feis/plants/shrub/coralt/all.html#5

Freyman, W. A., Masters, L. A., & Packard, S. (2016). The Universal Floristic Quality Assessment (FQA) Calculator: An Online Tool for Ecological Assessment and Monitoring. *Methods in Ecology and Evolution* 7(3): 380–383.

Hutnick, R. J., & Yawney, H. W. (1961). Silvical Characteristics of Red Maple (*Acer rubrum*). USDA Forest Service, Station Paper 142. Northeastern Forest Experiment Station, Upper Darby, PA. 18 p. Retrieved from https://www.srs.fs.usda.gov/pubs/misc/ag_654/volume_2/acer/rubrum.htm#:~:text=Growth%20and%20Yield%2D%20Red%20maple, in%20 70%20to%2080%20years

Mendel, E., & Nagraj, C. (2019). *Glenstone*. Landscape Performance Series. Landscape Architecture Foundation. https://doi.org/10.31353/cs1560

Minckler, L. S. (1965). Pin Oak (*Quercus palustris* Muenchh.). In *Silvics of Forest Trees of the United States*, pp. 603–606. H. A. Fowells, comp. U.S. Department of Agriculture, Agriculture Handbook 271. Washington, DC. Retrieved from https://www.srs.fs.usda.gov/pubs/misc/ag_654/volume_2/quercus/palustris.htm#:~:text=Pin%20oak%20is%20a%20 short, age%20(6%2C22)

12   Shade Envelopes

As areas become more developed and remove natural resources and vegetation, urban heat islands begin to negatively impact the thermal comfort of outdoor spaces. The increase in temperatures from impervious and low albedo surfaces absorb solar radiation and slowly release that accumulated heat during the evening, preventing spaces to cool during the night. Although there are a lot of factors and contributors to urban heat islands, introducing trees and pavilions can shade surfaces for less heat retention during the day.

The other significant impact shading can have on a space is the number of hours an area is cooled and when it is being cooled. It can be assumed that as a surface is shaded for an extensive amount of uninterrupted time, it will significantly cool more drastically than surrounding sunlit surfaces. Another consideration when shading spaces is the time of day and year that is most significant to outdoor comfort. With data resources being more readily available and incorporated into modeling methods, ideal times can be analyzed for shading potential during those peak high temperatures.

These are the elements that this chapter aims to analyze in the development of shade envelopes that are strategically modeled to output shade hours during high temperature at different time periods. The outcomes of this model can then be utilized to inform design interventions for ideal outdoor comfort.

GOALS

The purpose of this script is to establish a foundation for outdoor comfort analysis, since there are a multitude of direct and indirect variables contributing to urban heat islands that can be mitigated by performative landscapes. One of those variables that has a significant impact on outdoor comfort is the shading of surface materials. This model will not only demonstrate shadow projections of different elements but also the number of hours shade may exist within certain areas in a project site and how that can be strategically conducted during the hottest times of the day and year.

The following are the evaluation and assessment goals from the modeled shade envelopes:

1. Determine appropriate times and dates for when shade is most desirable – *hottest times*
2. Compare the percentage of tree canopy versus the percentage of shade

3. Measure the number of hours a set grid of surfaces is shaded for
4. Evaluate how areas with the majority of shade (>50% of daily hours) compare to minimally shaded areas

CASE STUDY SHADING

Before any modeling can be done, it is important to reference existing landscape performance projects that prioritize the reduction of surface temperatures. These case studies can inform and reduce assumptions about surface temperatures while also providing methods and outcomes from their extensive investigation of a project site for integration into the modeling workflow.

The table in Figure 12.1 shows several case studies from the Landscape Performance Series (LPS) website that cover a wide range of temperature reduction using an array of shading conditions on different surface materials with different coverage densities. With the table of unique temperature reduction outcomes, similar strategies can be applied to the development and refinement of a project to reference and compare temperature differences.

These case studies provide guidance for potential temperature reduction that are of similar quality and location. In the 'Outcomes' section of this chapter,

Project Name	Location	Shade Condition	Shade Coverage	Surface Shaded	Average Temperature Difference (°F)
AT&T Performing Arts Center: Sammons Park	Dallas, TX	Shade Structure	67%	Concrete	17°
Bagby Street Reconstruction	Houston, TX	Trees	70%	Pavement	21.6°
Belo Center for New Media	Austin, TX	Trees	N/A	Seating Area	19.8°
George "Doc" Cavalliere Park	Scottsdale, AZ	Trees/Shade Structure	N/A	Concrete and Stabilized D.G.	30°/45°
Phoenix Civic Space Park	Phoenix, AZ	Trees and Shade Structure	N/A	Turf/Concrete	12.4°/23.4°
Sundance Square Plaza	Fort Worth, TX	Structural Umbrellas	22%	Pavement	22°
Thomas Jefferson Visitor Center and Smith Education Center	Charlottesville, VA	N/A/ Trees and Shade Structure	N/A	Green Roof/ Courtyard	26.5°/29.7°

Figure 12.1
Matrix of different project sites demonstrating their reduction on surface temperatures by different means.
Source: Cho & Graham, 2014; Martin & Colter, 2014a, 2014b; Ozdil et al., 2014a, 2014b; Shearer & Tierney, 2015a, 2015b.

Figure 12.2
Shade structures at the Sundance Square Plaza for reduced surface temperatures.
Courtesy of Michael Vergason Landscape Architects, Ltd.

instructions are shown on how to incorporate the industry standard methods into the parametric model operations.

The Sundance Square Plaza project shown in Figure 12.2 is one example of how adequate shading can provide a significant temperature reduction on the surrounding surfaces (Ozdil et al., 2014b).

MODELING METHODS

The first step in conducting this shade study is using the Ladybug plugin to determine a specific time or day of the year that is applicable to a project goal. For this demonstration, the goal is to utilize shade for outdoor comfort. Although the summer solstice (June 21) is the longest day of the year in the northern hemisphere and would have the shortest shade projection, it doesn't necessarily mean it will be the hottest day of the year; in fact, in some areas that may already be a comfortable condition.

A more significant outcome would be to analyze the hottest time(s) of the year and use that as an evidence-based framework to conduct a surface shade study. In Figure 12.3, the *LB Analysis Period* component is set to July 3 (*7 _start_ month_ and _end_month_ inputs; 3 _start_day_ and _end_day_ inputs respectively*) since that is the overall hottest day with four out of the top six hottest hours being on that date.

The process of using the *LB Deconstruct Data* component for sorting values specific to the project needs is an effective way to do so. In this demonstration, *dry_bulb_temperature*, in Fahrenheit, is reordered from hottest to coldest temperatures using the *Sort List* component with reversed outputs. This will establish a specific date in relation to the hottest time of the year as the *first goal* of this script. The inputted *Keys* (*K*) can be used to sort all other data associated with that input such as other weather conditions, dates, and comfort levels.

Importing Climate Data for Shade Analysis

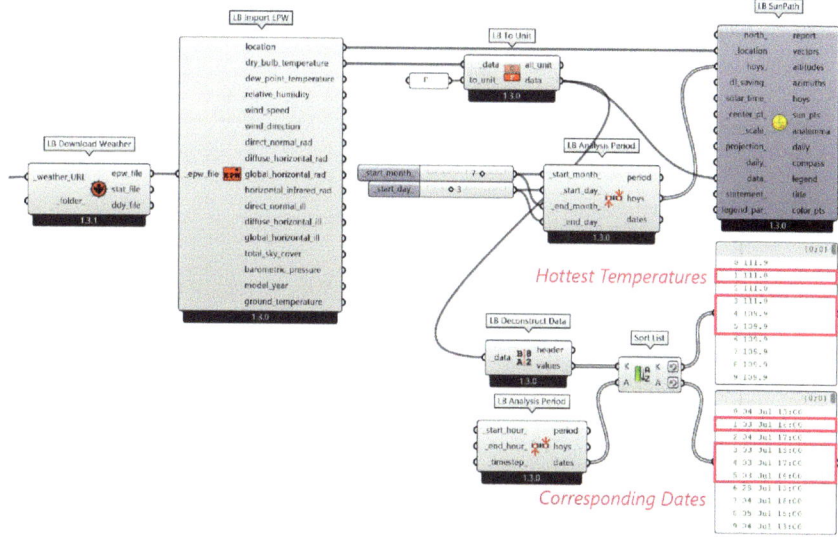

Figure 12.3
Using Ladybug climate date to establish hottest times of year for analysis.

With that date now determined from this process, another *LB Analysis Period* component is used with those parameters to isolate sun positions to just that date in the *LB SunPath* component. Another option, if a single date is not critical, is to analyze the top percentage of the entire year so that other dates and times are included as well. This is shown as part of the *Future Potential* of this chapter.

Surface Analysis

From the *vectors* generated from the *LB SunPath* component, the first and last ones (sunrise and sunset) are omitted with the *Cull Index* component, since graphically the shadows were too extreme and often those projections and temperatures are unsubstantial to shading conditions, as shown in Figure 12.4. The newly filtered sun vectors are now cast against select objects to create an array of shadow projections using the *Mesh Shadow* component.

Any three-dimensional objects such as trees, buildings, shade structures, and others can be used for this next step; however, it is beneficial to keep the geometry simple. Between the complexity of the geometry and number of sun rays used to create shadow projections, the computing power necessary to make all these calculations can overwhelm and even crash the program. The *Mesh Shadow* component suggests that only this can be used; however, *Breps* will also work, as shown in Figure 12.4.

When using trees, simple spheres will work well for the shadow projection while graphically representing more detailed trees since these two are not synonymous with each other. The resulting curves from *Outlines (O)* output from the component will need to be *flattened* in a separate *Curve* component for the next step. This will help in the organization of data branches in later operations of this script.

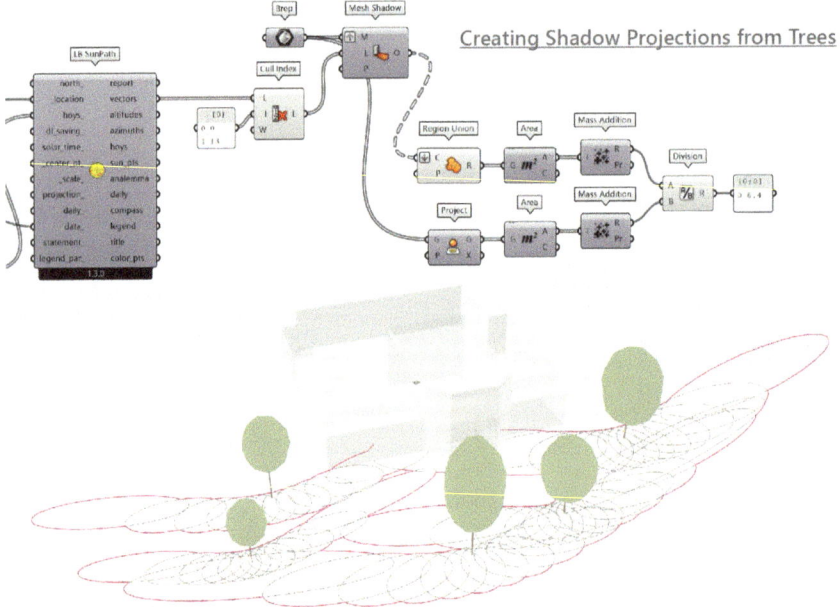

Merged Shadow Projection Boundary

Figure 12.4
Using sun path model to create shadow projections from trees for select hours of the day.

Figure 12.5
Generating a grid of
cells and points to
evaluate time within
shade envelopes.

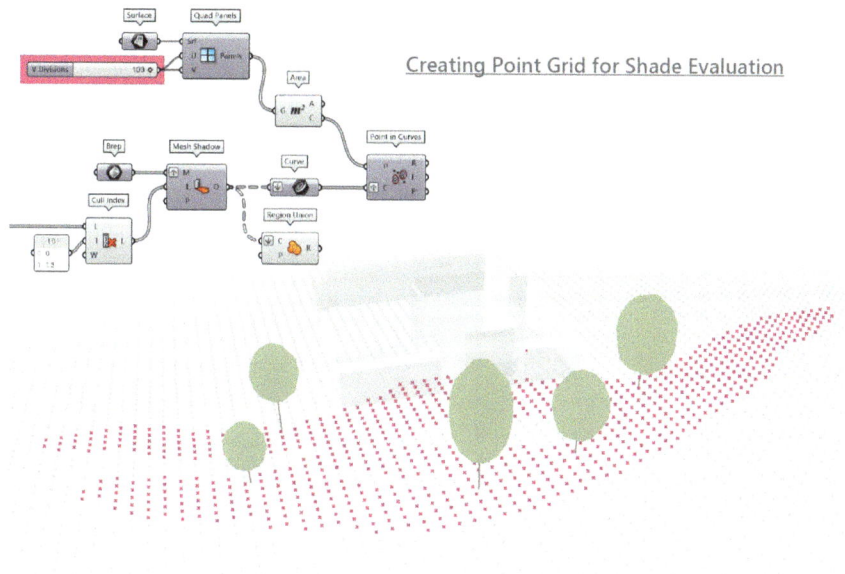

Creating Point Grid for Shade Evaluation

For communication purposes, all the *fattened* curves can be merged to represent an outline of all the sun vectors using the *Region Union* component.

The *second goal* of the script can now be achieved by comparing the total shade coverage throughout the day with the tree canopy area. The total shade coverage can be calculated from the *Region Union* component by extracting the *Area* and combining with the *Mass Addition* component. This value can then be divided by the tree canopy size using the same operation from a *Project* component of the referenced Rhino tree geometry. As a result, from the *Division* component, the trees provide almost 6.5 times the shade as compared to their canopy coverage. This value is still relative, but serves as a baseline for comparison until other iterations of tree placement, size, and quantity are modeled.

The next step in Figure 12.5 will begin to assess the number of overlapping shadow projections that occur within different areas of a project site. The level of detail and precision is contingent on the point grid density being used. There are a variety of ways of creating a grid of points; however, for this demonstration, this grid of points will also influence the graphic representation of this analysis, so a surface grid is first created with points extracted from each individual cell. For this, the Lunchbox *Quad Panels* component is used to create a grid of square surface from one large surface. The surface being referenced and inputted into the component can be of any size as long as it covers the extent of all the shadow projections being used.

The points created using the *Area* component will then be tested for inclusion using the *Point in Curves* component. The flattened curves being inputted into this component can then be *grafted* as part of the rebranching of the data. The grafting will treat each shadow curve independently from each other so that the resulting output will show how many curves contain the same point(s). For example, if six shadow projections have the same point contained within their curve, then that means that point or other corresponding geometry (surface) has received six hours of shade. This is what will be computed in the next operation of the script.

The resulting Relationship (R) output from the *Point in Curves* component determines how many points are contained within a curve. Although this may be useful for other purposes, in order to calculate how many curves a point(s) is be shaded by the data tree needs to be restructured using the *Flip Matrix* component to output the number of curves a point is inside of. The output is still using the criteria of 0, 1, and 2 being associated with whether a point is outside, coincident, or inside a curve, as demonstrated in Chapter 3, so the same operation from that chapter is being used to filter the data, as shown in Figure 12.6. The newly culled data can then be measured for the number of curves a point is inside of using the *List Length* component. Since points with zero curves is already understood as having now shade those points can be omitted using the *Larger Than* component with a *0* inputted so that only points with at least one hour of shade, or one curve that a point is inside of, are analyzed.

The resulting Boolean pattern from this component can then be used to filter out the Quad Panels, points, and number of curves from the *List Length* component using a *Cull Pattern* component for each individually. The bottom image of the figure demonstrates the geometric data being isolated to just the tree shadows, while the bottom *Cull Pattern* component of the script image contains the numeric data that represents the hours of shade for each point and surface being displayed from the above *Cull Pattern* components, satisfying the *third goal* of this model.

The different sets of culled data can now be used to visually communicate the impact shade has on surface conditions. The first form of transformation made to the geometry is changing each cell's size by the hours of shade using the *Scale* component, as shown in Figure 12.7. The cells will be the geometry being scaled

Filtering Enclosed Points and Panels from Shade Projections

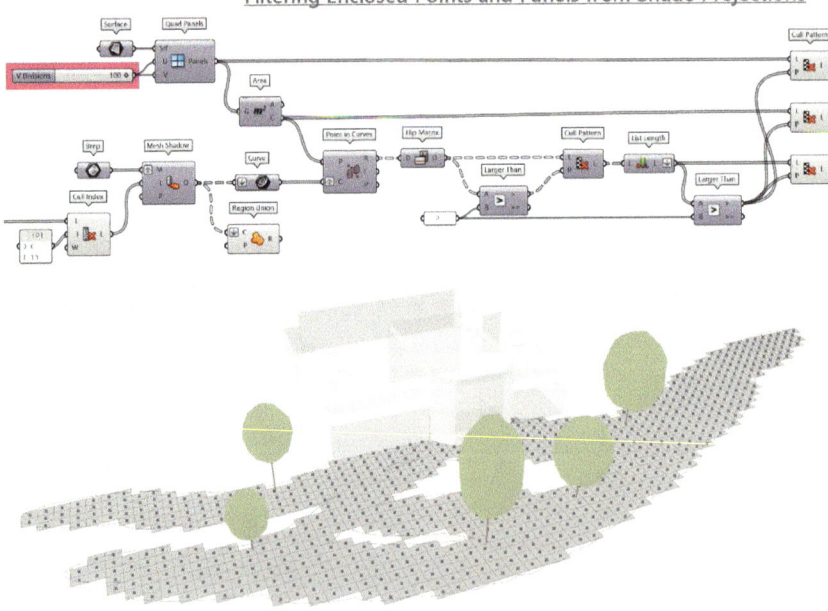

Figure 12.6
Using the evaluation process of points in shadow projections to isolate and partition data.

using their points as the center of scaling. The scale *Factor* (*F*) will be altered a little to enforce a better level of contrast between majority of shade versus minimal shading using the *Graph Mapper* component. After remapping the number of hours and connecting to this component, a *Bezier* graphing option is selected and given an *s-curve-like* graph. The intention behind this type of graph is that the left-hand side (minimal shade hours) will drastically increase at the halfway point toward the right-hand side (majority shade hours). For this type of operation in scaling geometry, it is also best to ensure geometry does not scale to *0* since that will create an error. That is achieved by making sure the smallest values (left-hand side) are above the *0* mark on the *y*-axis of the graph.

The scaled cells can also be colorized, as previously explained in Chapter 3 when using the *Gradient* component. Both the scaled geometry and colors can be connected to the *Custom Preview* component to visualize those unique shading conditions, as shown in the bottom image of the figure.

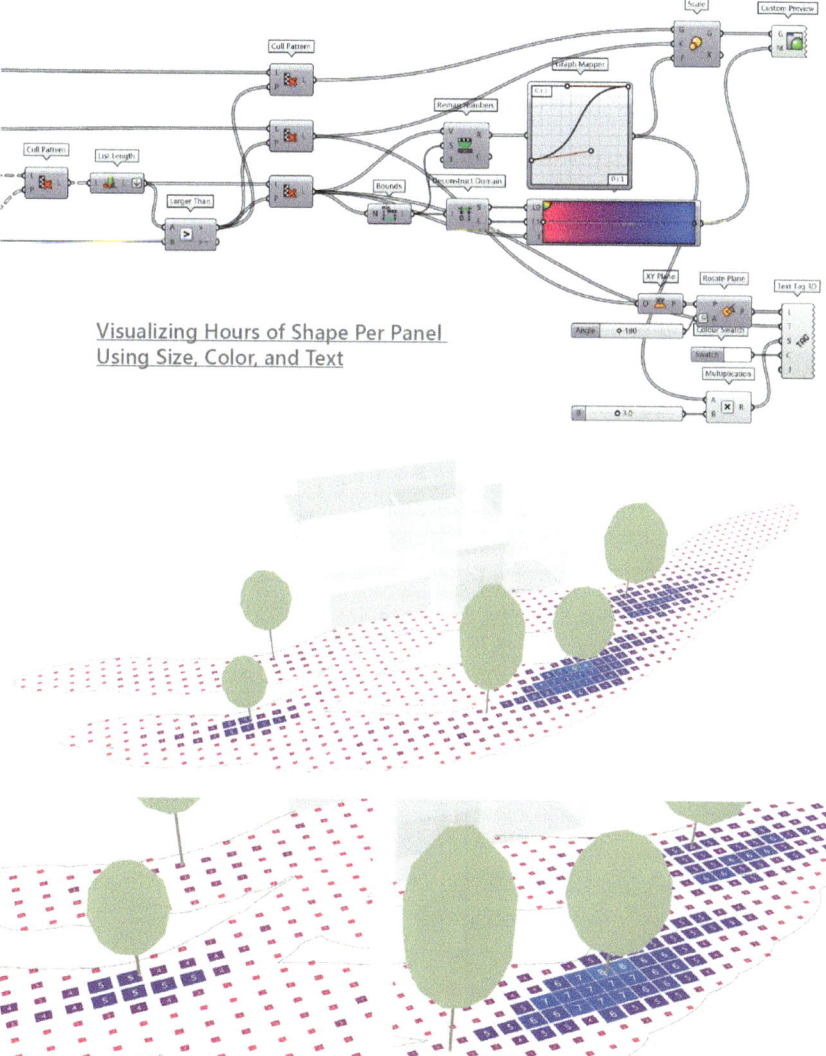

Visualizing Hours of Shape Per Panel
Using Size, Color, and Text

Figure 12.7
Modeling the shaded cells with color, size, and text to communicate their hours of shade.

The last element to add to the graphic communication of shade envelopes is the annotation of pertinent data. In this figure, the number of hours each cell receives shade from the trees is indicated with a labeling operation shown in the script. Although most of the parameters and operations remain the same, as previously demonstrated, a few adjustments are made for this model. Because text labels using an *XY Plane* will orient text in the positive direction, for best demonstration of shade patterns it was important to demonstrate the model from the north orientation. As a result, the text needs to be rotated *180 degrees* since the view is now facing in the negative direction of the *XY plane*. This will not always be the case, but rather contingent on the best orientation and view of the parametric model.

The other adjustment made with the text not already demonstrated is the ability to have the size of the text be reflected in the number of hours as well, like that of the scaling of the cells. By taking the same data from the *Graph Mapper* and *multiplying* it by a value for legibility purposes, the cells with the most hours of shade will contain larger-sized text than the cells with the least number of hours. This can also be seen in the bottom image of Figure 12.7.

Cumulative Analysis

In addition to the surface analysis output, different analysis can also be performed that outputs cumulative data from the shade envelopes. In the first example of Figure 12.8, temperature reductions from cases studies, as referred to the beginning of this chapter, can be integrated to the shade performance on different surface conditions. For this project site, it takes place in a similar location of the Phoenix Civic Space Park where concrete hardscape material is being shaded. So, based on the metrics from that case study, it can be suggested that similar temperature readings could be expected for surfaces that are shaded for the majority of the day. This does assume a lot of variables; however, it can be very informative when attempting to predict or simulate design proposals to achieve certain temperature reduction goals.

To begin this integration, the data first needs to be filtered once again to account for only surfaces, or cells, that are shaded for the majority of the day. Since the upper values for this script outputted seven to eight hours as the highs, a *Larger Than* component with a value of *4* is used to determine which set of data meets that criterion. The cells can then be reduced to only those values using the *Cull Pattern* component to merge into profile curves of those shaded areas.

Next, from those shaded areas, labels for the expected temperature reduction can be placed once rotated to the proper orientation for the *Text Tag 3D* component. The Text (T) that will labeled is the value referenced from the case studies table in Figure 12.1, being an average temperature difference of 23.4 degrees Fahrenheit for hardscape surfaces.

Another analysis that can be outputted from this model is the percentage of surface that have been shaded for the majority of the day. That output can be achieved by dividing the number of shaded surfaces receiving four or more hours of shade, in this case, by the total number of shaded surfaces. *List Length* components for both the new *Cull Pattern* components introduced in the previous

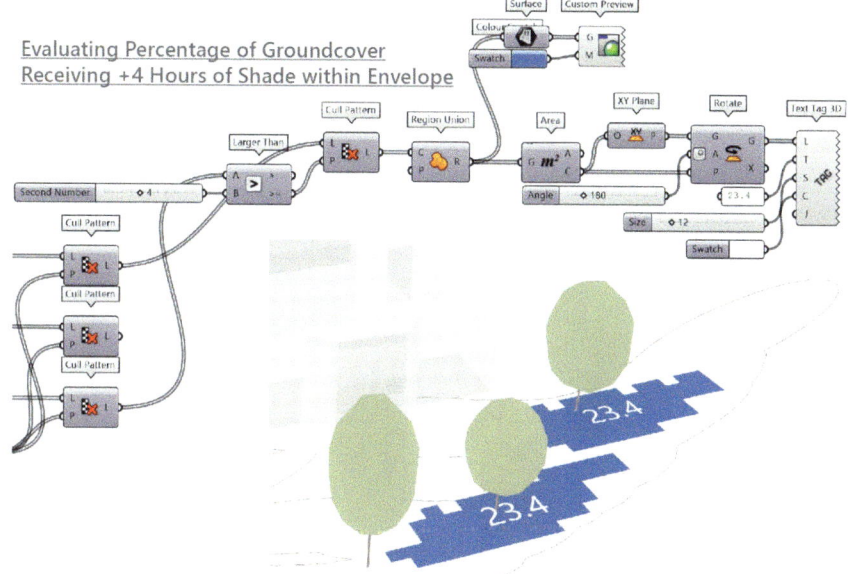

Figure 12.8
Evaluating
the potential
temperature
reduction from four
or more hours of
shade.

operation with the *Cull Pattern* component that it was connected to are used with a *Division* component to find a decimal percentage, as shown in the second script image of Figure 12.8. This value is then multiplied by *100* to convert it to percentage and joined with a *%* symbol using the *Text Join* component for proper labeling. The resulting text can then be connected to the *Text Tag 3D* component to annotate the graphic.

OUTCOMES

The data that can be extracted from this model can be analyzed and utilized in a variety of ways to begin the process of understanding outdoor comfort for pedestrians. Shade from trees, pavilions, and buildings all contribute to cooler surfaces, and from this model, surfaces can be analyzed for how long they've been shaded to further evaluate a project site.

In many of case studies showcased at the beginning of this chapter, the percentage of shade canopy is inventoried. This information or data can then be cross-referenced with the percentage of shade throughout the day or hottest times of the year to determine if those shading tactics are performing effectively. That becomes a valuable tool in the assessment of existing conditions as well as informing design decisions during the development phase of a project. Since the shading geometries (trees and structures) can be Rhino objects referenced into Grasshopper, they can be moved, resized, and multiplied for the Grasshopper model to feedback new data and assessment models based on those refinements.

Also, as stated in the previous section, the percentage or number of hours of shade can be measured to determine if an area is shaded for an extensive amount of time. Although not all case studies go into this level of detail, it can be assumed that surfaces receiving seven to eight hours of shade are going to be cooler than surfaces with only one to two hours of shade. This should also account for the air temperatures during those specific hours; however, it should still be substantially cooler, based on the results cataloged in the case studies. With the *Average* component, an overall average number of hours of shade can be measured for the entire project site so that it accounts for both minor and major shaded areas.

Lastly, the shade profiles throughout the day, which are always changing, can also inform patterns of shade to suggest pathways or occupiable outdoor spaces during hotter times of the year. This can be an interdependent process where desirable circulation works in tandem with shading conditions to output shaded and comfortable spaces.

Surface temperature is just one component when designing for outdoor comfort and does not have a direct correlation with other variables such as air temperature, wind, and humidity, but can be modeled with future chapters to give robust and comprehensive conditions that positively impact outdoor comfort.

Baseline Comparison

The resulting model of shade envelopes depicting surface temperature reduction can then be cross-referenced with a field survey as described in the 'Landscape Performance Guidebook's Temperature & Urban Heat Island' section. Within this section, there are potential metrics to consider worth using in reducing surface temperature by measuring throughout the site in relation to either a before condition or a similar space without adequate shading for a baseline comparison. The metrics for comparison can be in either temperature degrees or percent reduction. Temperature readings can be conducted with the use of handheld thermometers on site during the hottest months of the year when it is the peak time of day.

FUTURE POTENTIAL

As the first chapter modeling outdoor comfort conditions, this model serves as the foundation to make additional considerations with shade along with other variables such as air temperature, evapotranspiration, and windbreaks. With this model specifically, additional data can be extracted, as shown in Figure 12.9, or consider future script operations and considerations. Some of those considerations include measuring the percentage of hours that are shaded for the majority of the day, analyzing all the hottest hours throughout the year, and also cross comparing the shade conditions over different surface material.

Percentage of Time

In Figure 12.10, the *Cull Pattern* used to filter out the number of hours each cell is shaded for is used to determine the percentage of time surfaces are shaded for the majority of the day. This is using the same criterion operation of filtering out only surfaces that receive four hours of shade or more. Since this new *Cull Pattern* component is filtering out the hours of shade data, the resulting data can be connected to the *Mass Addition* component to sum the total hours for those surfaces. That value can then be divided by the total hours of shade by the trees to determine a percentage of surfaces that are shaded for the majority of the day, achieving the fourth outlined goal.

Figure 12.9
Percentage of area
shaded for four or
more hours.

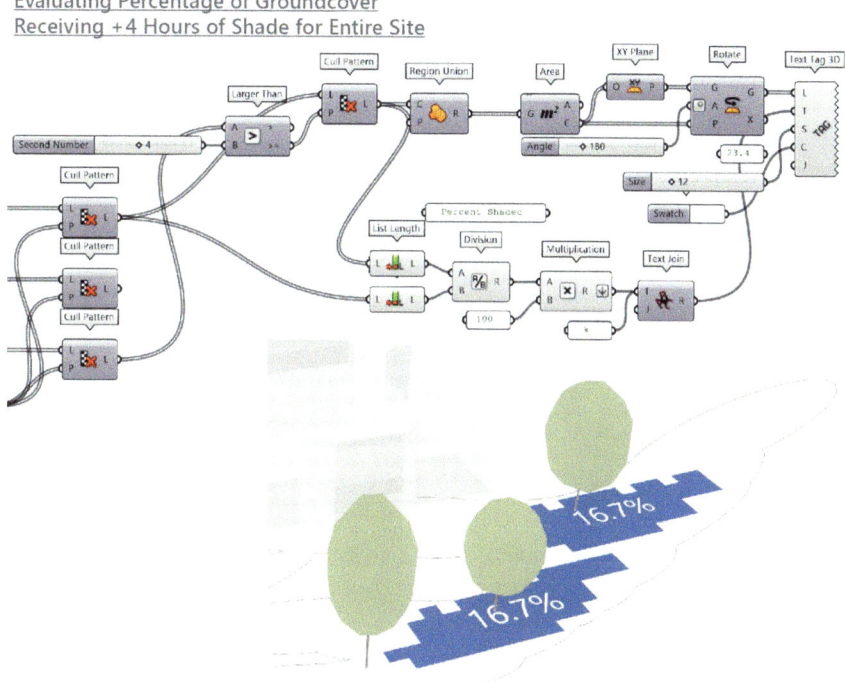

Figure 12.10
Percentage of total
hours shade during
evaluation period.

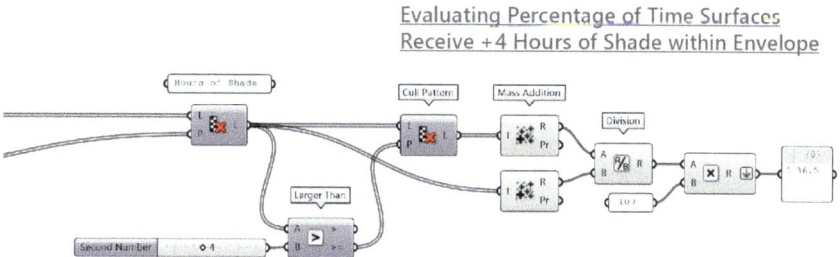

The resulting outcome is significant since Figure 12.9 showed that these areas only shaded 16.7% of the surface area; yet those same surfaces accounted for over a third of all time that was shaded at 36.5%. This concludes areas that there is a drastic contrast between hours of shade. Essentially, there is a small percentage of surface area that receive long hours of shade for cool microclimates versus most of the surface area receiving little to no shade for hotter uncomfortable conditions.

Hottest Annual HOYs

Another way to analyze the shade envelope of a site is to run the model for only the hottest hours throughout the entire year instead of the daily hours for a specific date. This will account for multiple dates but give a more specific outcome to only the hottest times when shade is most desirable. To perform this conditional analysis, the *LB SunPath* component's *hoys_* input will be replaced with this operation's output.

The *LB Deconstruct Data, LB Analysis Period,* and *Sort List* components are still the same as the initial script in Figure 12.3. Instead of using the sorted list output to guide the manual setting of an analysis period, the data will now be used directly to generate sun path vectors for those hottest times (HOYs) of the year. Since there are 8,760 HOYs with a corresponding temperature, finding just the top 1% will still result in about 88 HOYs for sun path vectors by *multiplying* the *List Length* component by *0.01*. Another option that can be used instead of a percentage is to still use the *Split List* component by simply connecting a number slider at the *Index (i)* input to find the top number of HOYs. In other words, a number slider value of *20* will output the top 20 hottest times of the year. The *A* output from the *Split List* component can now be connected to the *hoys_* input of the *LB SunPath* component, resulting in the shaded conditions illustrated in Figure 12.11.

Surface Material

Another option worth considering as part of the performative model is to analyze the shading impact of different surface materials that may be on an existing site or part of a design proposal. This is not demonstrated in this script; however, like the stormwater runoff scripts, the cells representing the base surface condition can be partitioned into their respective surface type such as pavement, turf, D.G., and others. Then, once the model has determined the number of hours of shade for the different surface types, the data for each will still be structured accordingly. So just like in Figure 12.8, when the referenced case study's temperature was inserted in the text tag, the same can be done with each different material type to show a more robust temperature fluctuation and percentage dependent on the material composition of the site.

These future incorporations along with the already existing performance script for shade envelopes establishes a foundation on the principle of shade

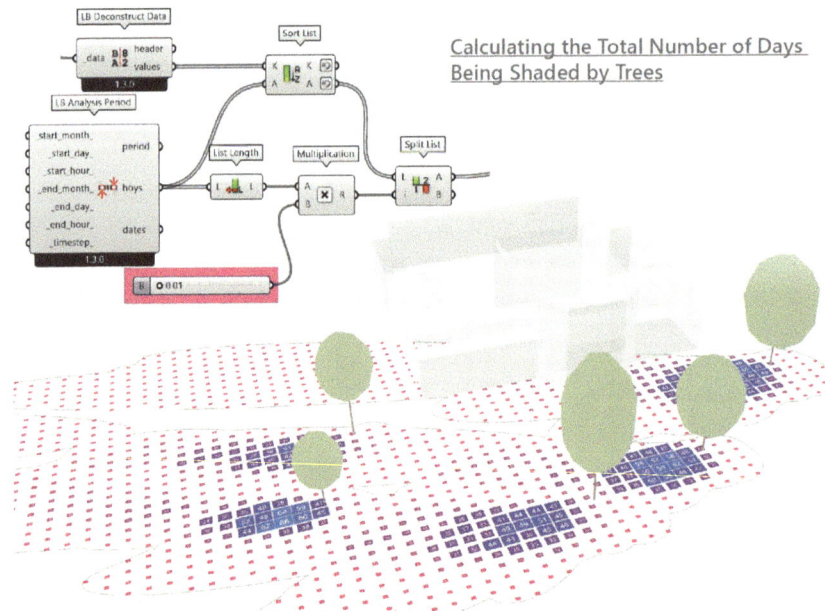

Calculating the Total Number of Days Being Shaded by Trees

Figure 12.11
Total number of hours being shaded throughout the hottest times of the entire year.

analysis – what is the shading condition when it is needed the most? The primary purpose of doing a shade analysis is to determine what the shading is like when temperatures are the hottest and most uncomfortable, since shade is the most effective way in cooling an outdoor space. Arbitrary dates and times should not be used when data can be extracted to inform proper design decisions. This is one of the powers of modeling with abstract data sets to influence the physical space of a site.

REFERENCES

Cho, L., & Graham, M. (2014). *Thomas Jefferson Visitor Center and Smith Education Center at Monticello*. Landscape Performance Series. Landscape Architecture Foundation. https://doi.org/10.31353/cs0700

Martin, C. A., & Colter, K. R. (2014a). *George "Doc" Cavalliere Park*. Landscape Performance Series. Landscape Architecture Foundation. https://doi.org/10.31353/cs0730

Martin, C. A., & Colter, K. R. (2014b). *Phoenix Civic Space Park*. Landscape Performance Series. Landscape Architecture Foundation. https://doi.org/10.31353/cs0750

Ozdil, T. R., Richards, J., Brown, R., Earl, J., & Stewart, D. (2014a). *AT&T Performing Arts Center: Sammons Park*. Landscape Performance Series. Landscape Architecture Foundation. https://doi.org/10.31353/cs0790

Ozdil, T. R., Richards, J., Brown, R., Earl, J., & Stewart, D. (2014b). *Sundance Square Plaza*. Landscape Performance Series. Landscape Architecture Foundation. https://doi.org/10.31353/cs0800

Shearer, A. W., & Tierney, N. (2015a). *Bagby Street Reconstruction*. Landscape Performance Series. Landscape Architecture Foundation. https://doi.org/10.31353/cs1000

Shearer, A. W., & Tierney, N. (2015b). *Belo Center for New Media*. Landscape Performance Series. Landscape Architecture Foundation. https://doi.org/10.31353/cs1010

13 Outdoor Comfort

In the previous chapter, it was discussed how the impact of shading ground surfaces can be valuable in combating the urban heat island effect (HIE) by creating comfortable outdoor spaces. Another important component in mitigating this urban issue is by evaluating how the air temperature can be reduced simultaneously within an outdoor space. The two methods of reducing urban heat islands use different metrics and cannot directly correspond with the HIE; however, it is understood that by reducing the two temperatures, conditions will improve (Environmental Protection Agency, 2021).

This chapter will demonstrate how both number of hours in the shade, like Chapter 12, along with the proximity of air temperature measurements to the ground can be factors for outdoor comfort. The performance script also relies on existing case study outcomes from Landscape Architecture Foundation (LAF) to give precedent on specific metrics and tree planting characteristics. These can serve as guides and precedence for projecting potential landscape performance goals on either other existing sites or future design intentions.

GOALS

When dealing with modeling and analyzing outdoor comfort for heat stress, it is important to determine both when those stressful times are throughout the year and how can shade mitigate those extreme temperatures. This performance model will provide a critical analysis of outdoor heat stress that can be mitigated with different planting and shade parameters based on existing case studies:

1. Evaluate the hottest time(s) of the year to analyze for outdoor comfort through air temperature
2. Incorporate the addition of humidity and wind speed to further refine heat stress
3. Model tree plantings or shade structures to measure the extent of shade envelopes
4. Cross-reference case study outcomes with performance model to assess air temperature mitigation from shade

DOI: 10.4324/9781003208020-16

PERFORMANCE METRICS

Measuring air temperature on a project site can occur at a specific focal point under unique shade conditions or be done throughout the site where there may be a lot of variances in shade. This is usually done with monitors and handheld temperature devices, but it is best to select one based on the level of accuracy appropriate for the project site and intended detail of information.

In Figure 13.1, differing shade conditions and their impact on temperature are shown as down from LAF's Landscape Performance Series' Case Study Briefs. Since there were slight differences in the cataloging of site characteristics, specifically tree planting density, the project size in acres along with the number of trees existing on the site were included to gauge how air temperatures were impacted.

These are snapshot analyses of the project site and they all vary in design, material use, composition, and location, so it is important to further investigate these projects and other background information to interpolate into the performance model. In Figures 13.2 and 13.3, there are examples of different shading strategies from designed trellises to dense tree planting.

Figure 13.1
Different reductions to air temperature from trees on project site.
Source: Chanse & Salazar, 2012; Ozdil et al., 2013; Cho & Graham, 2014; Martin & Colter, 2014; Özer & Stanford, 2014; Özer et al., 2014; Aman & Yildirim, 2019.

Project Name	Location	Project Size (Acres)	Number of Trees	Average Air Temperature Difference (°F)
Central Wharf Plaza	Boston, MA	0.3	25	10.4°
Klyde Warren Park	Dallas, TX	5.2	322	5.5°
1100 Block of Lincoln Road Mall	Miami Beach, FL	1.1	71	1.4°
SoundScape Park	Miami Beach, FL	2.5	355	4.7°
Phoenix Civic Space Park	Phoenix, AZ	2.5	111	1.8°
Yanaguana Garden at Hemisfair	San Antonio, TX	4.1	16	7-12°
Thomas Jefferson Visitor Center and Smith Education Center	Charlottesville, VA	13	146	1.4°

Figure 13.2
Shade structures with vegetation for dense shading.
Courtesy of Landscape Architecture Foundation and Skyler Stanford, 2016.

Figure 13.3
Dense tree planting
within an urban
plaza.
Courtesy of Charles
Mayer Photography.

MODELING METHOD

Like with the other comfort performance models, this one also begins with the use of the Ladybug plugin to establish parameters and benchmarks within the outdoor comfort model for heat stress. There are two directions to take in establishing strategies for outdoor comfort with this model by either determining the stress temperatures for a specific date or by using the hottest temperatures of the year to determine a specific date. This modeling method will demonstrate the latter, since that will, in turn, generate an analysis to inform responsive strategies against extreme conditions throughout the year instead of for a specific date.

Finding those extreme air temperatures will begin with the *Sort List* component with *reversed* outputs, as shown in Figure 13.4. The values from the *dry_ bulb_temperature* first need to be converted to Fahrenheit with an *F* input for the *LB To Unit* component and then have data isolated with the *LB Deconstruct Data* component. The *Sort List* component will also need an *A* input of all the hours of the year (HOYs) that can be generated from a *Series* component having a *Count* (*C*) input of *8,760*, representing every hour of the year from January 1 at midnight to December 31 at 11:59 pm.

From the *A* output of that component, the *LB HOY to DateTime* component can be used to convert the HOYs to the associated months, days, hours, and date. Based on the hottest sorted values, ranging from nearly 112 to 110 degrees, the date that can be used is July (*7*) third (*3*) since that date contains three out of the top five values. This will inform a future step and has flexibility to adjust the date for a different or multiple outcomes. These specific outputs will not be connected to any future components but serve as reference, as will the next step's outputs.

Figure 13.5 shows how to integrate the next set of parameters to aid in establishing benchmarks for the tree shade against hot air temperatures. This will follow a sequence of steps and visual outputs that will dictate the next new parameters, so it is important to follow the creation of the script operation incrementally.

The first visualized output will be from the *LB Psychrometric Chart* and *LB PMV Polygon* component to generate the bottom left-side image. This component

Figure 13.4
Importing climate
data to analyze
hottest air
temperatures.

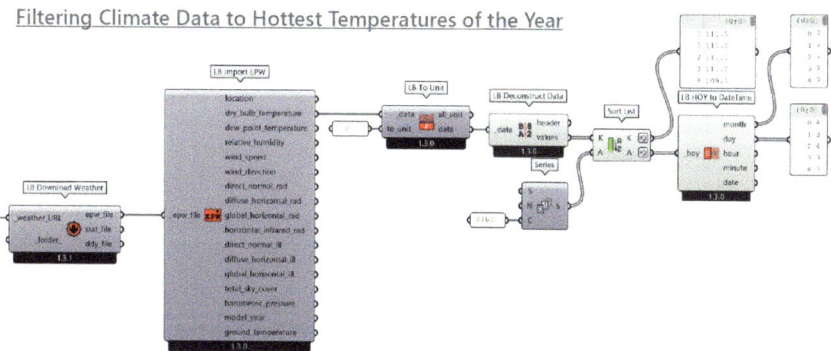

Figure 13.5
Comparing the
impact wind has on
comfort zones for
its context.

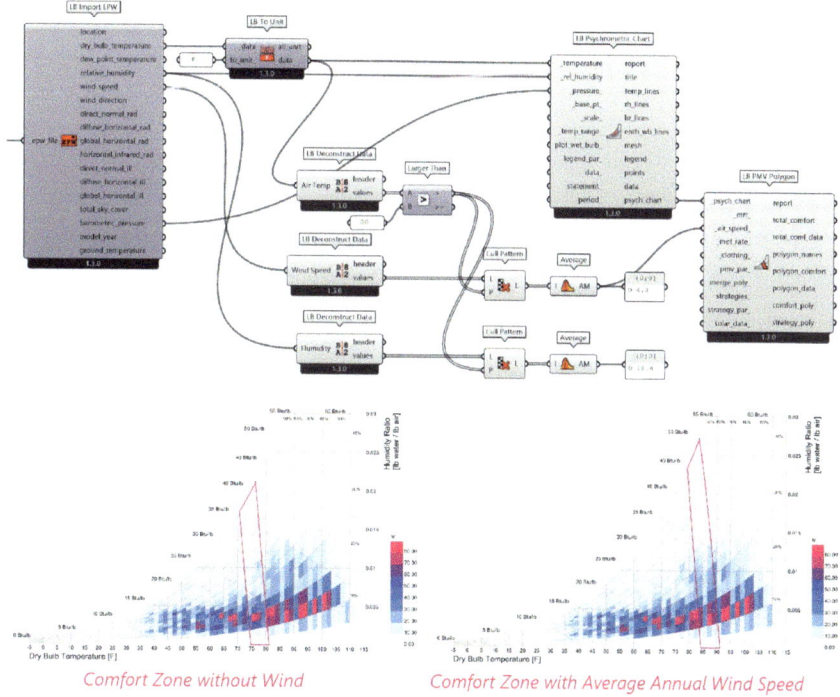

Comfort Zone without Wind   Comfort Zone with Average Annual Wind Speed

requires the *dry_bulb_temperature*, converted to Fahrenheit, *relative_humidity*, and *barometric_pressure* from the *LB Import EPW* component to be connected to their respective psychrometric chart inputs.

Before connecting *_air_speed_* to the last component, the psychrometric chart shows that without considering wind as part of the analysis, the outdoor comfort of space is around *75–80* degrees with low humidity and 71–77 degrees with high humidity, according to the polygon. It is important to note as well that this chart shows that the majority of hours, 50–80, are within the low humidity range of 20% or less. The next operation of this script sequence will show how the introduction of the average wind speeds and humidity for specific conditions will further refine this outdoor comfort model.

Since wind speed can have a significant impact on the perception and feeling of comfort in extreme air temperatures, the next step will demonstrate how to incorporate that parameter for a specific condition. This is important because the overall average wind speed can be significantly different than the average wind speed during different extreme conditions, whether it be hot or cold. As shown in the first chart image, the hottest air temperature could be while still considered comfortable was around 80 degrees, so climate conditions need to be filtered to anything about that threshold.

The *LB Deconstruct Data* component will first be used on the converted *dry_ bulb_temperature* data to filter only those temperatures above 80 degrees by using the *Larger Than* component with an evaluation input set to *80*. The Boolean pattern created from *Larger than* output will then be used to filter out the deconstructed wind speed and relative humidity values. If the project site is in imperial units, it would be logical to think wind speed, measured in meters per second by default, would also need to be converted. That is not the case when calculating outdoor comfort with the Ladybug psychrometric chart since it requires this parameter to remain in meters.

A *Cull Pattern* component is used for both the deconstructed wind speed and relative humidity values, each connected to an *Average* component. From these two outputs, the average wind speed when air temperatures are above 80 degrees is *4.8* meters per second and humidity is *18.4%*. The average wind speed can now be connected to the *_air_speed_* input of the LB PMV Polygon component. With that final parameter, the seemingly outdoor comfort with this wind speeds now ranges from *85* to *90* degrees where most of the hours are within that average outputted humidity value. This temperate range will now serve as the benchmark to analyze against in order to reduce air temperature so that the hours and spatial coverage of outdoor comfort can be expanded by shade.

From those parameters in the last two steps, the *LB SunPath* component and the analysis of outdoor comfort can be appropriately refined as it focuses on only those extreme heat stress periods throughout the year. In Figure 13.6, an *LB Analysis Period* component is set to July 3 to generate those *hoys* to input into the sun path component. The other parameter for this component is a statement of *a>90*, so that only the hours with an air temperature (dry bulb) greater than that for this specific date are outputted. The resulting sun path model generates 12 out of a potential 13 sun vectors, each an hour of the day, indicating that these hours all have a temperature above 90 degrees. The *vectors* output of this component needs to be *flattened* to simplify the data tree structure for the next steps.

Tree geometry can now be introduced to the analysis process, where for this demonstration the trees are modeled and referenced as meshes opposed to Breps. The reason for this is ease of computational processing power since meshes are easier to analyze and operate for this script. If trees are already represented as polysurfaces or Breps, they can always be converted to meshes within Grasshopper by connecting their referenced component to a mesh component.

In the left-side image of Figure 13.7, the trees are densely placed for ideal comfort conditions, as demonstrated by the case study brief's matrix in Figure 13.1. From the referenced tree meshes, their *Centroid (C)* can be created

Figure 13.6
Modeling the
hottest day of the
year depicted from
analysis.

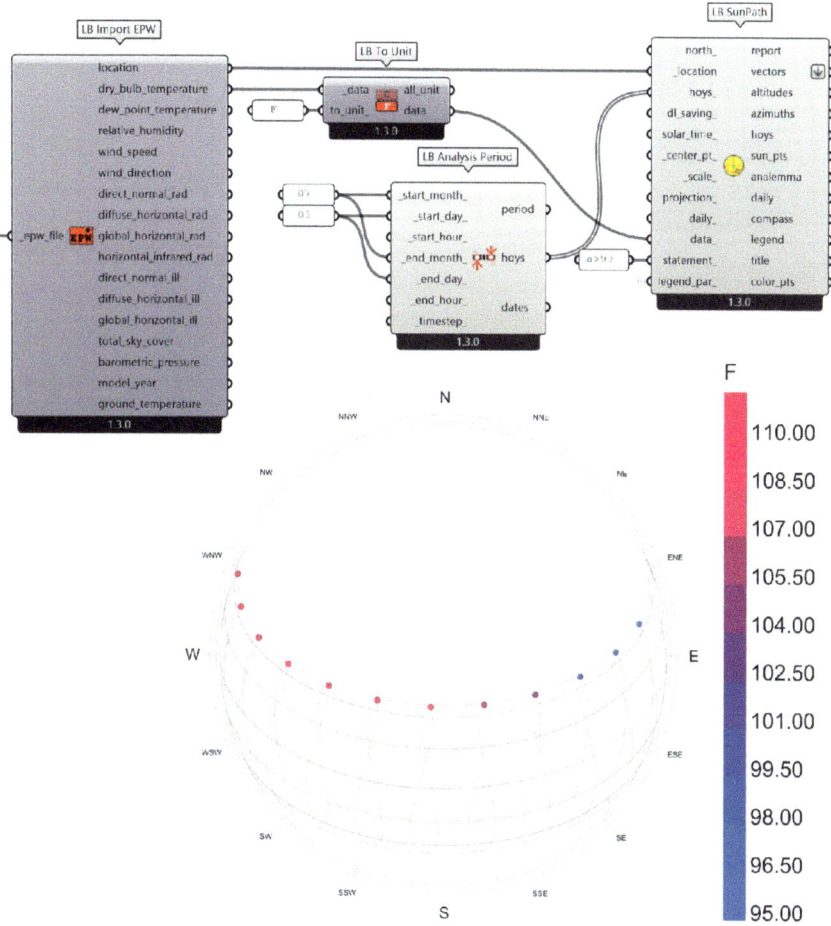

Using Comfort Model to Dictate Conditional Statement for Sun Study

from the *Area* component and used as the *Origin* (*O*) for the *Plane Normal* component. Although the vectors from the previous step were *flattened*, they will now be *grafted* at the *Z* input of this component. This is so every sun direction is treated separately for the group of trees instead of analyzing one sun vector per tree. Based on the tree placements and sun vectors, an intersecting curve can be generated with the *Mesh|Plane* component to represent its shade/shadow. After this curve is cleaned up with the *Trim Tree* component to match the *grafted* branching structure of the sun vectors for the *Project Along* component, it can then be connected to the *Geometry* (*G*) input of that component as well. The cleaned interesting and projected curves can now be used to create a closed shadow envelope for testing by connecting them to a *Ruled Surface* and *Cap Holes* component.

The output from this final component should be flattened to simplify the data tree structure once again for the next steps. The right-side image of the model of this figure shows how each tree now has a generated closed surface representing a shadow envelope for all 12 sun vectors throughout the hottest hours of the day.

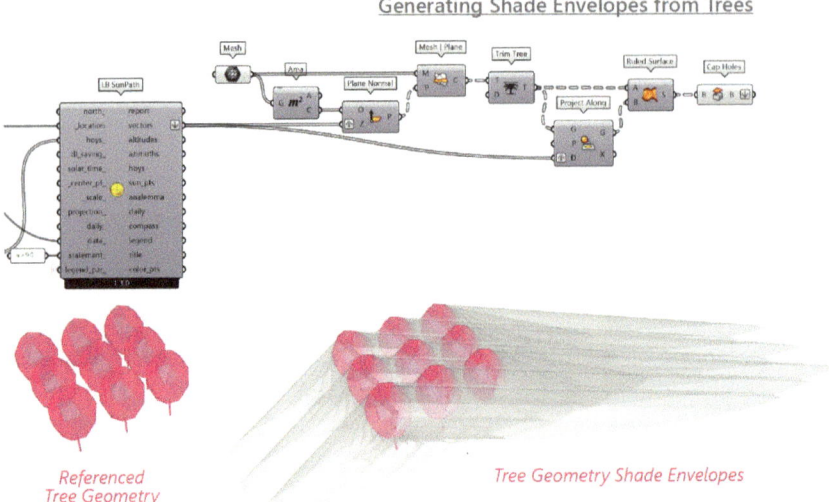

Generating Shade Envelopes from Trees

Figure 13.7
Creating shade
envelopes of
referenced Rhino
tree geometry
from the sun path
vectors.

*Referenced
Tree Geometry*

Tree Geometry Shade Envelopes

Similar to the operation in the previous chapter, 12 Shade Envelopes, these shadow geometries will be evaluated to determine how much of the space and for how many hours are they being shaded. Figure 13.8 shows how the *flattened* output from the *Cap Holes* component is connected to a *grafted* input of the *Point in Breps* component with a collection of random points to test for the *Point* (*P*) input. For this example, the points are representing air in space in a random configuration; however, in the 'Future Potential' section of this chapter, this can be modeled in a more logical configuration, a grid, to analyze.

The random points are being generated from a *Populate 3D* component where the space filled with points is a closed surface, referenced *Brep*, that covers the extent of the shadow projections. The density or *Number* (*N*) can be any value since that is not the specific parameter being measured for performance in this model but rather a visualization of the model. This can be modified like the previous chapter to treat these points as a performance measurement, as also shown at the end of this chapter, but for variety and other new opportunities, the remaining steps of this script will treat them as only a visual indicator.

The next operation is the same as the previous chapter as well in determining how many of the hours within a specific three-dimensional space of the model is being shaded. It is achieved by the sequencing of the *Flip Matrix, Cull Pattern*, and *List Length* component, with the last one having a *flattened* output since the data branching of the points is no longer needed.

This output can now be used to measure the number of hours being shaded using the *Larger Than* component with an evaluation of 0 so only the points being shaded are considered. The resulting Boolean pattern from this component is used to filter both the data from the *List Length* component (hours) along with all the random points. The culled points are then Deconstructed to find their *Z* value (elevation/height) as part of the weighted rating process in the next step.

Figure 13.8
Evaluating random
collection of points
within shade
envelopes.

Modeling Random Points to Represent Air Particles

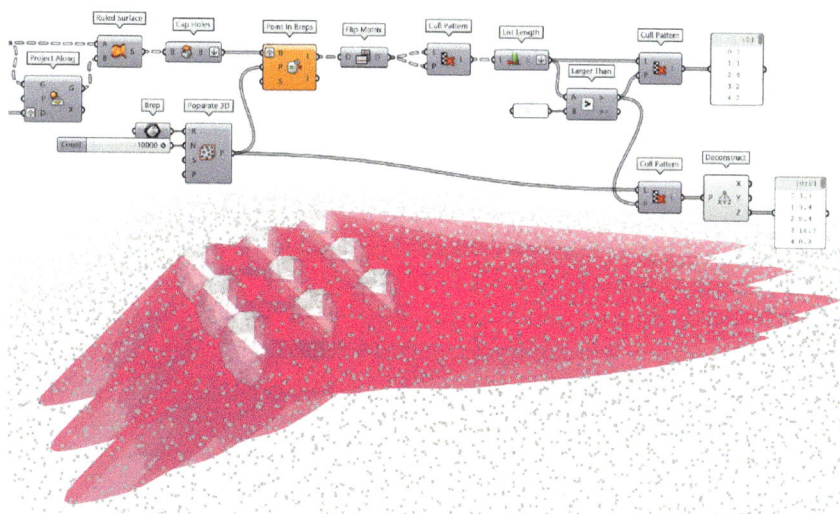

Figure 13.9
Creating rating
criteria for hours
of shade and
proximity to
ground.

Creating Criteria for Air Temperature Ratings

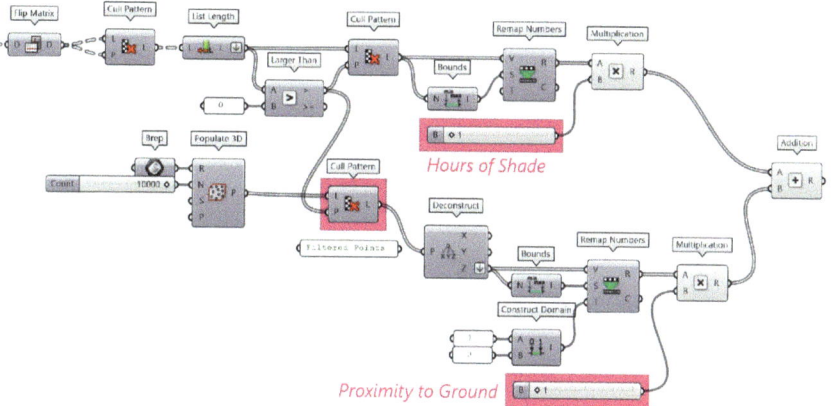

Now that both point parameters have been filtered out, they can be ranked and weighted for preference over one another or treated equally, as will be demonstrated in Figure 13.9. Since one output is number of hours while the other output is the *z* coordinate, they need to be recalibrated to have a shared rating system. Both will use their range of low to high values with the *Bounds* component and connected to the *Remap Numbers* component. The number of hours is a linear relationship, since more hours means cooler conditions or higher rating. The *z* coordinate is an inverted relationship, since the lower *z* values (closer to the ground) should have a higher rating since cooler air from the shade falls while warm air rises. This inverted relationship can be created by using a *Construct Domain* component that has a *1* for the *A* input and a *0* for the *B* input to flip the values.

Both sets of remapped values can be weighted now with a *Multiplication* component and number slider to assign preference to one parameter over the other. If the intention is to give preference over the number of hours of shade, then that slider can be increased or vice versa for the *z* coordinate parameter. These values can be *added* for a cumulative output that will be visualized in the next step and have the different weighted outcomes demonstrated.

For this visualization of the warm air mitigation from tree shade, in Figure 13.10, both point parameters will be weighted equally. From the *Addition* component, a *Bounds* and *Deconstruct Domain* component is used to determine the lower (*L0*) and upper (*L1*) limits of the *Gradient* component. The cumulative values will also be connected to the Parameter (T) input. The gradient is done in a way to show lower 'warm' values with a warm color while the higher 'cool' values receive a cool color; however, this is to the discretion of the visualization style of this performance model.

An added element to this colorization output is an operation to filter out the top cool conditions from the tree shade. The function of this operation utilizes *E* output of the *Deconstruct Domain* component to get the highest generated value from the previous calculation. This output can then be *Multiplied* against a decimal value between *0.00* and *1.00*, representing percentage, to isolate the top percentages

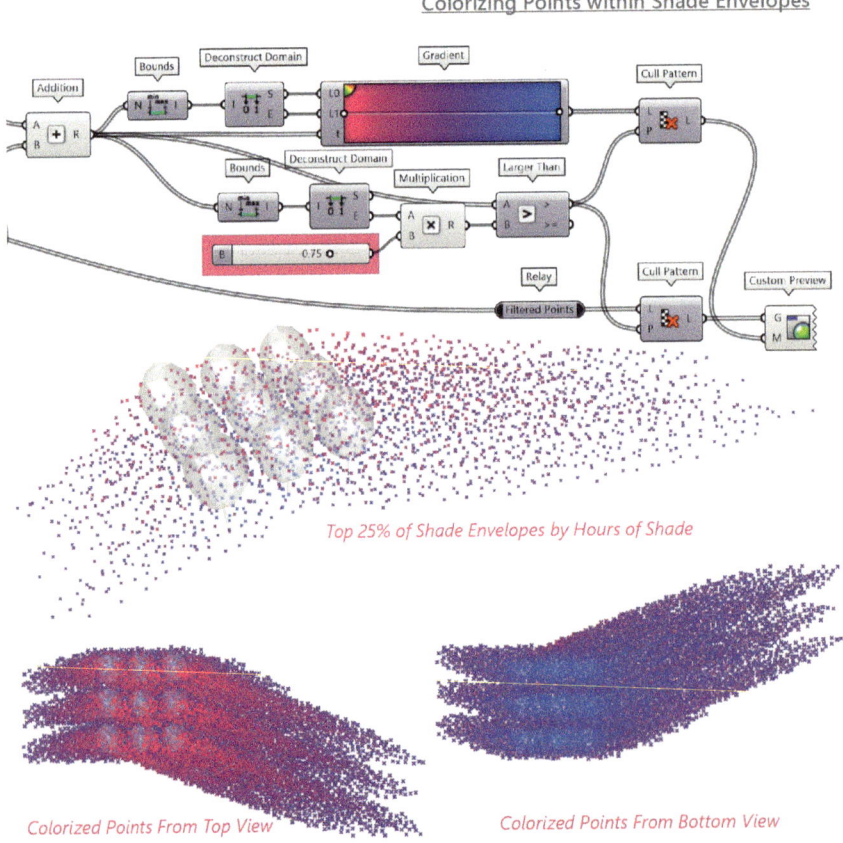

Colorizing Points within Shade Envelopes

Top 25% of Shade Envelopes by Hours of Shade

Colorized Points From Top View

Colorized Points From Bottom View

Figure 13.10
Colorizing point
cloud by number of
hours in the shade.

of the model. A number slider set to *0.75* will create a benchmark that only considers values at 75% of the highest value and greater, for example. This computed value will then serve as the criteria for a *Larger Than* component that evaluates the cumulative input from the *Addition* component.

The Boolean pattern from that component can then be used to filter out both the colors from the *Gradient* component as well as for the geometry points being evaluated. Both will be connected to a *Custom Preview* component showcasing the different potential outputs, shown in the model images at the bottom of the above figure. The bottom left-side image shows the points from the top where the warmer shaded air exists within, while the right-side image shows the colorized points from the bottom where all the cooler air begins to settle within the tree shade.

The multiple images in Figure 13.11 demonstrate different colorized model outputs depending on how the different weighted rankings are set or what percentage is filtered out, generating a variety of iterations and comprehension of this robust performance model.

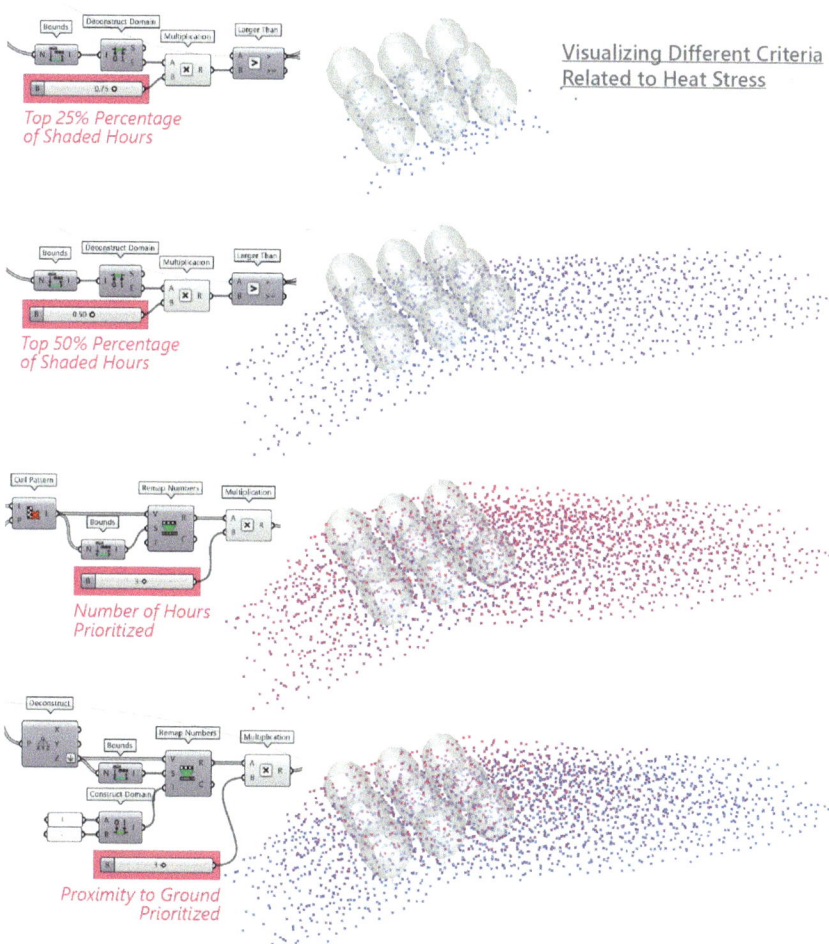

Figure 13.11 Demonstrating how different priorities to the rating criteria affect the impact of tree shade.

The dynamics of this portion of the model and script provide a variety of options to analyze and assess how different shade conditions and tree densities can be comprehended for specific project goals. It allows for the control of a few parameters to output an excessive amount of unique data sets for outdoor comfort through reducing air temperatures in the shade.

ADDITIONAL CONSIDERATIONS

Evaluating the shading conditions can be performed in a variety of ways as mentioned earlier by a shift from visual exploration to a more logical procedure by using defined measurements of space, both horizontally and vertically. This first requires the shift of redefining the points for evaluation to a grid pattern with set dimensions to reference throughout the process; so although it may not matter what the density of the grid is, it will be important to maintain that dimension and measurement throughout consistently. This same approach can also be applied to vertical layers of space since warm air rises and cooler air falls, so this principle can be used as an assessment parameter.

Gridded Points

As mentioned earlier in the evaluation of points within the shade envelopes, it may be more conducive and logical when formulating outcomes to use a gridded system so the points can be measurable. In Figure 13.12, a grid of points is used where they are spaced five feet from each other and are elevated one foot up to eight feet. This new grid set can now be evaluated for a percentage of shade containment at each elevation level. Measuring this percentage first requires the creation of a set of domains with the *Consecutive Domains* component. The *Numbers* (*N*) for this component can be derived from a *Series* component that starts below the lowest point, in this case that would be −0.5 since the lowest point is zero. The count for the series should result in the number of levels, and since this site starts at zero feet and ends at eight, a value of *9* is used.

This component can then be connected to a *grafted Domain* (*D*) input of the *Includes* component to create a Boolean pattern with the *Z* values of the points. The generated values will be used in a *Cull Pattern* component to parse out the number

Generating a Grid of Points Opposed to Randomized Points

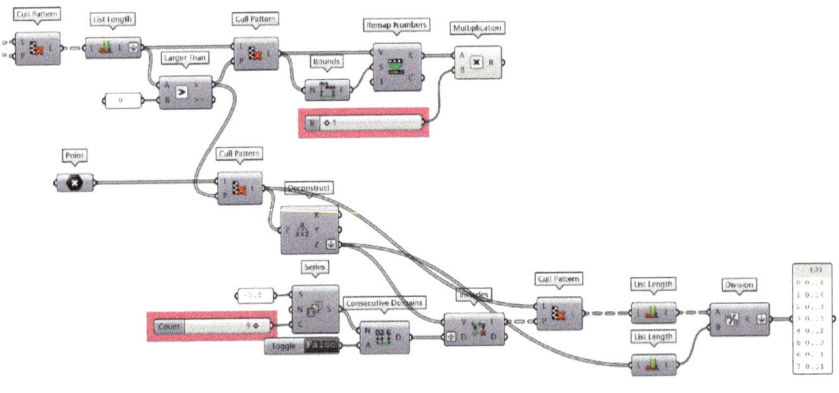

Figure 13.12 Applying a grid of points in lieu of random points for evaluation.

Figure 13.13
Creating layers
of shade based
on proximity to
ground.

Generating Layers of Shade by Hours and Proximity to Ground

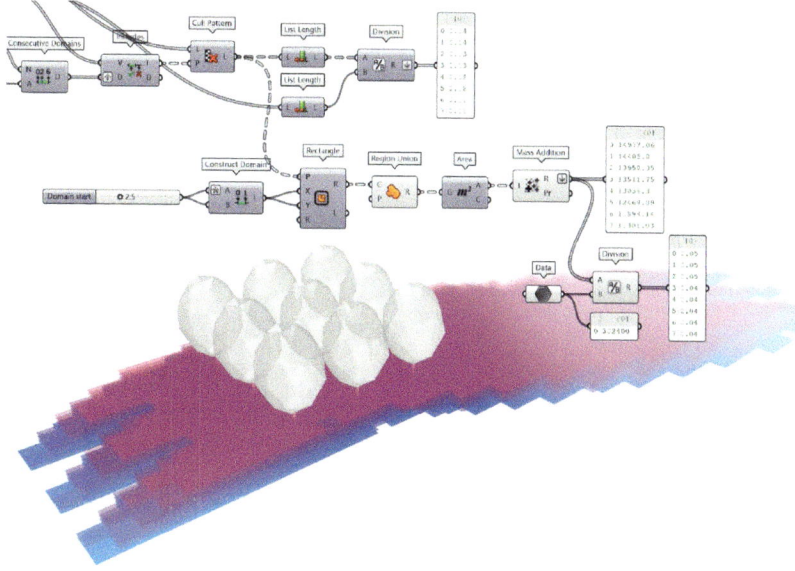

of points per elevation, or Z coordinate. These points can then be counted with a *List Length* component and *divided* with a *List Length* of all the points to calculate the percentage for each elevation level.

Percentage of Different Height Levels

As a continuation of the previous step, the percentage of the overall project site can be calculated by the different levels of shade. Since there will be both a geometric and numeric outcome from this operation, the points are assigned a physical geometry instead of just a data association of 25 square feet, with them all being five feet from each other in either direction. The script in Figure 13.13 shows how a square curve is assigned to each parsed out set of points with a *Construct Domain* component value going *2.5* in both a negative *expression* of −*x* and positive direction. This domain is used for *Rectangle* component that has the curves on each level joined with the *Region Union* component to calculate their *Areas*.

Some levels may contain several joined shapes, so a *Mass Addition* component is used to calculate multiple closed curves on the same level. This can now be *divided* by the total project area to find the percentage of shade at each level.

OUTCOMES

Reverting back to the climate data from the Ladybug plugin component, the site conditions can now be further evaluated based on the qualitative aspects of the tree configurations. This operation is not directly linked to the previous modeling process; however, it is influenced by both the modeling of trees and contextual data from the matrix in Figure 13.1.

This step, in Figure 13.14, will determine what percentage of the uncomfortable hot conditions throughout the day can actually be reduced into the comfortable

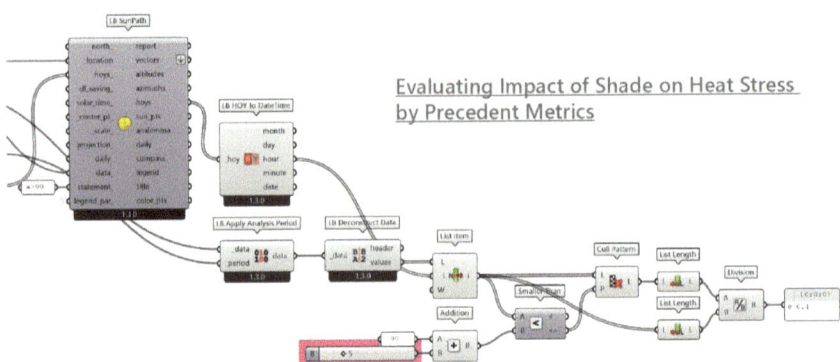

Figure 13.14
Assessing the
impact trees have
on increasing
time of shade for
outdoor comfort.

parameters from the beginning of this chapter. The first step is to determine which HOYs are to be measured from the *LB HOY to DateTime* component. These hours will serve as the *Index* (*i*) of the *List Item* component that has the deconstructed *dry_bulb_temperature* values inputted.

The outputted temperatures will be evaluated with a *Smaller Than* component to calculate what percentage of the day has been reduced to a comfortable state from the shade on the project site. The criteria for this evaluation will be the hottest comfortable condition from the wind-aided psychrometric chart (*90*) plus the expected air temperature reduction for this regional context and tree planting density. This model has a 100% planting density in a location that results in 5 degrees air temperature reduction, so an addition of *5* from a number slider is inputted into the *Addition* component as well. Since the previous steps of the model does not calculate the temperature reduction directly, this operation essentially formulates that temperatures up to 95 degrees will actually feel comfortable, since that extreme will be reduced by 5 degrees and fit within the comfort model. So depending on the tree density and location, the number slider can be adjusted to match the performance of air temperature reduction from existing case studies within that region.

The Boolean pattern created from this evaluation criteria can then be used to filter out the daily hours with the *Cull Pattern* component and *divides* the *List Length* of that value with a *List Length* of all the hot hours of that day. This model outputs a value of *0.10*, or 10%, of that air being reduced to a comfortable outdoor condition within those shade envelopes.

Baseline Comparison

Like with the previous chapter on shade envelopes, it is crucial to not treat this digital model as an end-all analysis model but as a tool to help synthesize, refine, and develop to depict the dynamic conditions of a project site. This can only happen by cross-referencing field work, ground truthing, and external research to accurately portray current and projected outcomes.

Determining the performative outcome for air temperature reduction will require a comparison between either a before condition of the project site. Some project either did not include this part of the initial site investigation or lack sufficient data for comparison, so an alternative is taking temperature readings of a nearby or adjacent property that resemble the pre-designed space or contains similar conventionally designed elements and materials. Weighted averages for different spatial elements, such as proximity to turf, water features, or other microclimates, can also be a valuable measurement for comparison.

FUTURE POTENTIAL

Since the performance model can currently measure the hours of shade for air temperature, the outputs can be further developed to account for consecutive hours of shade and how this may impact or be weighted in relation to the proximity to ground. Since warm air rises, the air temperature closer to the ground will be relatively cooler, so it may be valuable to rate the air temperature in a similar weighted ranking, like the speculation already made with the model.

Weighted Values: Hours of Shade versus Proximity to Ground

There are currently no conclusive studies that have determined that areas in shade for longer periods of time will result in relatively cooler temperatures; however, there have been studies conducted on the differences in proximity to the ground. In one study (https://treefund.org/wp-content/uploads/2018/04/Rahman-et-al.-2018.pdf), it was found that the average air temperature difference between a measuring point at 4.5 meters and 1.0 meters from the ground was between one and half a degree Celsius cooler. The reason for this was that the deep shade closer to the tree canopy was significantly more impactful on cooling air temperature than evaporative cooling from vegetated ground covers. This was especially the case during extreme hot conditions where those same ground covers acted more like hardscape materials.

These findings along with the utilization of this model can result in very robust and meaningful outcomes as part of the analysis of existing conditions or projecting positive impacts on the environment from landscape performance design.

REFERENCES

Aman, A., & Yildirim, Y. (2019). *Yanaguana Garden*. Landscape Performance Series. Landscape Architecture Foundation. https://doi.org/10.31353/cs1570

Chanse, V., & Salazar, J. (2012). *Central Wharf Plaza*. Landscape Performance Series. Landscape Architecture Foundation. https://doi.org/10.31353/cs0270

Cho, L., & Graham, M. (2014). *Thomas Jefferson Visitor Center and Smith Education Center at Monticello*. Landscape Performance Series. Landscape Architecture Foundation. https://doi.org/10.31353/cs0700

Environmental Protection Agency. (2021, September 15). *Heat Islands*. EPA. Retrieved from https://www.epa.gov/heatislands/learn-about-heat-islands#_ftnref1

Martin, C. A., & Colter, K. R. (2014). *Phoenix Civic Space Park*. Landscape Performance Series. Landscape Architecture Foundation. https://doi.org/10.31353/cs0750

Ozdil, T. R., Modi, S., & Stewart, D. (2013). *Klyde Warren Park*. Landscape Performance Series. Landscape Architecture Foundation. https://doi.org/10.31353/cs0590

Özer, E., Alvarez, V., & Gonzalez, G. (2014). *1100 Block of Lincoln Road Mall*. Landscape Performance Series. Landscape Architecture Foundation. https://doi.org/10.31353/cs0810

Özer, E., & Stanford, S. (2016). *SoundScape Park*. Landscape Performance Series. Landscape Architecture Foundation. https://doi.org/10.31353/cs1180

14 Windbreaks

Windbreaks are effective microclimate controllers and conservation strategies for areas with high exposure to winter winds (Kuhns, 2012). These interventions usually consist of trees and shrubs to reduce and redirect wind, but some reports have even shown that perennials and tall grasses have also been effective. Factors when considering the design of windbreaks should include height, density, species, length, as well as different varying number of rows of trees (Brandle & Finch, n.d.). Appropriate design of these windbreaks can yield multiple benefits beyond microclimates and conservation including reduced soil erosion, crop and livestock protection, and wildlife habitat. Those types of considerations as well as the incorporation of site program can lead vistas and screening to the surrounding context (Kuhns, 2012).

This chapter will focus on the height, species, and density of tree plantings as windbreaks using industry standard metrics for the redirecting and reduction of cold wind conditions. Although these are only a few of the design factors for windbreaks, the outcome of the model can still align with the many listed benefits beyond microclimate control.

GOALS

Like the other outdoor comfort models, this example focuses on how to create comfort for harsh cold conditions by using trees or other barriers to shield outdoor spaces from the wind. The Grasshopper script demonstrates the process of determining cold stress throughout the year and utilizing that data to dictate wind conditions for either the analysis of existing wind barriers or to develop strategic tree placement:

1. Determine the coldest times of the year using a comfort model
2. Parameterize both wind speed and direction for the coldest times
3. Use trees or other barriers to simulate windbreaks
4. Evaluate the percentage of wind being redirected from the windbreaks for multiple conditions

PERFORMANCE METRICS

One of the first factors when considering the performance of windbreaks from trees is their height. When referring to different windbreak resources and metrics, the reduced downwind or leeward side of trees is a multiplier of its height. For

DOI: 10.4324/9781003208020-17

example, when a table shows a tree has a 10H for a specific wind speed, then that means the wind will be reduced up to ten times a tree's height. Most tables from Forest Service agencies will show a variation to this *H* multiplier depending on the wind speed, density, and tree species. For much of the modeling demonstration, a 3H factor is used; however, it operates from a number slider so that value can fluctuate depending on those other tree considerations. Tree height will most likely vary along the windbreak, but the wind displacement H factor will be contingent on the tallest trees, not the shortest or median height.

Density

Regardless of the density of tree planting, there will always be some wind that will be let through, with the exception of walls and other solid rigid objects. It is, however, recommended that dense windbreaks are used to have greatest reduction in wind speed. The density of the windbreak is affected by the number and type of rows, species foliage, and spacing of trees which usually has a direct relationship in letting wind through – essentially zero density has no reduction in wind speed, whereas a 100% density lets minimal wind through.

Although this model is unable to simulate this, when creating planting density, it is recommended to be in the range of *60–80%*. This is because when all the wind is redirected above the planting from fully dense planting, the resulting low pressure on the leeward side will draw wind back into that protected zone. By allowing some wind through, the strength of the low-pressure system can be reduced.

Length

Like tree height, the length of a windbreak can have similar outcomes in redirecting and reducing wind speed. The presence of low pressures can also exist from the horizontal movement of wind, so it is often recommended the length of windbreaks is at least *ten times* the height of the tallest trees.

Species Selection

Specific tree selection is also important when considering effective ways to shield a project site from cold wind. Aside from the hardiness zones, soil suitability, and watering needs, tree selections can vary from evergreens or conifers, deciduous, and even shrubs. Some of the most effective evergreen trees include juniper and pine species as well as spruces. These types of trees can serve as a more permanent windbreak or be planted amongst deciduous so during the dormant periods there is still adequate foliage to protect an area.

Due to their size and slight range in dormancy during the wintertime, the following deciduous trees can be utilized while maintaining diverse planting:

1. Populus species (poplars and cottonwoods)
2. Salix species (willows)
3. Fraxinus species (ashes)
4. Ulmus species (elms)
5. Morus species (mulberries)

Other specific species can also include *Gleditsia triacanthos* (honey locust), *Robinia pseudoacacia* (black locust), and *Celtis occidentalis* (hackberry). With shrubs, there are a lot more varieties of species that can be used as long as they meet some of the previous parameters but should be hardy and native to the site.

Windbreak Rows

The layout of tree planting in single to maximize limited available space or the use of multiple rows to incorporate wildlife habitat and larger protection can all have a significant impact on diverting cold winds around a project site. Projects that have little space to plant where wildlife habitat is not a priority, such as dense urban settings, a single row of dense junipers, spruces, or other evergreens can effectively work. If deciduous trees are used, it is recommended that ones with a narrow crown are used.

A two-row windbreak will still need dense planting like a single row, but can be done so as either a standard or alternating lining of trees. A standard row can have either two rows of dense evergreens or one row of evergreen and the other row composed of large shrubs and deciduous trees. An alternating planting, also known as a twin-row, will have a space in one row filled with a tree in the other row, like a checkerboard arrangement. This approach can still be done when little space is available in an urban setting as well as begin to provide some wildlife habitat to the design. Expanding to either a three-row or four-row windbreak follows the same process of having one row of evergreen trees while the others have more flexibility, while simultaneously improving the quality of wildlife habitat. Each additional row will require more space, so that will be a limiting factor to larger project sites (Kuhns, 2012).

Additional Microclimates

Although the priority of this performance model is to address windbreaks against extreme cold and windy conditions, the resulting planting strategy can also be used to create microclimates for cooling spaces as well as for other comfort purposes. Making spaces cooler during the hotter times of year with windbreaks can be achieved by keeping the understory clear of dense branching or shrubs to ensure ventilation. This will require more analysis to determine the direction of those warmer winds so that the planting strategies can accommodate both extremes (Bentrup, 2008).

By reducing the wind speed on the leeward side of a windbreak, microclimate factors such as air temperature and humidity will increase which decreases the amount of evaporation and water loss from plants. Air temperatures, up to ten times a tree's height, can be increased by several degrees versus areas unprotected from the wind. This will be the case for both day and nighttime temperatures. Humidity in these protected leeward sides have been found to be 2–4% higher than in the surrounding unprotected areas. With higher percentage of moisture in the air, agriculture and other plants for harvesting will use less water and be more productive. This should be consistently monitored, because if humidity levels become too high from very dense planting, diseases and other issues may become more prevalent in crops.

Like air temperature, soil temperatures will also tend to be warmer in these sheltered areas, allowing for planting and germination to occur earlier in season when the growing season may be shorter for specific species. When windbreak rows are planted in a north-south orientation against easterly or westerly winds, the soil temperatures can also be warmer along the southern edge because of heat reflection from the sun. This will, however, inversely lead to cooler soil temperatures from shade on the north side of the windbreak (Brandle & Finch, n.d.).

MODELING METHODS

As with most modeling of weather and climate conditions, the Grasshopper Ladybug plugin is used to collect annual hourly data. The same setup is used as demonstrated in the plugin and inventory of climate data as shown in the script image of Figure 14.1. For the modeling of windbreaks, there are several ways to evaluate the discomfort of cold stress with Ladybug to analyze and respond with tree placement. In this demonstration, the *LB UTCI Comfort* component is used to classify cold stress numerically with negative values ranging from −5 to −1. Another option for measuring cold stress is with Ladybug's *Psychrometric Chart* component, also demonstrated in previous chapters.

This comfort modeling component requires the inclusion of *dry_bulb_temperature, relative_humidity*, and *wind_speed* in their respective inputs from the *LB Import EPW* component. The *_mrt_* input for this component is optional, but can be attained from the *LB Outdoor Solar MRT* component. Once that component is connected with its respective inputs, it can be run with the *Boolean Toggle* set to *True*. This will now allow for the comfort model component to operate once it is also connected with a *Boolean Toggle* set to *True*.

The *category* output from the comfort model component can then be used to parse out all the comfort levels to just the cold stress values, as shown in Figure 14.2. The *LB Deconstruct Data* component is then used to separate the values from the header data so that the *Smaller Than* component can filter out all the values greater than *0* with the *Cull Pattern* component. This pattern created from this operation will be used multiple times with different data sets.

Importing Climate Data for Cold Stress

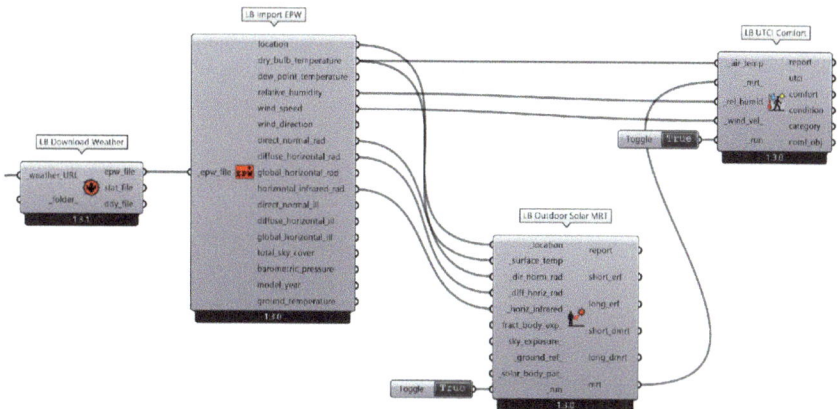

Figure 14.1 Importing climate data for modeling periods of cold stress throughout the year.

The rest of this script demonstrates the variety of other data that can utilize the Boolean pattern with a *Cull Pattern* component that can be applied necessary parameters in later steps. To attain dates and times for specific cold stresses and wind conditions, the *LB HOY to DateTime* component is used to generate every hour of the year (HOY) and its associated date. This component requires the input of all the HOYs which can be created with a *Series* component that has a count of *8,760*. The date outputted from this component will be the recommended data to filter; however, the other outputs can also be of value depending on the desired project narrative. Next, both the wind speed and direction outputs from the *LB Import EPW* component, each illustrated with a renamed *Relay*, are deconstructed for their values and filtered out.

Now that all the values have been filtered, they can now be inputted into a *Sort List* component with the initial values from the *LB UTCI Comfort* component being the sorting factor. As a result, the dates, wind speed, and direction will all be sorted, starting with the coldest conditions, as shown with the panel outputs.

Another data output that may be of use is the *Average* of both the wind speed and direction if it is intended to find overall conditions to the project site. For this to function properly in the future steps, the output from this component needs to be *flattened*.

Now that there are a variety of outputs from the previous operation, multiple options can be explored with this model as it relates to data for a specific date, condition, or averages. To find a specific value starting with the coldest conditions, Figure 14.3 illustrates how a *List Item* component for both the wind speed and direction is used with a number slider to scroll through the list of different outputs. As stated before, another option utilizes the sites overall or average conditions, which is what will be demonstrated for the development of this script.

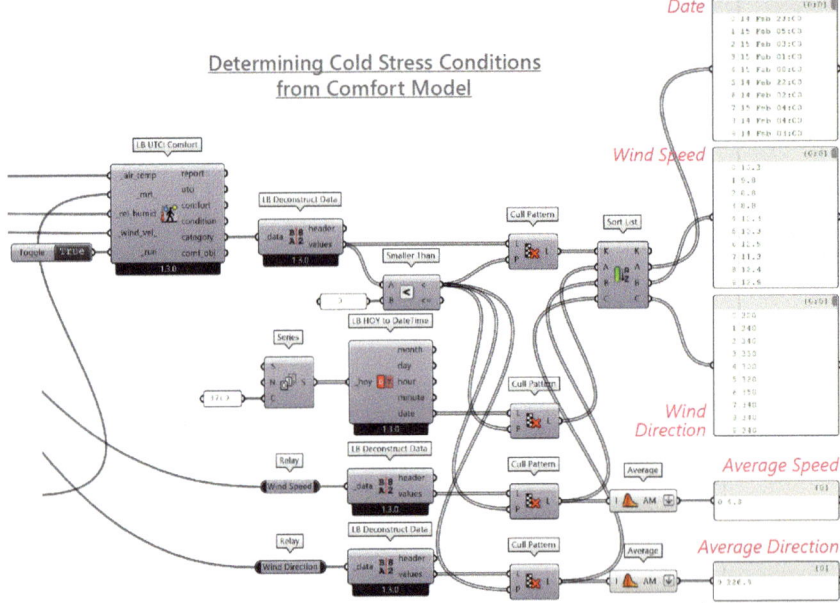

Figure 14.2 Generating different output parameters to evaluate windbreaks from trees.

The modeling of these chilled wind conditions begins with the creation of a wind silo to isolate the visual to a digital container. The size of this silo is contingent on the size of the project site, but can be first simulated with a circle created in Rhino and referenced with a *Curve* component. It is recommended to use a circle to maintain uniform spacing of the wind points, whereas irregular shapes may cause inconsistent wind lines at the corners for certain wind directions. From the referenced circle curve, it can be converted into a closed volume with a *Boundary Surfaces* and *Extrude* component set to a given height with a *Unit Z* value. The generated volume from this curve is illustrated in the model image of this figure.

The next step for this wind silo is the parameterizing of wind speed and direction. Wind speed is modeled using a *Line SDL* component where the *Start* (*S*) can simply be the center of the silo curve extracted with an *Area* component. The initial direction of this output line should start along the x-axis by inputting the *Unit X* component. The reason for this is because when assigning wind direction, a zero value is representative of due east which is the x-axis and progresses counterclockwise. The *Length* (*L*) of the curve utilizes the wind speed value; however, for the functionality of this model it is required to be measured at a tenth of its value by giving an *expression* of *x*0.10*.

This generated curve can now be oriented to the appropriate direction using the *Rotate* component with the *Angle* (*A*) set to *Degrees* while using the same center point as the *Plane* (*P*). The correctly oriented curve can be connected to the *Line* (*L*) input of the *Vector Force* component with the extruded volume serving as the *Bounds* (*B*).

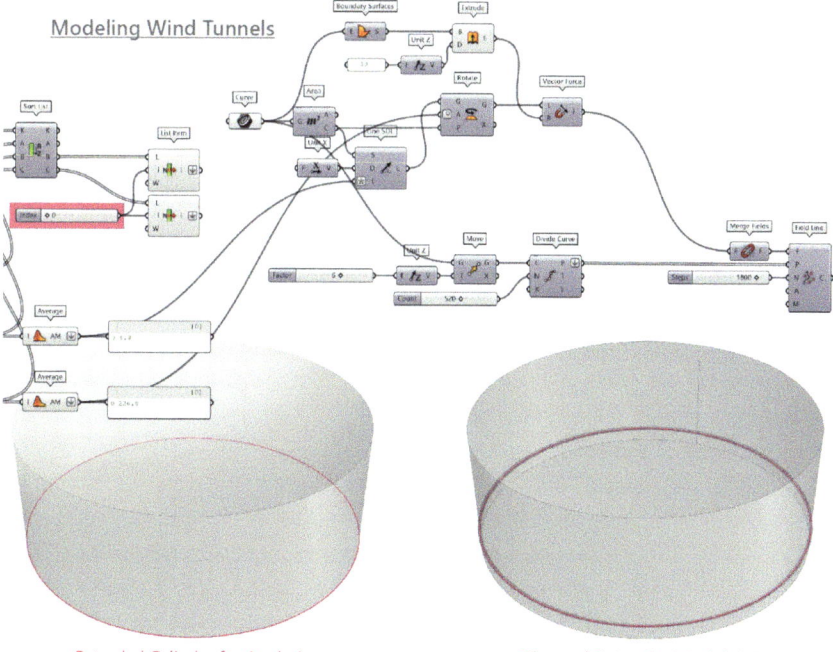

Figure 14.3
Wind tunnels
to evaluate the
impact trees
have on different
wind speeds and
direction.

Extruded Cylinder for Analysis   *Elevated Points for Wind Origin*

Before wind lines and their characteristics is defined, the elevation and density of the wind source needs to be set. The initial referenced curve for this wind silo should be raised to a height of around six feet using the *Move* component with a *Unit Z* value set to *6*. This elevated curve can now have a set density of points to serve as the wind source using the *Divide Curve* component, in this case, a value of *520* is assigned for relatively high density. The points from this component will then need to be flattened before inputted into the *Field Line* component.

For best performance of the script, it is recommended to connect all the necessary parameters first before inputting the *Field* (*F*) value, since this requires substantial computer processing power. The *Number of Samples* (*N*) input should also be greater than the default value of *1,000* for accurate line generation. This demonstration uses a value of *1,800*. Although there is currently only one field value, the *Field* (*F*) output from the *Vector Force* component should be connected to a *Merge Fields* component and then connected to the *Field Line* component.

Now that wind parameters have been incorporated into the wind silo, trees and their ability to serve as windbreaks can be simulated based on their size. The first step is to use circle curves made in Rhino to represent the trees, as they are referenced in Figure 14.4. These references curves will be used to extract different characteristics of the trees to inform windbreak parameters such as the size and length of wind trail. The size of the trees is first used to create a *Point Charge* from the center of each tree circle using the *Area* component. The *Charge* (*C*) input of this component utilizes the size of the circle extracted with the *Dimensions* component, where either the *U* or *V* output can be inputted, since a true circle will have equal sizes at either dimension. The *Bounds* (*B*) input will once again use the *Extrude* component geometry to contain the field lines within the designated area. The outputted *Field* (*F*) can now be combined at the *Merge Fields* component by dragging the wire while holding *Shift*.

The next step is to create a line charge to simulate the wind trail after it comes in contact with the tree or represented point charge. A *Line SDL* component is first used to create a curve at the center of each tree. The rotated curve representing wind direction is then used to generate a vector with the *End Points* component to create a *Base Point* (*A*) and *Tip Point* (*B*) for the *Vector 2Pt* component. And lastly, the length of this curve uses an operation of finding the referenced tree curve's diameter by doubling the radius from a *Deconstruct Arc* component. A tree's windbreak potential is generally three times its diameter (Kuhns, 2012), so to calculate this, the *Radius* (*R*) is given an *expression* of $x*2$ and then increased with the *Multiplication* component to a power of *3*. Since this wind trail curve is beginning at the center of the circle, it also needs an *Addition* of the *Radius* (*R*) value to shift the starting length at the edge of the tree curve. This can now be connected to the *Length* (*L*) input of the *Line SDL* component.

The final step is to convert this curve into a *Line Charge* with the *Charge* (*C*) value being similar to the tree diameter. This is achieved by using the *Multiplication* component for the *U* output of the *Dimensions* component with a value of around *0.95* to make it slightly smaller for functional purposes. This is a number slider because it may need to be adjusted depending on the scale of the project site. The *C* input will also require an *expression* of *−x* to invert the charge direction. Once again, the *Line Charge* component will use the *Extrude* geometry as the *Bounds* (*B*) input one last time. This can now be combined with the other charge values.

As shown in the model images of Figure 14.4, scrolling through the different wind speeds and directions from the *List Item* component or using the average values, a range of curve lengths and orientation for wind can be simulated.

The visualization of these windbreaks is very similar to previous examples of taking a specific parameter from these field lines and colorizing them as demonstrated in Figure 14.5. Since all the field lines initially begin at the same height, it makes sense to use a point's *Z* value to delineate the change of direction by the trees. This can be achieved by first using the *Divide Curve* component to generate a specific density of points. The model image of the below figure demonstrates different visual outcomes depending on the point density. The points from this component are then *flattened* at the input of the *Deconstruct* component, so that each can be evaluated relative to each other for proper colorization.

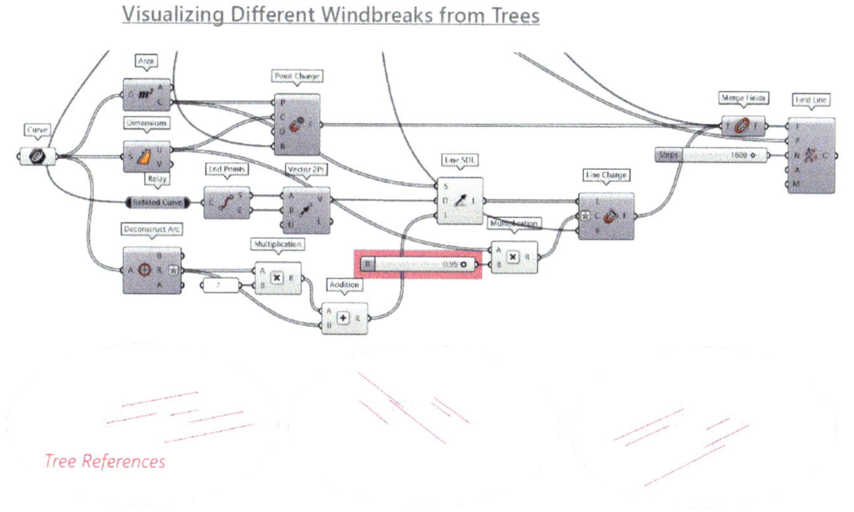

Visualizing Different Windbreaks from Trees

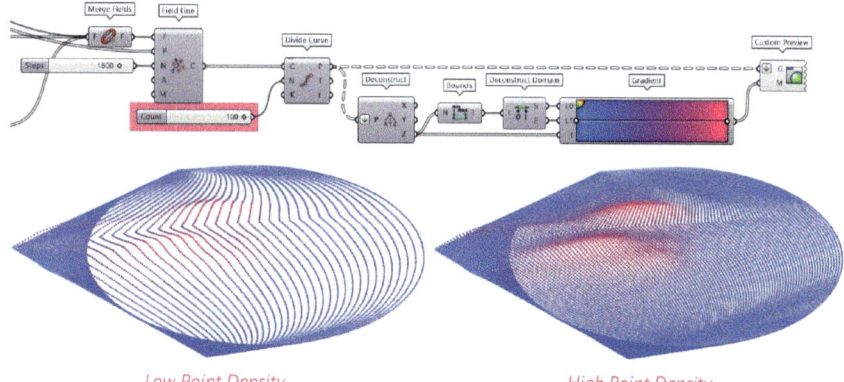

Figure 14.4
Referencing circles as tree canopies to generate wind dispersion from different directions.

Tree References

Modeling Windbreaks with Colorized Points

Figure 14.5
Visualizing different densities of windbreaks through colorized points.

Low Point Density   *High Point Density*

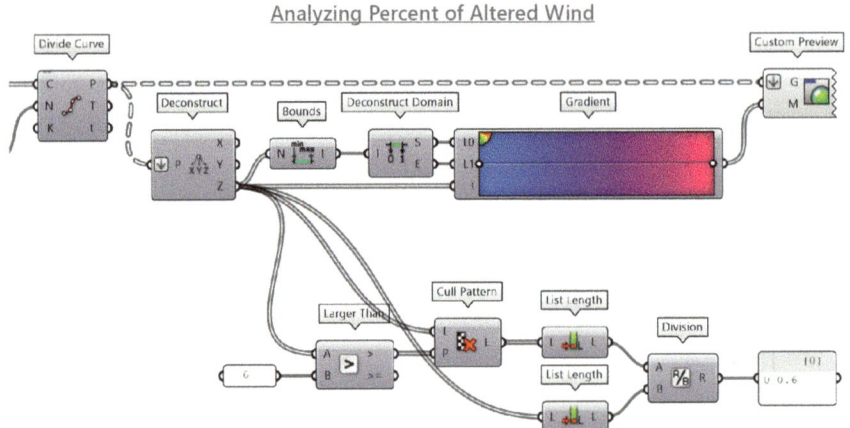

Figure 14.6
Assessing the percentage of wind within project site being diverted from tree windbreaks.

The *Z* output is connected to the *Parameter* (*t*) input of the *Gradient* component while the lowest and highest *Z* values are determined with the operation from the *Bounds* and *Deconstruct Domain* components. The *flattened* points inputted into the *Custom Preview* component can now be assigned the gradient color values.

The visualization of these windbreaks does not need to be performed through points but could also be simply field lines; however, from the extracted points and their *Z* value, the trees can be further evaluated for their windbreak potential.

This technique also pulls the *Z* values from the *Deconstruct* component to determine what percentage of points have been moved by the windbreaks, as shown in Figure 14.6. The *Z* values are first inputted into the *Larger Than* component and compared with their initial starting height of *6*. All the points or, in this case, values that have shifted up due to the windbreaks will be filtered out with the *Cull Pattern* component.

Now a *List Length* component can count the number of values that have changed and are calculated against the *List Length* of the original count number with the *Division* component to find the percentage in decimal format.

OUTCOMES

This model specifically was able to demonstrate a significant impact on shielding space from harsh cold winds. In comparison to the test space (wind silo), the trees only occupied 2% of the total area; however, through the different wind conditions the trees would shield between 40% and 60% of the wind. This was a significant finding from this model considering trees are able to shield three times their height on their leeward side.

The modeling process for this topic is still highly speculative, since there are very few operations or plugins specific to modeling wind conditions around objects that can generate numeric outputs. Most existing plugins provide a visualization of windbreaks; however, the evaluation of their performance and effectiveness is challenging to extract from the model. The other limitation to the other available models is that the outputted wind directions are limited to a single horizontal plane, whereas for windbreaks from trees, the primary intention is to demonstrate the vertical displacement of wind, as demonstrated in the key resources showcased at the beginning of the chapter.

FUTURE POTENTIAL

There are still several factors when it comes to windbreaks that this model has not explored, yet due to some limitations, however, can be further explored and tested with revisions to the script. Tree types, such as evergreens and deciduous, and their specific shape can have significant differences in simulating windbreaks. Other elements in the landscape such as walls and screens also have the potential to shield from cold winds.

Evergreens versus Deciduous

Evergreens are typically the preferred type of tree for windbreaks since they will have foliage year-round and are typically dense, thus acting as a solid barrier. This, however, can be a design or planting challenge to avoid monocultures, poor aesthetics, or other considerations when it comes to planting diversity. So when it comes to the modeling of windbreaks from trees, the model has the potential to categorize trees by these two types and generate different outputs depending on seasonal change or tree dormancy.

Tree Shape

The point charge used in this model essentially creates a sphere of influence on the wind conditions; however, the overall shape of trees does not typically grow like that but is cone or oval shaped. Either adjusting the point charge height to represent the different height of a tree or using a series of point charges in the vertical direction of one tree can begin to consider different tree shapes.

Another option is to divert from the point charge method altogether and focus on simulating the shape of the windbreak form instead. This method is a bit more cumbersome for the modeling process, since the focus shifts to creating geometric forms and having that dictate the wind patterns versus having the current vice versa occurring.

Walls and Screens

Like the previous consideration, the addition of other windbreak elements such as walls and screens could make this model more comprehensive to the design process. This approach, however, falls to the same limitations as before since the use of point charges are limited to spheres, whereas walls and screens will be rigid and linear.

REFERENCES

Bentrup, G. (2008). *Conservation Buffers—Design Guidelines for Buffers, Corridors, and Greenways*. General Technical Report SRS–109. Asheville, NC: U.S. Department of Agriculture, Forest Service, Southern Research Station. 110 p. Retrieved from https://www.fs.usda.gov/nac/buffers/docs/conservation_buffers.pdf

Brandle, J., & Finch, S. (n.d.). *How Windbreaks Work*. University of Nebraska Extension. Retrieved from https://nfs.unl.edu/publications/downloads/ec1763_0.pdf

Kuhns, M. (2012). *Windbreak Benefits and Design*. Utah State University Cooperative Extension. Retrieved from https://forestry.usu.edu/news/utah-forest-facts/windbreak-benefits-and-design

15 Walkability

There are many factors for pedestrians that determine the walkability of route that can scale in both detail and complexity. For some, the visible directness to a destination may impact their decision to take a specific route, whereas for others the quality of path's material may be more important to avoid tripping hazards or comfort on their feet. These metrics that measure walkability vary amongst different professions along with the intended outcomes of these output ratings. So, although there may not be a universal method to measure and model walkability (Brashear & Khawarzad, 2011), one of the consensuses among all is that it requires multiple factors to measure walkability because a single preference is rarely sufficient in dictating what is a walkable path.

This idea also reinforces that the shortest path is not necessarily the best path. A short and fast route can also be trapped in close proximity to heavy vehicular traffic with no option of refuge if the route is too uncomfortable or unsafe. Sometimes, pedestrians will go out of their way to stay within a shaded path when environmental and weather conditions are too harsh when exposed to the elements. That is why and how this method will demonstrate how a multitude of factors and considerations can be implemented within a model to give variety and choice driven by known outcomes of walkability.

GOALS

The intention of this modeling process is to measure and evaluate how different walkability factors impact a pedestrian's preference to choose a specific route to their destination. The shortest distance is often a singular consideration in most site analyses, so this model aims to demonstrate the incorporation of other factors for a more comprehensive depiction of walkability:

1. Determine the shortest route to a destination point along a street network
2. Incorporate walkability factors to output a comparable rating score
3. Output multiple walkability conditions to evaluate different site constraints

PERFORMANCE METRICS

There may be no standardized method for measuring walkability, considering all the different variables of users and scale; however, one credible resource that incorporates walkability elements important for landscape architecture is the *Kansas*

Figure 15.1
Different factors
that impact
walkability.

Walkability Factors

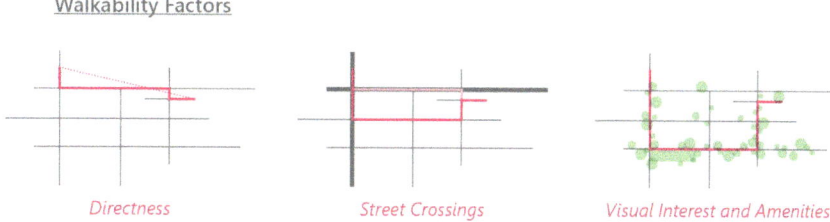

Directness　　　　　　*Street Crossings*　　　　　　*Visual Interest and Amenities*

City Walkability Plan (LSA Associates, Inc., 2003). The value of this resource is that it uses assessment metrics to evaluate the needs of a pedestrian system. These metrics are scalable from a citywide network to the neighborhood level. There is a total of five metrics that this method uses; however, for the ability to model walkability the three that are included are directness, street crossings, and visual interest and amenities.

The directness of a path evaluates the shortest possible route in comparison to the actual distance, as shown in the left image of Figure 15.1. This can be measured as a ratio between the measured actual distance with the path distance to determine the rating for this metric. A good evaluation will have a ratio of less than 1:1.2, a moderate rating between 1.2 and 1.6, and a poor level of service will be larger than 1.8. Street crossings consider numerous elements which require a high level of detail in the modeling process, so for this rating the width of a crossing based on the street classification is used, as depicted in the middle image of Figure 15.1. These widths vary from 72' to over 96'. The last metric, visual interest and amenities, is another category that can incorporate different levels of detail. For the purposes of this demonstration, tree canopy coverage is used as it varies from full (100%) coverage to sparse (<10%) coverage, illustrated in the right-side image of Figure 15.1.

When it comes to directness, the complexity or number of turns to get to a route is generally measured as undesirable for most pedestrians. However, depending on other conditions that affect the directness, it can counter that preference which is why the accumulation of multiple factors will provide a more comprehensive outcome. For example, in the *Street Crossings* image of the figure, pedestrians may forgo the shortest and more direct route in order to avoid wide street crossings and heavy traffic. Another instance may be for a more shaded pathway that significantly detours a pedestrian for a more aesthetically and comfortable experience.

MODELING METHODS

Evaluating the walkability metrics of a project site is entirely contingent upon the model's level of detail regarding street classifications and network connection, crossings, field of vision, and aesthetics or tree canopy. These modeling methods can be even further refined with qualitative measurements of sidewalk material, crossing signals, and light posts. For the purposes of this demonstration, the rating system for street crossings or classifications, directness, and protection from the elements (tree canopy) from the previously mentioned 'Performance Metrics' section will be used.

To best prepare the Grasshopper script, it is important to have curves of the road network already in place, preferably in Rhino. This road network can be created from the previously mentioned plugin *@it* for shapefiles, an existing computer-aided design (CAD) or Rhino file of streets, or other mapping plugins such as Elk that imports layers from *OpenStreetMap*. From this road network, streets should be designated by layer as having:

- Layer 1: Three or fewer lanes (less than 72 feet wide crossing)
- Layer 2: Four to five lanes (less than 84 feet wide crossing)
- Layer 3: Six or more lanes (crossing no wider than 96 feet)

Layers 4 and 5 account for the qualitative metrics mentioned before and are only optional for this example to account for future potential of this script. Figure 15.2 shows a *Curve* component representing each layer (street classification), including the optional layers.

These five curve components will be merged with the *Entwine* component to maintain the classifications within an ordered branching structure for the future steps of rating the street crossing conditions. The *Shortest Walk* component (plugin) will require all these curves to be *flattened* so that a continuous route can be formulated but will only need to be done for this operation. Next, either a referenced surface or point, or even a mix of both, needs to be referenced to represent a starting and end point for the walkable route. In this example, a referenced point is used as the starting point representing a housing parcel and a surface is referenced with its *Centroid (C)* from the *Area* component representing a neighborhood park as the end point.

To account for the potential to add multiple starting points, a *Closest Point* component is used and connected with the starting point using a *Line* component

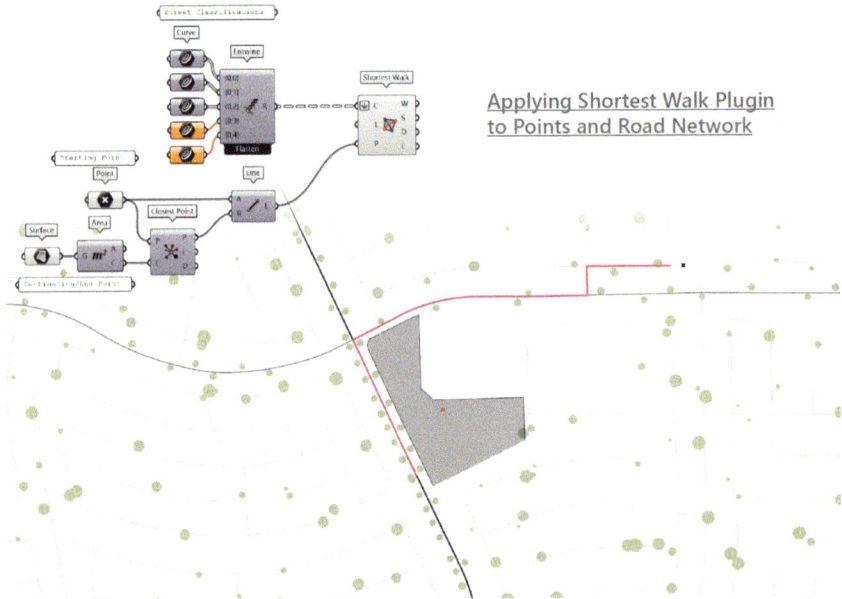

Applying Shortest Walk Plugin
to Points and Road Network

Figure 15.2
Using different
street classifications
for the shortest walk
network path.

for the *Wanted Path* (*P*) input of the *Shortest Walk* component. This will result in a curve representing the shortest possible route between the two points using the street network.

The outputted curve can now be evaluated using the three different walkability metrics to combine for an overall rating assessment.

The key to keeping the road classification branches intact to evaluate for their width or number of lanes throughout the shortest route uses the same operation of utilizing data indices, as demonstrated in previous chapters. In addition to the *Shortest Walk* component generating a curve, it also outputs the indices of each curve used from the street network, known as the *Succession* (*S*), as shown in Figure 15.3. The indices are first parsed out in the same structure as the curves from the *Entwine* component using an *Item Index* component with the *List* (*L*) input flattened while the *Index* (*I*) input maintaining the curves' branching structure. The limit of each branch is then extracted from newly parsed indices using the *Bounds* component. The values from the *S* output can now be determined if they reside within one of the previously created limits using the *Includes* component.

This operation will result in a Boolean pattern that can be used with the *Cull Pattern* component to isolate the indices to only the ones being used as the shortest path curve while maintaining the branch structure of street classifications. This pattern of indices can now be used to group all the individual curves within those same classifications using the *List Item* component with a *flattened* input from the *Entwine* component.

The total length of each street classification is then measured using the *Mass Addition* component as well as the total length of the shortest path curve. The total length can also be pulled from the *Length* (*L*) output of the *Shortest Walk* component as well. These two massed addition values can now be divided to find their percentage or weighted value. This is a similar process to the weighted coefficients for calculating stormwater runoff.

Whenever this model does not include one or more of the street classifications, in this case, the optional layers 4 and 5, a null will be generated from this operation. This will affect the outcome of this calculation in the next step, so it is important to remove any null using the *Replace Nulls* component with a *0* value used as the *Replacements* (*R*) input. That way, when these values are multiplied by their respective rating value, it will result in a *0* and not the value it is multiplied by (*5–1*). This descending rating list from *5* to *1* will be the standard evaluation metric for all the path classifications like shown in the walkability methods resource. All the values for each street classification can now be added to give a total rating for

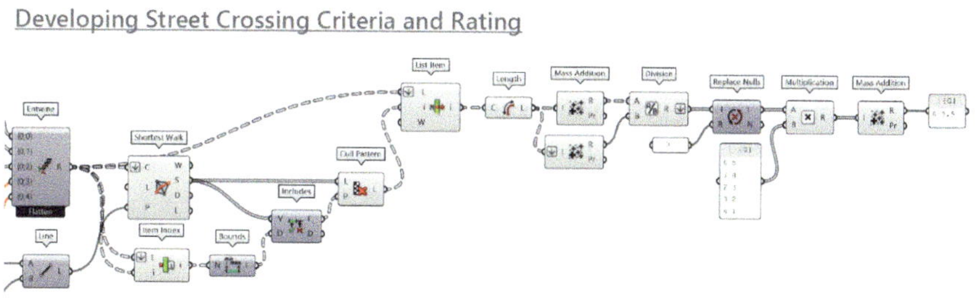

Developing Street Crossing Criteria and Rating

Figure 15.3
Creating a rating criterion for the different street classifications and crossing widths.

the street classification of a path's walkability performance. The resulting weighted rating for this path's street classification is a *3.9*.

The next rating to evaluate the shortest path with is the directness of the route. The less complex or fewer number of turns a path has, the more preferable it can be for pedestrians and walkability. The first step in measuring this in Figure 15.4 is by *dividing* the length of network path (*Shortest Walk* component) by the direct path length. Like the previous image's operation, an *Includes* component is used to determine which ratio classification this path falls within. The given values from the referenced methods document are connected to the prior *Consecutive Domains* component. This component also needs a *False Boolean Toggle* as part of the operation to ensure that the domains do not add toward a sum total, but rather stay sequential.

When creating consecutive domains, the number of values to construct these limits will always be one more than the number of domains. For example, with this demonstration, there are five rating classifications for directness, so therefore a sixth value is needed to complete the extent of the last domain. This value can be exaggerated in a way that if a project site has a maximum distance of two miles, a value of four or greater can be used to ensure all possible distances are included. The value used will not affect the outcome of the rating as long as it fulfills this criterion.

The Boolean pattern created can now be used on a *Cull Pattern* component with the same *5* to *1* rating list as before. The resulting rating for this path's directness is a *2*.

Evaluating the path's visual interest and amenity can be performed with various street elements; however, for this example, the street's tree canopy coverage will be measured. Since the script will be measuring canopy coverage, trees need to be modeled as curves at their respective canopy size. This can be achieved by including an aerial image to plot and model in Rhino. The modeled curves will then be referenced and measured for their diameter using the *Deconstruct Arc* component's *Radius* (*R*) output with an *expression* of $x*2$, as shown in Figure 15.5.

Since the model uses street centerlines, the following operation will be used to determine if a tree is shading the street and adjacent sidewalk. First, the curve's *Centroid* (*C*) from the *Area* component is connected to the *Curve Closest Point* component along with the generated path. This will output distance values to

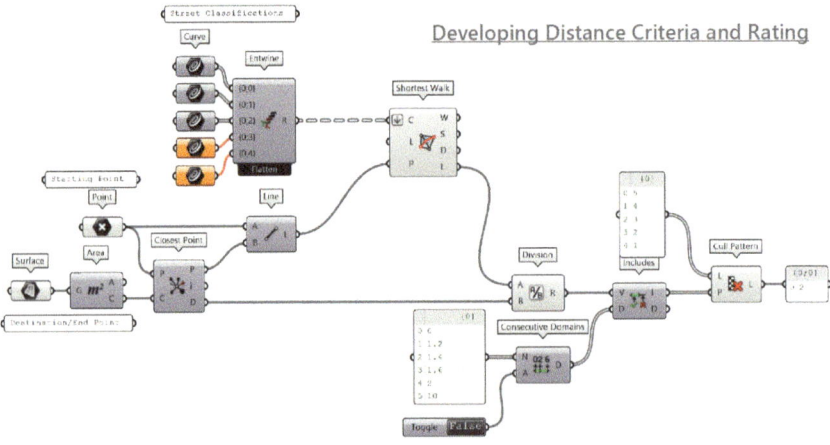

Figure 15.4
Creating a rating criterion for directness of route.

determine if they are within close proximity to the shortest path by using the *Smaller Than* and *Cull Pattern* components to filter only values within the distance of each respective tree canopy.

The culled diameter of each tree is added together to calculate what percentage of the shortest path is being covered after using the *Mass Addition* and *Division* components. The resulting percentage is then classified with the same given rating system.

To evaluate full coverage of the path, a domain between 2 (200%) and 1 (100%) is used and then sequentially to 0.50 (50%), 0.25 (25%), and 0.10 (10%) using the *Consecutive Domain* component operation. The same process of operations is then used as before to incorporate the walkability rating system that results in a tree coverage value of *3*.

Now that all the walkability ratings have been measured for the path, the values can now be combined and averaged out using a *flattened* input for the *Mass Addition* component and then *divided* by *3*, since there are three rating categories, as demonstrated in Figure 15.6.

With the walkability ratings calculated, the data can now be visualized along the path for a more comprehensive communication of this performance topic. Since all the rating scores are out of five, a *Text Join* component is used with a */5* text added after the rating score as been inputted. The first step will be labeling the rating score on the path using the *Text Tag 3D* component. The placement of the text can be done with a *Point On Curve* component and adjusted as necessary to ensure proper legibility of the text. This parameter, along with the others used for this operation, will most likely need to be changed once all the settings are inputted. The rating score is then used for the *Text* (*T*), a number slider for the *Size* (*S*), and a swatch for the *Colour* (*C*).

Next, a circle is used where the size and color parameters are dynamic to the rating score, where a large pink size indicates a high rating, whereas a small blue one represents a low rating. To perform this responsive visualization, the rating score is first adjusted to reflect an appropriate *Radius* (*R*) for the *Circle* component using the *Remap Numbers* component and two *Construct Domain* components for the *Source* (*S*) and *Target* (*T*) inputs. The source domain will be between *1* and *5* reflecting the walkability rating and the target domain will use values between *200* and *500*. These

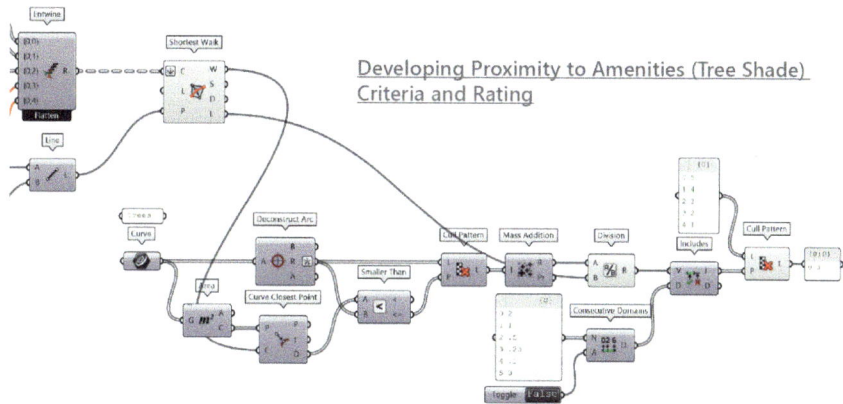

Developing Proximity to Amenities (Tree Shade) Criteria and Rating

Figure 15.5
Creating a rating criterion for proximity to route amenities, specifically shade trees.

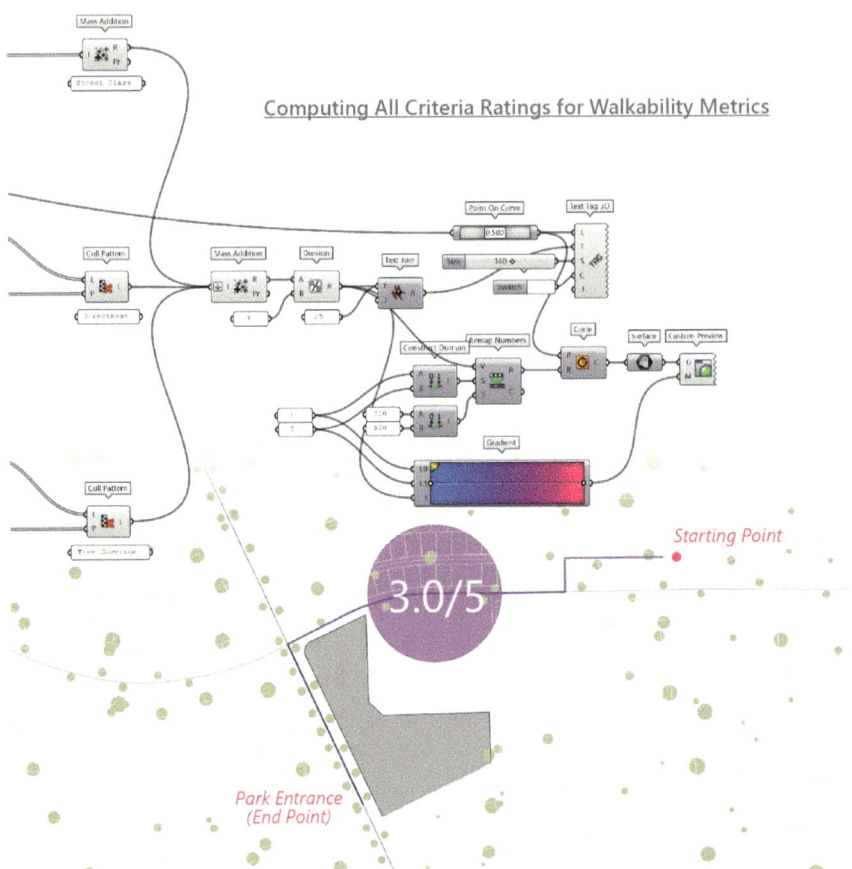

Figure 15.6
Aggregating all
the rating criterion
to visualize the
shortest route
through color, size,
and text.

values will require some testing to determine the most legible smallest and largest sizes for the circle. Again, this will be contingent on the scale of the project but can be easily determined using a number slider with a high range of values.

The circle can now be colored after being converted with a *Surface* component to respond using the same parametric logic. A *Gradient* component is used with a *Lower limit* (*L0*) and *Upper limit* (*L1*) using the same values as the source domain. The *Parameter* (*t*) input will then use the rating value to determine where it falls within the color spectrum. The resulting color value can then be connected to the *Material* (*M*) input of the *Custom Preview* component along with the converted surface.

An alternative to displaying the walkability score as a fluctuating-colored circle is to apply these settings to the destination point of the path, in this case a neighborhood park, as shown in Figure 15.7. The operation is simpler than the previous method since there will be no size adjustment, only text and color change. The label *Location* (*L*) for the text will be from the extracted point from the neighborhood park surface geometry, which was also used as the end point for the desired path of the *Shortest Walk* component. The color gradient operation will remain the same; however, the *Custom Preview* component will be connected to the park surface geometry from the beginning of the script.

Figure 15.7
Comparing different
routes that seem
closer yet have
lower walkability
and vice versa.

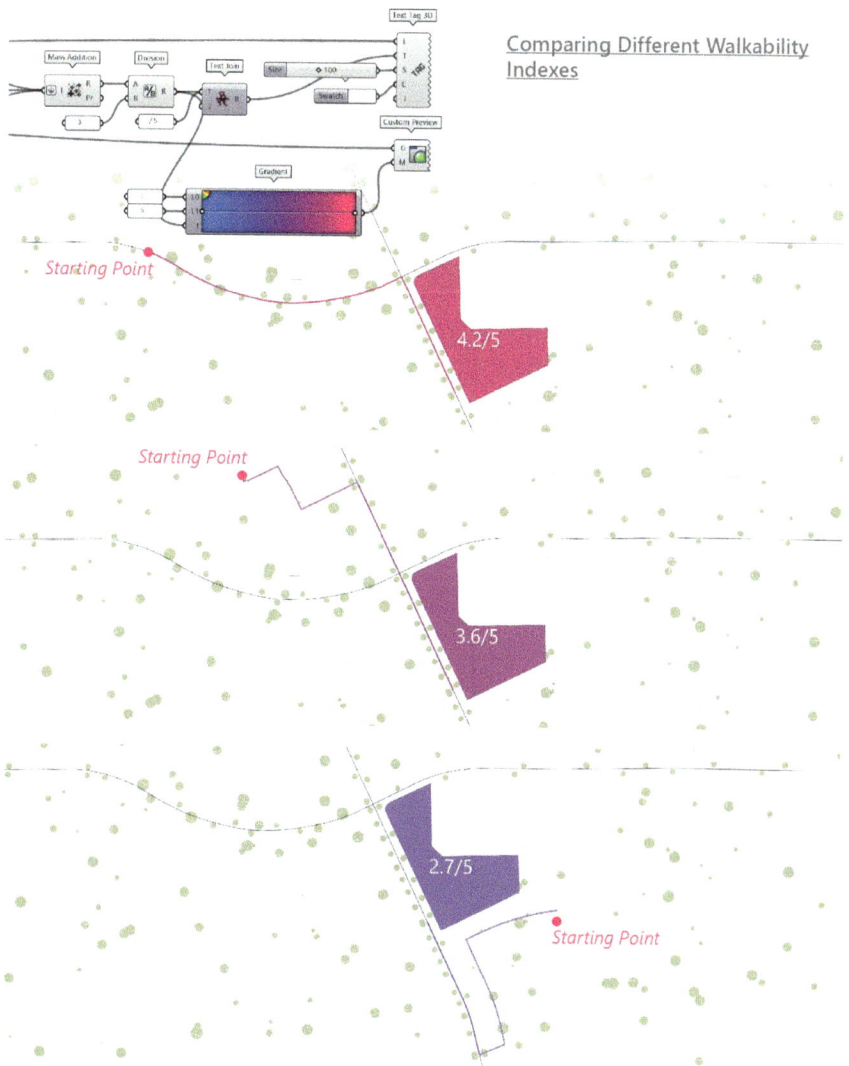

Figure 15.7 also shows three image examples of different walkability conditions utilizing this new label method. The images starting from the top demonstrate a decreasing value in the walkability score, even though the starting point indicated with a *black X* seemingly moves closer to the neighborhood park. This reinforces that the objective of this model is compiling multiple metrics to determine a more accurate analysis of walkability where distance alone is not the main predicator for a desirable path, since route complexity, street types, and shade coverage will equally influence this landscape performance.

The overall outcome of a path's walkability can be difficult to comprehend without seeing the individual rankings itemized and displayed as well. Figure 15.8 shows how that can be achieved by adding a subset of operations for those individual metrics. The operations are the same for the road classifications and directness, whereas the tree canopy coverage requires substantially less components.

The first grouped operation utilizes the curve outputted from the *Shortest Path* component, represented in the *Relay* of the script. The sequence of using the five components from that curve for the *Location* (*L*) of the *Text Tag 3D* component dictates the location and orientation of the label. The value rating from the road classification (street class) operation is derived from the component in Figure 15.6 and joined after the panel labeled *Road Classification:* is inputted into the *Text Join* component. This component is then used as the *Text* (*T*) input. A number slider can then be used to determine the most appropriate size for this label along with the other two ratings to keep everything consistent. This operation is then copied and replaces the shortest walk curve with a *Line* component created from the start and end points, as shown in Figure 15.2. The panel's text is then changed to *Directness:* that connects its respective rating.

For the canopy coverage label, a *Cull Pattern* component can be used with the pattern from the *Smaller Than* component in Figure 15.5 that connects the *Area* centroids from that same figure. With a *List Item* component and number slider

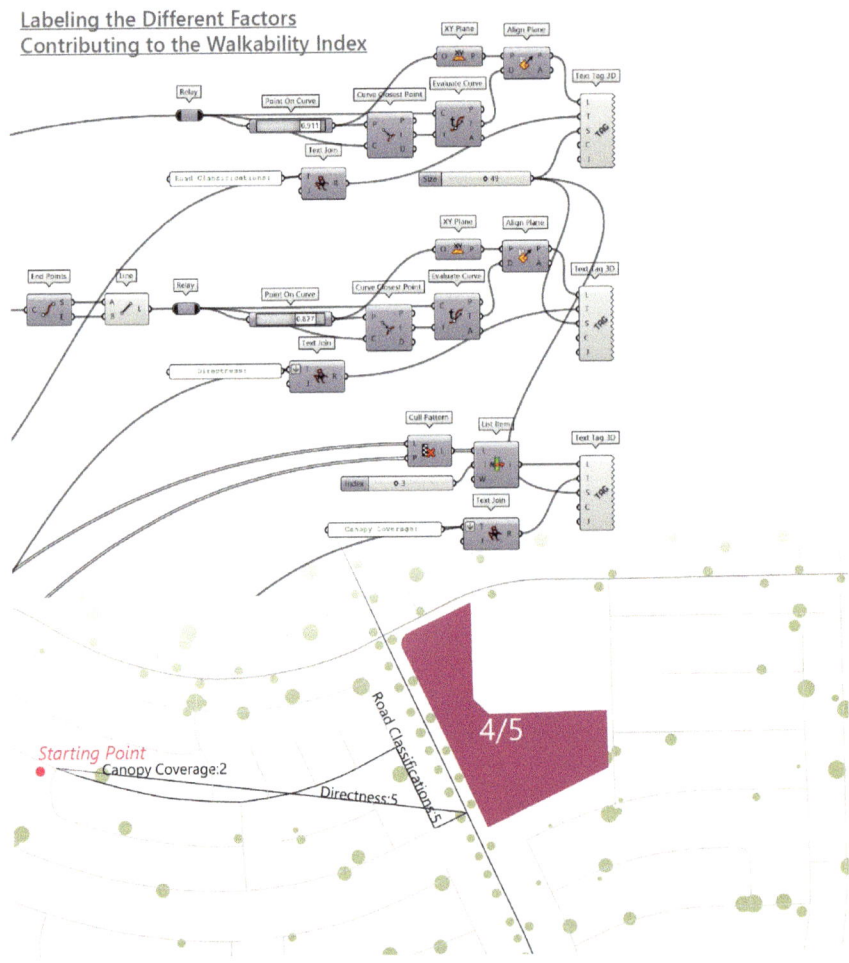

Figure 15.8
Including the individual ratings for walkability to assess their contribution to the overall rating.

for the *Index (i)* input, an appropriate point can be selected to choose a location for that rating label. Lastly, a *Text Join* component is used again with a panel labeled *Canopy Coverage* with that same rating system.

OUTCOMES

From this process of creating paths from start and end points, a walkability performance is evaluated to determine the quality of the path and how the different factors impact it. The path itself is the shortest possible distance between points but necessarily the best and mentioned at the beginning of the chapter. However, shorter distances are often preferable, but still need to consider the other factors. By knowing which factor is having the largest negative impact on a path's walkability, design tactics can be guided to improve the path conditions.

This can be a fast iterative process of testing different locations such as residents, businesses, and public recreational spaces that generate different ratings for a more comprehensive scope of a project site and surrounding context.

FUTURE POTENTIAL

The modeling of walkability performance has a lot of potential to incorporate other dynamics such as additional ratings, multiple starting points, alternative routes, and path improvements.

Additional Ratings
Since this model is currently speculative map, other factors that consider qualitative conditions such as crosswalks, lighting, and path material condition can later be incorporated once an on-field survey assessment has been conducted. The addition of these conditions can provide a much more detailed and thoughtful perspective on what and how a path is defined as walkable. This can be further elaborated on once environmental and climate factors are incorporated as well.

Multiple Starting Points
Another option with this script is modeling multiple starting points to a desired endpoint. The value of this technique is for a side-by-side comparison of routes which can also indicate which factors may affect them differently or if it's a shared factor to guide design tactics, as shown in the first image of Figure 15.9.

Alternative Routes
Similar in functionality of the script for the previous option, a single starting point can be given alternative routes to the shortest distance. These alternative routes can be dictated by needing to include an intermediate point of interest, preference to avoid wide streets, or priority to travel along the most shaded path, as seen in the second image of Figure 15.9. This can be achieved by either including additional points for the desired path or by isolating the street path network to either cull wide street classes or to only consider paths near trees.

Figure 15.9
Comparing multiple
route options
between a start
and end point with
additional stops.

Multiple Staring Points and Multi-stop Routing

Path Improvements

In turn, these alternative considerations can also serve as metrics to guide design decisions and strategies for improved street conditions and better walkability. Additionally, if the field analysis indicates poor street conditions significantly impact the walkability, then this can lead to design interventions at specific locations along desired routes. And as suggested before, if the preference is to take routes under dense tree shading, this can then suggest where future tree planting would be most appropriate.

It is this type of informed and dynamic modeling of walkable routes that can easily transition from the analysis of space to conceptual development that follows performative metrics.

REFERENCES

Brashear, P., & Khawarzad, A. (2011, February 15). What Is Walkability? How Do You Measure It? Take-Aways from This Year's TRB Meeting. Project for Public Spaces. Retrieved from https://www.pps.org/article/what-is-walkability-how-do-you-measure-it-take-aways-from-this-years-trb-meeting

LSA Associates, Inc. (2003). Kansas City Walkability Plan. Retrieved from https://www.kcmo.gov/home/showpublisheddocument/583/636953455908800000

16   Viewsheds

When traversing a landscape or urban setting, it is important to acknowledge significant vistas or, inversely, obstructions that may lead to harm or safety concerns. These viewsheds, therefore, become a valuable tool and component to evaluate locations on a project site that afford these types of opportunities, good or bad. When viewsheds become obstructed by barriers such as builder corners, trees, or other visual obstacles, safety and access are diminished leading to uncomfortable situations.

Although it is highly preferred to introduce a dense tree canopy in urban settings due to their extensive benefits, these assets can quickly impair business signage resulting in reduced patronage. Other site amenities such as signage or other vertical elements can be contradictory to best practices for safety design. All these different performance metrics can be implemented to this model demonstration where it will focus on defining viewshed parameters for project goals and objectives.

GOALS

The primary output of this performance model is a visualization of a pedestrian or vehicular viewshed from a defined vantage point. The generated geometry can significantly represent how the quality of a viewshed varies as the vantage point traverses a space and becomes obstructed by surrounding elements:

1. Visualize viewsheds from a series of vantage points
2. Incorporate elements of the built environment, such as trees and buildings, to evaluate visual barriers
3. Determine the percentage of impaired viewsheds by site elements
4. Consider the amount of time a view of specific elements is impaired by

Although many of the goals are qualitative based, the additional potential of this model is elaborated on in the conclusion of this chapter to incorporate quantitative measurements for evaluation.

PERFORMANCE METRICS

Visual clarity from the human is usually restricted within a 60-degree horizontal field of view (FOV) and begins to diminish until around 120 degrees (Mass Transit

DOI: 10.4324/9781003208020-19

Railway, 2017). The clarity also relies on distances where the sharpness and focus of an object can vary significantly as well. Within the preferred 60-degree FOV, objects can be seen clearly up to 300 feet or further, whereas in the 120-degree FOV, objects become less defined (Tara et al., 2020). Since there are significant variations to the different degrees of view and related distance, this chapter will maintain parameters of 250 feet within a 60-degree FOV and 150 feet within a 120-degree FOV. But like the suggestions in the many previous chapters, referring to external resources will be valuable for credible implementation of these different parameters and metrics.

MODELING METHODS

The viewsheds will build off a curve centerline that acts as a pedestrian or vehicular pathway. For best results, once the curve is drawn in Rhino, it should be moved vertically between five and five and a half feet to simulate the eyeline of a pedestrian. This curve is then referenced into Grasshopper where the *Divide Curve* component extracts a set number of points. For this demonstration, as shown in Figure 16.1, a number slider set to *60* is used; however, depending on the length of the Rhino curve, this value may need to be smaller or larger for a preferred spacing of points. The points from this component are then inputted into a *List Item* component where another number slider is used to select a specific point from the *Index (i)* input. The maximum value for this slider should be equal to the set value of the previous number slider, in this case, 60. This slider will be used to scroll through different points or locations, yielding different viewsheds in the final output of this script.

Next, the output from that component is connected to the *Point (P)* input of the *Curve Closest Point* component while the pathway curve is connected to the *Curve (C)* input. The *Parameter (t)* output from this component is used to orientate a plane tangent to the curve, regardless of its location by connecting it to the *Parameter (t)* input of the *Horizontal Frame* curve with the pathway curve connected as well. This orientated plane is demonstrated in the right-side image of Figure 16.1.

The viewsheds that will eventually stem from this operation will also need visual barriers or obstacles to demonstrate what may impair someone's view along a path. To do so, a *Brep* component is used to contain a set of different geometries that serve as buildings, indicated as grey boxes, and trees, shown as green spheres in the right-side image of the figure above. Although the demonstration graphics are predominantly shown in plan, these objects are all three-dimensional.

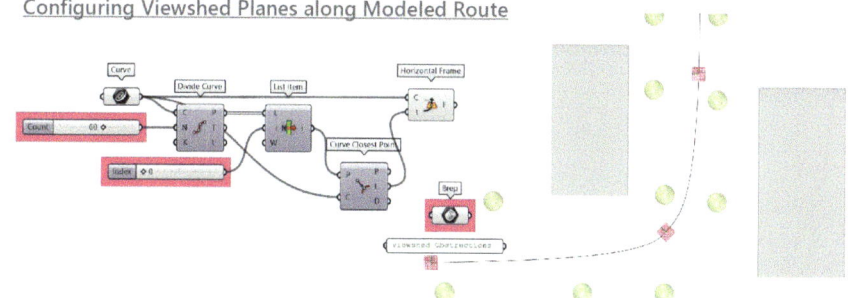

Configuring Viewshed Planes along Modeled Route

Figure 16.1 Plotting viewshed points along a designated route from a curve in Rhino.

The *Horizontal Frame* is then connected to both the *IsoVist* component, shown in Figure 16.2, and a *Brep|Plane* component, where the *Brep* representing the buildings and trees is connected to as well to create curves from those intersections. The *IsoVist* component is responsible for detecting barriers in the viewsheds and can be refined with a set of additional parameters. The first parameter is the value *360* for the *Count (N)* input so that there is a point for all three 360 degrees of view. The *Radius (R)* parameter uses both *250* and *150* values to simulate view clarity respective to two different view angles. The initial 60-degree view angle of a pedestrian can clearly see distances up to 250 feet, whereas the peripheral view of 30 degrees on either side can only see distances clearly up to 150 feet.

With these parameters now set, one can see in the bottom image of Figure 16.2 all the intersection points for both distances created by the barriers from the horizontal plane viewpoint. The next series of operations will demonstrate how to represent the two different viewing angles.

From the colliding intersection points created by the buildings and trees, the view angle for 60 degrees at a distance of 250 feet is created in Figure 16.3. The *Points (P)* from the *IsoVist* component are restructured with the *Flip Matrix* component, so that the *List Item* component with the default index of *0* can first isolate the points to the 250-foot distance. Two *Split List* components are used to find the first 30 points to the left and the other 30 points to the right. To get the right-sided points, the *List (L)* input of the second component needs to be *reversed*

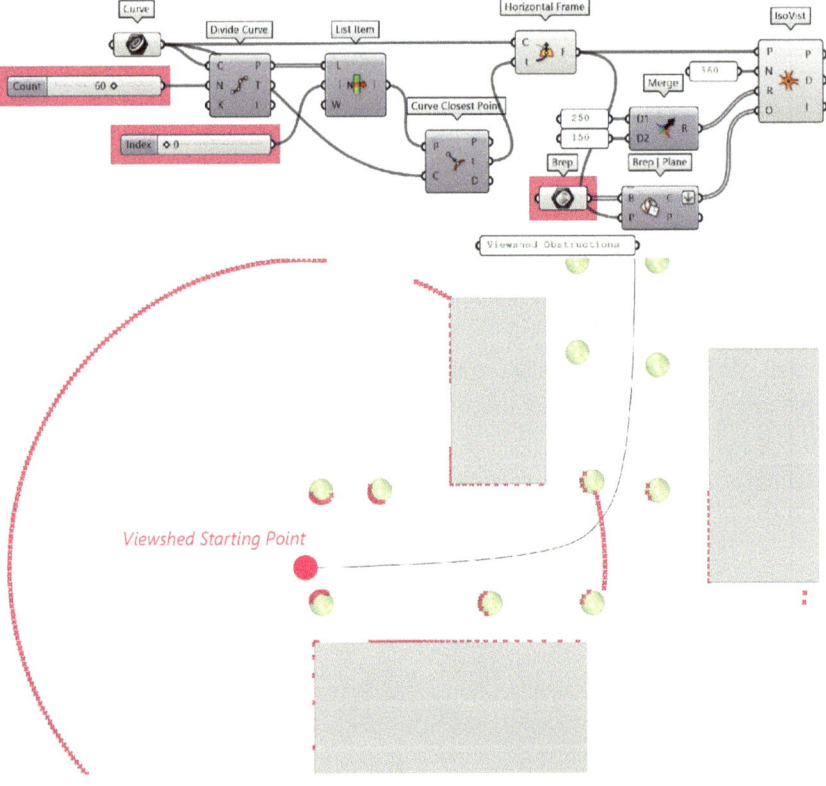

Applying Viewshed Distance Parameters

Figure 16.2
Determining different viewshed distance for different horizontal fields of view.

to rearrange the points in the opposite direction. A panel with a value set to *30* is used for both the component's *Index* (*i*) input as well. Lastly, the same component that had the input reversed will also need to have the output *reversed* once again to reorder the points to coincide with the other collection of points that will be joined with the *Merge* component.

It is important for the *Merge* component to join the inputs in the correct order. The first input for the component should be the *A* output of the first (non-reversed) *Split List* component, followed by the initial *List Item* component that is extracting a specific point from the divided curve. The final input should then be the reversed *A* output of the other *Split List* component. Each input of the of the *Merge* component should also be *flattened* to ensure that each list has the same number of branches and joins correctly. All the merged points can now be used to create a closed curve with the *PolyLine* component with the *Closed* (*C*) input set to *True* with a Boolean toggle. The bottom image of Figure 16.3 demonstrates how this newly created viewshed curve is created and changes, depending on the slider's value for the initial *List Item* component at the beginning of the script.

The same logic can be applied to the next set of operations to procure the wider 120-degree view angle, with 30 degrees on either side of the one just created, as shown in Figure 16.4. The data will first be isolated to the shorter 150 distance, using the *List Item* component with an index of *1* from the *Flip Matrix* component, since visibility and clarity is less defined at the peripherals. The output of this component will once again be *flattened* and *split* for both peripheral sides. With both *Split List* components, an expression of *x-1* is used for the *Index (i)* input with the same panel value of *30* so that these viewsheds share a point and therefore are tangent with the first view angle. If that expression is not applied, there will be a slight gap between them instead of being flush. Similar to the previous operation as well, the second *Split List* component's *List* (*L*) input needs to be *reversed*.

Creating a 60-Degree Viewshed

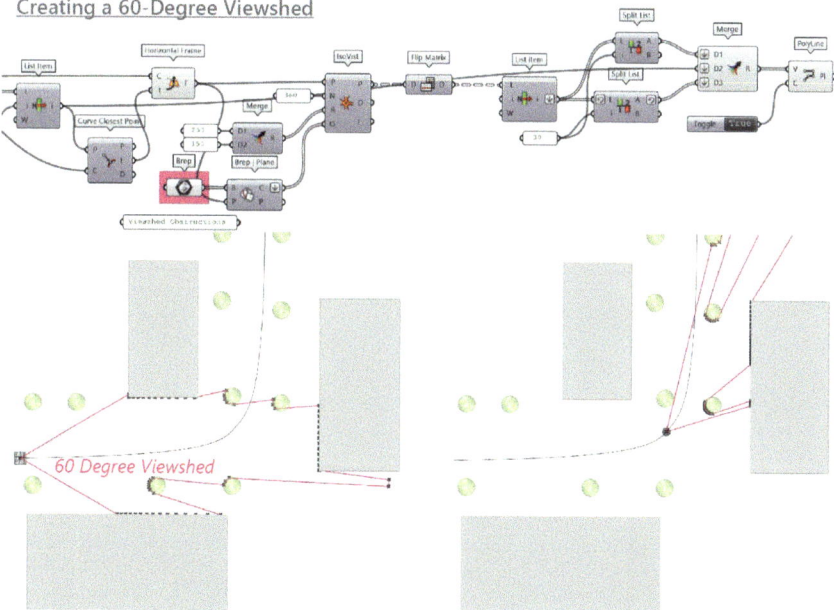

Figure 16.3
Creating closed curves of a 60-degree viewshed for a distance of 250 feet at different points.

From these split lists, they will be split again using the *B* output from the previous *Split List* components. The *A* output from this second set of components can now be combined with the initial *List Item* component at the beginning of the script with a specific order for each one. For the left-hand side peripheral, the *A* output of the first split list is used as the *D1* input of the *Merge* component and the initial *List Item* component as the *D2* input. For the right-hand side peripheral, the opposite order inputted into the *Merge* component is used between the same respective components.

Once again, all the inputs going into this component should also be *flattened* to maintain the same data branching. A *PolyLine* component is then used for each to create a closed curve representing the two peripheral viewsheds, as shown in the bottom image of Figure 16.4.

For additional visual aid to the model, the outputted polyline curves can be converted into a surface with the *Boundary Surface* component and assigned a color fill using the *Custom Preview* component, as shown in the script of Figure 16.5.

The next operation of this script is intended to add annotation of incremental distances within the different viewsheds using radial curves and labels. To do so, the initial *List Item* component from the beginning of the script is once again used to create a series of concentric circles with the *Circle* component. For multiple circles, a *Series* component is inputted for the *Radius* (*R*) of that component that starts (*S*) and steps (*N*) at a value of *50*. A total of *four* circles are generated with that as the number slider value connected to the *Count* (*C*) input. This can be seen in the left side model image of Figure 16.5.

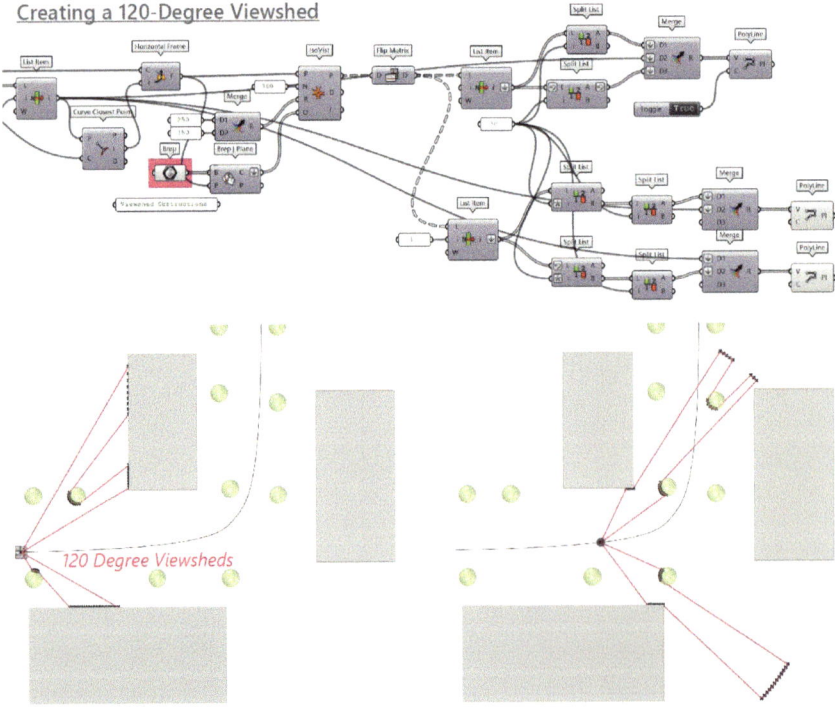

Figure 16.4
Creating curves for the 120-degree viewshed at a distance of 150 feet.

Instead of showing these as whole circles, it can be more effective to show them split and contained within the primary viewshed. This can be achieved by determining the points of intersection using the *Circle* component as the *A* input of the *Curve|Curve* component and the viewshed *PolyLine* component as the *B* input. These outputted intersection points (*tA*) can then be used as the *Parameter* (*t*) input of the *Shatter* component that is *grafting* the inputted *Circle* component.

The circles have now been shattered but all the parts still remain, so an operation is used to isolate the curves to just the ones located inside the primary viewshed. A center point of each curve can be used to determine if it sits within this viewsheds by extracting it using the *Curve Middle* component for each curve. This point is then tested for its inclusion using the *Point in Curves* component where the viewshed curve, *PolyLine* component, is the testing curve. The outputted values from this component can then be evaluated against an *Equality* logic of *2* to generate Boolean values for the *Cull Pattern* component. The *Shatter* curves will be the inputted list for this component which will output only the interior intersecting curves, as shown in the right-side model of Figure 16.5.

The newly split curves of their respective distances can now be annotated and further visualized using a variety of graphic characteristics that include color, text, and style, as shown in the complete script of Figure 16.6. A *Text Tag 3D* component is used to label the curves at their center using a *Curve Middle* component from the culled list.

The text that will be displayed will utilize the radial distances in addition to the unit of measurement – feet. Using the outputted values from the *Series* component, illustrated with a *Relay* element in the script image, these values are

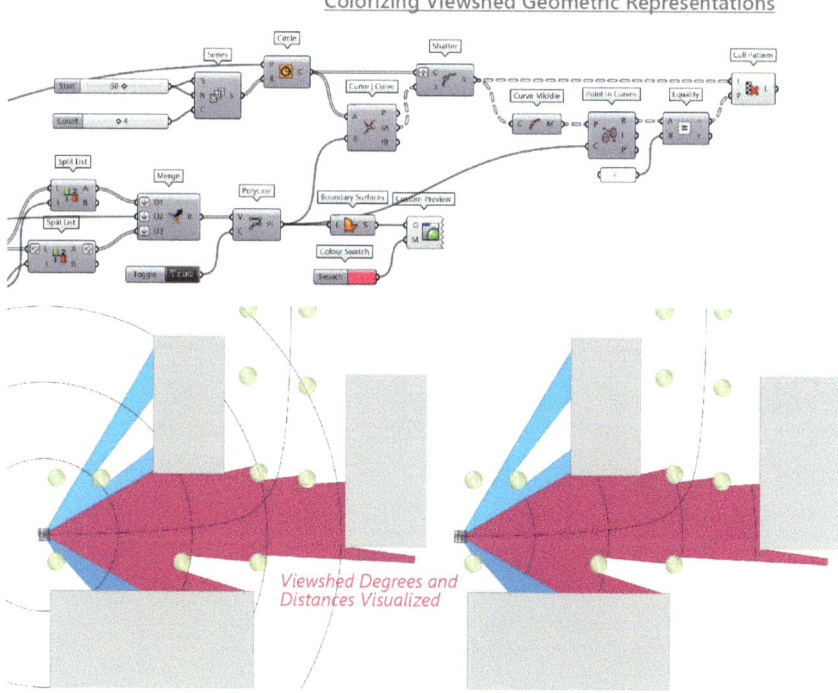

Figure 16.5
Colorizing the different horizontal fields of view with distance benchmarks.

converted to a text after *flattening* the input. The same relay will also be used to count the number of distances with the *List Length* component so that the *Repeat Data* component knows how many times the unit of measurement label, *ft*, needs to be duplicated. This will also be *flattened* and converted with another separate *Text* component. The values in the two *Text* components are then restructured with a *Graft Tree* component to isolate them for combining using the *Text Join* component. The resulting distance labels are then *grafted* and inputted into the text tag component. Size and color of the text can then be configured appropriately to the scale of the project site model.

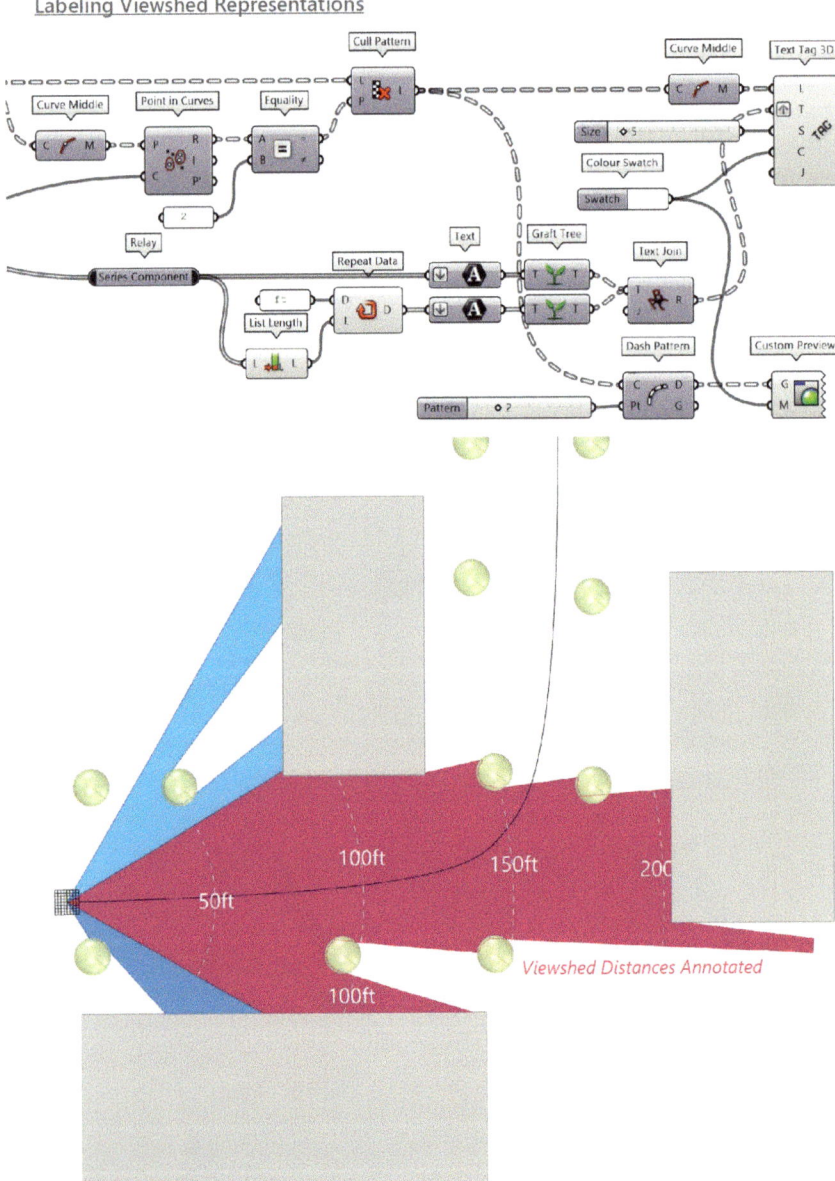

Figure 16.6
Labeling the different distance benchmarks for the viewsheds.

Changing the style of these radial curves can also be beneficial in deciphering them from the other visible curves and indicate that they are measurements instead of intended objects. For this demonstration, the curves are made into dashed lines with the *Dash Pattern* component where a number slider is used to determine the spacing and size of the dashes. These dashed lines can then be colorized with the *Custom Preview* component using the same color as before or with a different swatch.

As stated at the beginning of this chapter, the number slider inputted into the initial *List Item* component can control which vantage location is visualized for these viewsheds. The below series of images in Figure 16.7 illustrates a series of progressive locations along the path curve and the fluctuating conditions of the viewsheds. Graphically, this script and process can be used to show static snapshots of specified conditions or generated into a short video with the *Animate...* option for the number slider.

Figure 16.7
Demonstrating
the sequence
of viewshed
obstructions along
a route from trees
and buildings.

As the Rhino model develops with additional barriers, such as buildings and trees, and the path curve changes, all these characteristics of the viewsheds will respond accordingly for a dynamic output of information. This can be useful when it comes to evaluating the visibility and access to site amenities.

OUTCOMES

Although the major significance of this model is visually based with the generated viewshed geometry, the script still contains data that can be useful in evaluating the performance and condition of a pedestrian space. For example, in the demonstration of extracting data from the primary viewshed, the percentage of impaired or obstructed views was determined from a given vantage point. This can be useful data and potentially weighted to inform design strategies.

The same principle could also be applied to not only what percentage of a view angle is being impaired, but based on the distance to the vantage point, what percentage of space beyond these barriers is also being obstructed. Although this script does not demonstrate this measurement, it is something that can and should be considered as part of the performative outcome of this model.

ADDITIONAL POTENTIAL

Aside from what was already stated in the measurement of obstructed views, other potentials of the parametric model could include a three-dimensional viewshed, visibility of signage, and safety from vehicular traffic.

Three-Dimensional

The way space is perceived is not on a singular horizontal plane but within a three-dimensional cone that includes vertical planes. By creating a series of horizontal planes at the vantage points and incorporating the same logic of peripheral clarity, a three-dimensional viewshed can be generated for a more comprehensive and expanded perception of space. This becomes a valuable refinement as it could potentially consider tree and building height along with other obstacles below or above a single horizon line.

Signage

The addition of a three-dimensional aspect of a viewshed would better evaluate the visibility of storefronts and signage to assist business traffic. If there are too many impairments to the visual access of businesses along a street, then they can begin to face economic challenges in promoting sales. To help in this evaluation process, signage and storefronts could be represented with points that the viewshed can determine if it is within its field of vision. Additionally, it can further expand on that by determining what percentage of time it is visible from a vantage point along a curve.

Safety

Similar to the previous opportunity, and arguably most important, is the ability to evaluate safety concerns with vehicular traffic. Crosswalks and thoroughfares adjacent to streets can also take the same approach with representing points to serve as benchmarks for safety.

These types of potential can make this performance model much more robust beyond the qualitative visualizations of viewsheds by considering the importance of what should be frequently visible for both pedestrian and vehicular traffic.

REFERENCES

Mass Transit Railway. (2017). Proposed Comprehensive Residential and Commercial Development atop Siu Ho Wan Depot Final Environmental Impact Assessment Report. Retrieved from https://www.epd.gov.hk/eia/register/report/eiareport/eia_2522017/EIA/html/Appendix/Appendix%2011.1.pdf

Tara, A., Lawson, G., & Renata, A. (2020). *Measuring Magnitude of Change by High-Rise Buildings in Visual Amenity Conflicts in Brisbane*. Landscape and Urban Planning. 205. https://doi.org/10.1016/j.landurbplan.2020.103930.

SECTION 4

17 Conclusion

ENGAGEMENT – AUGMENTED REALITY

Modeling landscape performance serves as a dynamic decision-making tool for the accelerating global interest in evidence-based approaches and responsive design strategies. As natural landscapes and ecological design are being recognized for their robust services and improvement to our built environment, models can aid in evaluating their effectiveness and ability to do so with quantifiable measurables. Through the advancement of landscape performance and integration of quantitative metrics into design practices, landscape architects and allied professions can begin to evaluate the health of our environment for appropriate and sensible strategies to enhance the ecological, social, and economic services of a space.

As landscape architects, we need to become leaders and advocates for ecological processes and services in ways that make this knowledge tangible and accessible beyond our professional conventions. The modeling process with this specific platform affords the ability to communicate these complex systems through both digital interfaces and physical demonstration pieces to reach a universal audience. It is the intention of this book to provide an accessible and replicable workflow for others to learn, adapt, and integrate into the modeled outcomes into their specific environmental, social, and economic needs. The potential of communicating landscape performance with both digital interfaces and physical demonstration pieces is grounded on the foundation of parametric scripting to connect technology, design, and community engagement with landscape performance.

As already stated, these computational scripts are formulated from standardized metrics and calculators accessible through leading entities such as EPA's Green Infrastructure Guidelines, the Sustainable Sites Handbook, and LAF's Landscape Performance Series website and guidebook. The parametric modeling of landscape performance provides dynamic visualizations, immersion, and interactive experiences that conventional data tables and calculators are unable to provide independently. By connecting the data metrics of landscape performance to a variety of user interfaces, the user will gain a better understanding and significance to the role ecologies and nature-based solutions have to the vitality of future urban growth.

Due to the fundamental workflow of these performance models creating responsive outputs to the changing parameters of different factors, a dynamic feedback loop is provided for critical decision making. The inclusion of internal, external, and field survey data creates an adaptive feedback of analytical data to the modeling

DOI: 10.4324/9781003208020-21

process as well. It is essentially never a static or definitive modeling tool. The value of this is the dynamic criticism and evolution of how users understand information and the refinement that follows for discernable, informative, and accessible communication of ecological processes.

This digital and computational platform can also be formatted for web interfaces to host these performative dynamics in an interactive demonstration for a wide range of national and global audiences. The accessibility and dissemination of landscape performance metrics, benefits, and impact are crucial in the success of creating stewards and leaders in the preservation of our natural and built environment. One of the goals of introducing this digital tool into a web-based educational framework is to shift design thinking and methods to an adaptive and iterative approach that engages with students, community members, and decision-makers.

CROSS-PLATFORM COMPATIBILITY

The combination of a real-time interface allowing for quickly arranged option scenarios, the ability to co-create and engage with the community, and the real-time integration of data can fundamentally reroute design thinking methods. Abstract and formal ideas can merge into one process, helping designers innovate, increase options, and see the implications of those decisions simultaneously with their collaborators as a closed feedback loop.

Parametric modeling of landscape performance provides an advanced learning experience for the user to engage with innovative technology, create robust analytical modeling and assessments, and foster a participatory experience between community members in the advocation for healthy sustainable cities. This can come in the form of augmented and virtual reality, large-scale mapping procedures, parameterized policies, and even dynamic animations. These digital forms can also be hybridized with malleable model making, revealing a symbiotic connection between instrumental tools in design thinking. The cross-platform and mixed-media opportunities have tremendous potential to increase the capacity for designers to work with communities, advocate for healthy living environments, make design more efficient by linking tools that help speed up the process, and help everyone involved see the benefits of proposed designs.

The use of a real-time responsive interface allows for scenario building, the ability to co-create and engage with the community, and a seamless integration of data can fundamentally reroute the design thinking process. Abstract and formal ideas can merge into one workflow, helping the user engage, innovate, increase options, and see the implications of those decisions simultaneously with their collaborators as a closed feedback loop.

ENGAGEMENT – AUGMENTED REALITY

Augmented reality is not only gaining traction as an innovative representation tool but with the integration of parametric modeling and performance metrics, it can also serve as a decision-making process (Duenser et al., 2008). Students, community members, and stakeholder groups can rapidly generate scenarios that align with design goals and objectives that relate to social, economic, and environmental benefits for measurable outcomes. With the augmented interface, information and data becomes perceptual and responsive to real-time data, performance parameters, and user decision making. With the influx of real-time quantitative data that updates during this process, there is a profound opportunity to fundamentally shift design thinking and action based on these augmented outcomes. By embedding measurables and metrics, a new design process and methodology can potentially emerge that enables the respective parties to generate scenarios for accessing and evaluating with their goals and objectives.

With augmented reality, the understanding of space can come in the form of both tangible and intangible elements of a project site. It allows for the user to engage, interact, and participate with amenities of the built environment by refining space to their preference and suitability. The augmented reality interface is like a chessboard. Pieces such as landforms, trees, and built structures are loaded into a perceptual interface that can be seen simultaneously in the computer, the physical world, and a merger of the two in augmented reality, as shown in Figure 17.1. With this interface, students, designers, and community members can move pieces around, seeing perceptually what impacts different designs may have.

Within this same augmented workflow, landscape performance outputs can be conveyed to communicate quantitative data on the design iterations to assess the tradeoffs and outcomes of different scenario models, as shown in Figure 17.2. Now the design experience goes beyond the perception of space to include additional information aiding the process and development of a concept. The figure also demonstrates how with some augmented software programs, the model making from a smart device will simultaneously generate a working Rhino model for further refinement and detailing.

Regardless of how the augmented environment is engaged with, this platform affords the opportunity for the user to make real-time perceptual decisions of how a space should be through synchronized qualitative and quantitative metrics.

Figure 17.1
Full scale and real-time augmented reality interface for exploring concept designs on site.

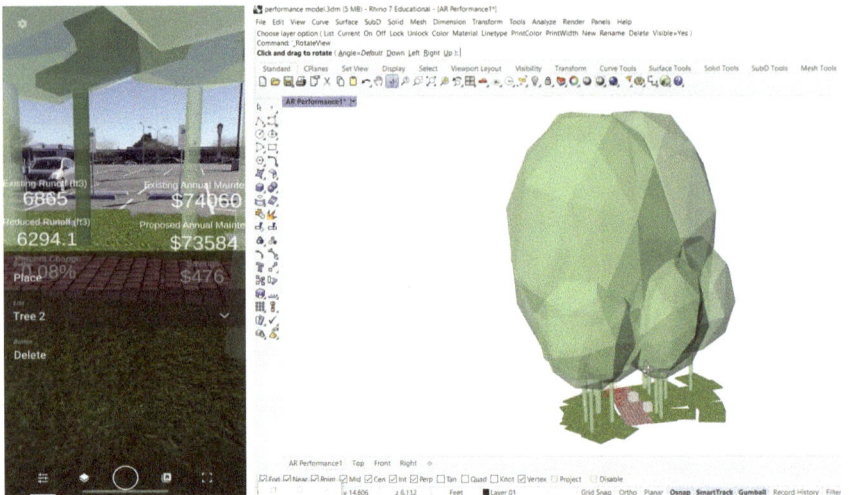

Figure 17.2
Synchronization
between on-site
design development
and computational
model workspace
demonstrating
performance
metrics.

Design Decision Making

The metadata that dictates these 'physical' and tangible objects can also be utilized in the form of intangible information to aid in the decision-making process. Data in the form of landscape performance metrics, as shown in Figure 17.3, can be integrated, so the user may understand and perceive a more comprehensive outlook to their decisions as it can produce environmental, social, and economic outputs depending on their design priorities. With augmented reality, this robust interface can be accessible to a larger audience beyond just designers. It provides opportunities for collaboration between stakeholder groups to service community accommodations in response to their needs as everyone involved can contribute to the design process. This builds advocacy and stewardship to preserve natural ecological systems and design new ones mimicking ecosystem services.

This mixed reality of augmented and virtual spaces provides an advanced learning experience for the user to engage with innovative technology, create robust analytical modeling and assessments, and create a participatory decision-making experience to advocate for healthy sustainable cities (Wang et al., 2013). The hybridization between malleable model making with augmented data reveals a symbiotic connection of instrumental tools in design thinking. It correlates with many STEM-related principles of evidence-based strategies where design can act as a healing response to problematic issues of urban heat island, water scarcity, public health, and equitable resources within the built environment. The inclusion of quantifiable data seamlessly integrates with performance metrics and methods as it operates as a real-time assessment tool to measure and evaluate defined project goals and objectives.

This augmented reality design process has tremendous potential to increase the capacity for users, designers, and the community to work together and advocate for healthy living environments, make design more efficient by linking tools that help speed up the process, and help everyone involved see the benefits of proposed designs, as shown with augmented performance models in Figure 17.4.

Figure 17.3
Conceptual
performance
models that can be
explored through
an augmented
workspace.

Performance Decision-Making

Shaded Comfort   *Hydrology*   *Accessibility*   *Shielded Comfort*   *Erodibility*

Figure 17.4
Student work by
Landyn Green
at the University
of Arkansas's
Landscape
Architecture
Department using
computational
performance
metrics for
impactful and
responsive decision
making. Part of the
curriculum's LARC
2335 studio course
co-taught with Ken
McCown.

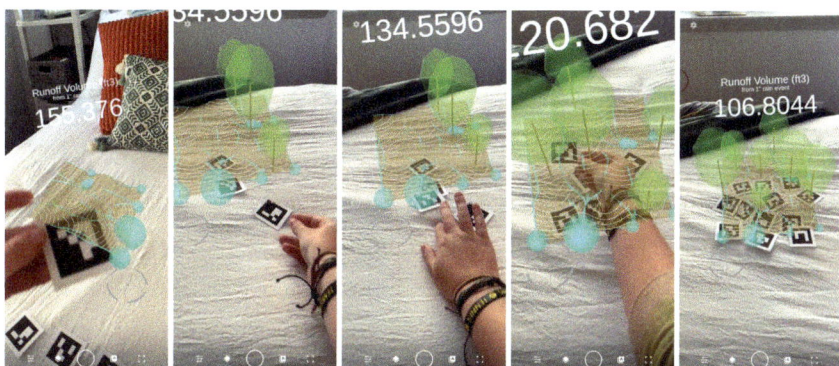

The interface can remain perceptual or conceptual, like in Figure 17.4, while the responsive real-time data helps the participant see quantitative impacts of qualitative design decisions. This method links smart devices, metadata, computational software, mixed media, and project sites to a physical and attainable world. By linking these virtual and real sites together, a new design process can potentially emerge that enables the user to make large gains to assess their generated iterations.

The p[AR]k Sandbox

Augmented reality sandboxes are becoming a common tool to interact with and understand terrain heightfields as a colored point cloud projection on sand media. Although beneficial to understand these fundamentals, the point cloud projection can be advanced further to measure other characteristics of slope, aspect, and drainage for the measurement of ecosystem services that include tree benefits, stormwater management, erosion control, outdoor comfort, and accessibility. Providing visualizations and interactive properties through an augmented reality interface gives the user ownership and agency behind their decisions for outdoor spaces (Figure 17.5).

The feedback collected from the point cloud of the sandscape is processed through a computational model to analyze and generate visuals and charted information based on various parameters. In addition to sand media, the Oculus Quest 2's controllers can be used to place various amenities for outdoor spaces that include trees, benches, water features, paths, and other elements that interact and respond to defined characteristics for community engagement during the process.

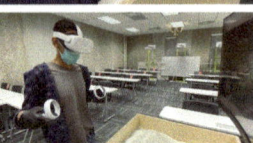

Figure 17.5
Students
participating in the
hybrid p[AR]k
experience to
evaluate landscape
performance within
a pocket park
design scenario.

STRATEGIC PLANNING – MAPPING PROCEDURES

As demonstrated in many of the landscape performance models, importing GIS shapefiles is a common and viable option with the modeling software, so its relevancy to large-scale mapping projects is certainly an option to apply different methods and metrics to. This should, however, be limited to performance methods that are appropriate for this type of scale but may include urban forest benefits, stormwater management, and walkability, to name a few. Another advantage to this workflow is that there are more opportunities to communicate the data and metrics as three-dimensional spaces rather than the common two-dimensional convention of most GIS platforms. The drawback to this is that information can get convoluted by inserting unnecessary forms of graphic representation or diluted due to scaling constraints.

One effective application of the parametric modeling of landscape performance metrics can be seen in Figure 17.6 where the process of utilizing shortest distance calculations can convey different accessibility outcomes to fresh food, often resulting in designated food deserts. In the left-side image of the figure, when measuring for access to fresh food, a half-mile radius is generated to determine which residents are classified as living within a food desert – outside the circle. Although this is helpful in identifying the number of residents outside an accessible distance from a grocery store, the results can be misleading because this portrays access as direct distance instead of the reality of the network distance.

The shortest walk operation, previously demonstrated in the book, portrays a different message of access to fresh food by utilizing the existing road and path network of a neighborhood. Since routes to grocery stores are rarely a direct line, as indicated by the other model, many residents can be at a disadvantage due to complex road networks. When applying this modeling method for residents in North Las Vegas neighborhoods, 40–70% of residents within the half-mile radius are actually further than that deemed accessible distance. Therefore, by the traditional method of radius distances to map food desert, these residents would not be classified as living within a food desert, even though they are physically outside those parameters. This is one of many examples where modeling landscape

Figure 17.6
Comparing
traditional radius
distances for
proximity maps to
network distances
using parametric
modeling.

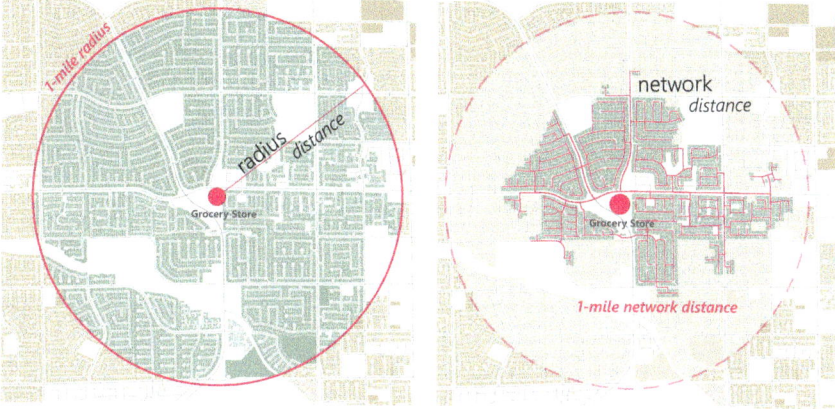

performance through these advanced computational models can provide accurate and compelling arguments to go beyond conventional and traditional approaches of improving the built environment.

Campus Tree Benefits

One example where this scale and method has been applied to is a university's tree inventory to calculate and assess different tree benefits. The positive effects of working urban forests have been well documented in US Forestry reports, including Greg McPherson's Desert Southwest Community Tree Guide (McPherson et al., 2004) and the Davey Resource Group's Urban Forest Resource Analysis of Inventoried Public Trees (Davey Resource Group, 2013). Urban tree canopies are particularly influential in harsh desert landscapes that lack surrounding forests to mitigate the effects of air pollution, urban heat islands, and monsoonal rain events. These benefits, however, can be easily diminished by proper planting or poorly structured municipal codes that prevent sufficient soil volume in tree planter designs, which will be elaborated on further in the next section on policies.

Qualitative and quantitative methods were both used in the assessment of existing tree health, as this would significantly influence the impact of tree benefits. When first collecting tree benefit data as part of the field surveying and ground truthing phase, healthy tree conditions were compared to unhealthy tree conditions and found that the struggling trees performed on average 45% less than their healthy counterparts. As a result of this finding, it was important to evaluate tree health based on the planting condition and the US Forest Service rating criterion. The rating criterion is the Tree Health Metrics (Hallett, 2018) developed by US Forest Service Forest Ecologist Rich Hallett. Using this method, tree health was assessed by documenting the crown health and measuring the trunk's diameter at breast height (DBH).

As part of the quantitative process, the campus tree inventory was extracted from an online resource that included tree coordinates, species name, and DBH, all of which are necessary to retrieve accurate tree benefit measurements. This data was then imported and cross-referenced with an external dataset of different tree species' annual benefits. The dataset included cost savings, stormwater interception, and carbon dioxide sequestration values based on the tree's DBH to compare with the existing tree inventory. In the below figures, different analyses were performed for their benefits to evaluate strategic planning to manage, preserve, and improve planting conditions for the university, as mapped out in Figure 17.7.

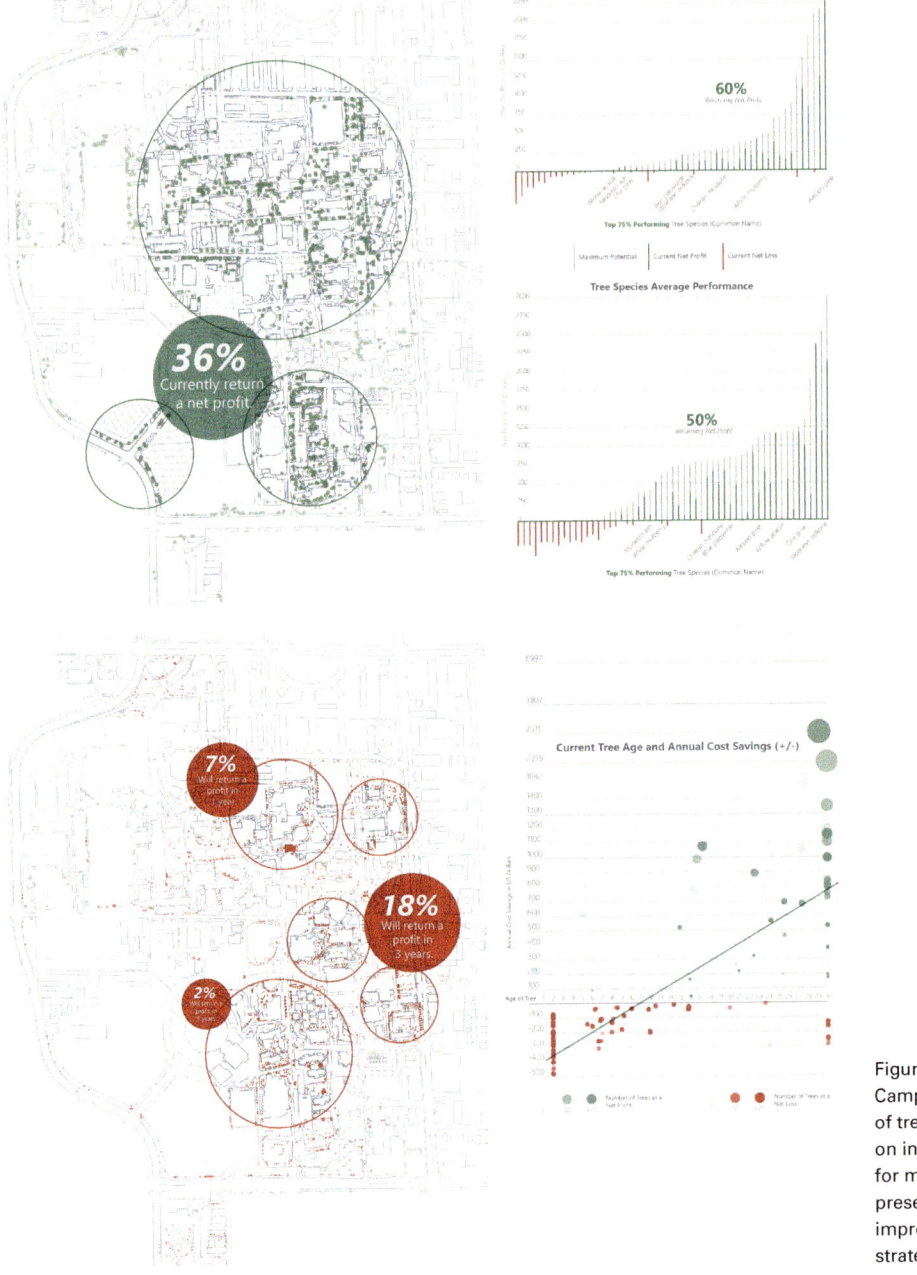

Figure 17.7
Campus mapping of tree's return on investment for management, preservation, and improvement strategies.

With the tree health and parameters collected, tree DBHs were inputted into online calculators to determine current and future economic and environmental benefits from individual trees. These calculators include i-Tree Design (USDA Forest Service, 2017) and the National Tree Benefit Calculator (Casey Trees and Davey Tree Expert Co., n.d.). These calculators were able to provide benefit measurements for cost savings, stormwater runoff interception, and carbon dioxide sequestration annually throughout a 30-year lifespan. The data collected for all the trees on campus was then used to begin the evaluation process of performance in comparison to baseline values, as shown in the below figures.

To ensure appropriate future planting strategies for the campus, the assessment process began by limiting tree species calculations to the Southern Nevada Water Authority's *Regional Plant List* (SNWA, 2021) and consulting with local practitioners, leading to 68 unique tree species. These species were then entered into different calculators to tabulate their annual DBH growth and performance outputs over a period of 100 years. According to the i-Tree Design (USDA Forest Service, 2017) calculator, the DBH growth of most trees slowed or ended by the 30-year mark, thus leading to that being the extent of annual calculations for this study.

POLICY AMENDING – PARAMETERIZING CODES

Many municipal and city codes rely on different quantifiable metrics for implementing parking regulations, setback requirements, and land use development that can be incorporated into a parametric model to evaluate the potential for implementing said strategies or even recommend amendments to. Codes are often disseminated as abstraction of the built environment that may not be able to effectively convey the implications to place making or community identity, environmental impact, or even economic viability. One of these municipal regulations that will be demonstrated is the impact parking lots have on stormwater runoff volume, outdoor comfort and accessibility, and economic investment.

Performing Parking Lots

The main goals of this study were to encourage municipal code change that promote green infrastructure in parking lots through revised configurations of parking stalls, planter size, and surface material that provide adequate ecosystem services. Much of the redesigning of the parking lots drew directly from county codes to parameterize and support these decisions. Discrepancies and limitations of the codes were also identified, but will not be elaborated on in the provided examples. The parameterizing of county codes for parking lots resulted in dense tree canopies and implementation of green infrastructure for an urban forest while minimizing parking stall loss or in some cases, increasing stall count without diminishing planting communities. Collectively, all these improvements to the often utility of parking lots created an effective performative landscape with a focus on stormwater management and tree benefits.

The first phase focuses on utilizing different aisle and stall dimensioning and orientation for safer circulation and reallocating areas for green space and soil volume, as shown in Figure 17.8.

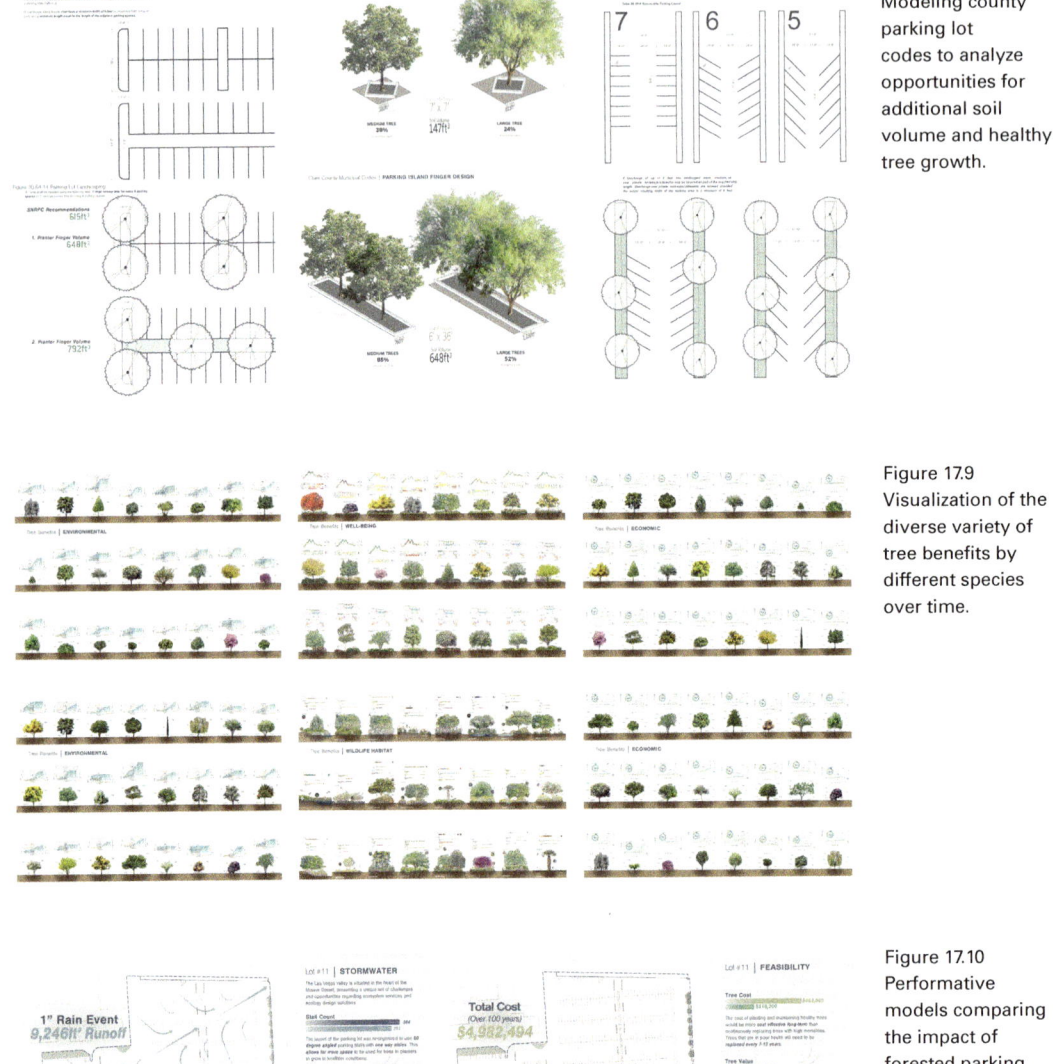

Figure 17.8
Modeling county
parking lot
codes to analyze
opportunities for
additional soil
volume and healthy
tree growth.

Figure 17.9
Visualization of the
diverse variety of
tree benefits by
different species
over time.

Figure 17.10
Performative
models comparing
the impact of
forested parking
lots to their
existing conditions
for stormwater
management and
long-term economic
savings.

Next, select trees species considered appropriate for parking lots by the valley water authority were measured for their benefits to determine which would be most suitable in fulfilling the intended project goals and objectives for reduced stormwater runoff, carbon dioxide sequestration, and long-term cost savings, as shown in Figure 17.9. Lastly, an alternative material to asphalt was chosen and strategically integrated into appropriate locations throughout the parking lot.

The comparative studies visualized in Figures 17.10 were components of the final deliverables that included before and after evaluations for stormwater management and cost savings. Other qualitative metrics were also included to assess potential for wildlife habitat and comfortable outdoor spaces from shade and evapotranspiration from dense plant communities.

TEMPORAL FACTORS – DYNAMIC ANIMATIONS

The complexity and compression of ecological processes and landscape performance can be attainable and inclusive to a broad audience through animated and interactive visualizations. Conventional static graphic representations often only provide a snapshot of the fluctuating dynamics of our ecological environment. Animated graphics are not a novel graphic convention, as software is becoming more accessible with shortened learning curves; however, they can be underutilized as walk-through visualizations that don't capitalize on the other variables of a 'living' graphic. The value of using either animated and interactive visualizations through modeling landscape performance is to demonstrate changes of time and scale to express the relationship more significantly between the natural and built environments. With time-based animations, hourly, daily, seasonally, and yearly patterns can be measured and visualized as a parameter in this modeling technique.

Because these models provide a more comprehensive visual and informative perspective of landscape performance topics, a universal audience can experience different layers and complexities of environmental systems and their symbiotic relationships demonstrated throughout this book and other landscape performance methods. Figure 17.11 showcases how traditional static approaches to the topic of stormwater and hydrological analysis compares with the transition of the same information to a model that visualizes data in a more dynamic format.

Although three-dimensional representations are a benefit of these computational models, this does not necessarily translate as better communication of a topic. It can be argued however that it simply affords the opportunity to explore the best option to communicate the necessary information effectively instead of the limiting capacity of most other software programs.

Arid Meadows

One example frequently demonstrated is how different rain events may impact our living environment due to rainfall intensities, urban development introducing more impervious surfaces, and regrading of the landscape. Although they are a series of snapshots from a video animation, the images in Figure 17.12 demonstrate how monthly rainfall averages range throughout a city and are affected by different land covers and slope conditions.

The relationship between different inches of rainfall throughout the city (data extracted from rain stations) with their comparative locations demonstrates how annual or overall precipitation values is not sufficient enough to understand how this type of data affects a region. It is seen in the images of previously listed figures how rain events vary significantly throughout the city and will have drastically different outcomes when compounded with different densities of impervious land cover

Figure 17.11
Exploring the
potential dynamics
of temporal
representation
of stormwater
performance
through animations
in the bottom image
in contrast to the
static conventional
representation of
the same topic.

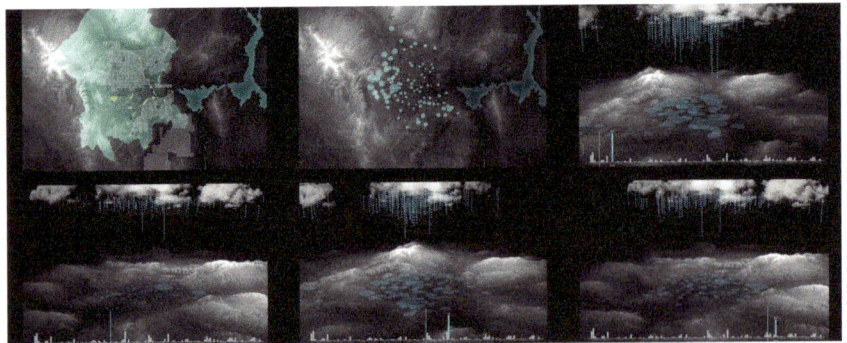

Figure 17.12
Snapshot images of
an animated model
for rain stations
and the impact
topography and land
cover have on urban
runoff throughout
a city's unique
conditions.

and topography. There is too much variability for projects of this scale to manage with conventional practices, whereas parametric modeling can make the process of incorporating large datasets and multiple factors into a robust and comprehensive model relatively easy.

This example and the many others throughout the book demonstrate the seemingly limitless potential of making landscape performance models instrumental tools to the comprehension and advocacy of ecosystem services throughout our built environment to mitigate issues of climate change, social equity, and economic viability. As a tool, these models can be valuable in aggregating tremendous amounts of information as spatial layers of data that comprise a project site or condition. They can perform basic or advanced calculations to analyze multiple factors through baselines comparisons and scenario building. Outputs from these performance models can then ultimately be utilized as guidelines for responsive design strategies that directly address the issue in a more fluid design workflow process.

REFERENCES

Casey Trees and Davey Tree Expert Co. (n.d.). National Tree Benefit Calculator. Retrieved from http://www.treebenefits.com/calculator/

Davey Resource Group. (2013). Urban Forest Resource Analysis of Inventoried Public Trees. Retrieved from http://forestry.nv.gov/wpcontent/uploads/2013/09/LasVegas_ResourceAnalysis_20130802.pdf?forcedefault=true

Duenser, A., Grasset, R., & Billinghurst, M. (2008). A Survey of Evaluation Techniques Used in Augmented Reality Studies. ACM SIGGRAPH ASIA 2008. https://doi.org/10.1145/1508044.1508049

Hallett, R. (2018). *Tree Health Metrics: One Page Guides*. The Nature Conservancy.

McPherson, G., Simpson, J. R., Peper, P. J., Maco, S. E., Xiao, Q., & Mulrean, E. (2004). Desert Southwest Community Tree Guide: Benefits, Costs, and Strategic Planting. Arizona State Land Department Natural Resources Division, Urban & Community Forestry Section & Arizona Community Tree Council, Inc. Retrieved from https://www.fs.fed.us/psw/publications/mcpherson/psw_2004_mcpherson002.pdf?forcedefault=true

Southern Nevada Water Authority (SNWA). (2021). Regional Plant List. Retrieved from https://www.snwa.com/assets/pdf/water-smart-plant-list.pdf

USDA Forest Service. (2017). I-Tree Design User's Manual. Retrieved from http://www.itreetools.org/applications.php

Wang, X., Kim, M. J., Love, P. E. D., & Kang, S.-C. (2013). Augmented Reality in Built Environment: Classification and Implications for Future Research. *Automation in Construction* 32: 1–13. ISSN 0926–5805, https://doi.org/10.1016/j.autcon.2012.11.021

Index

Note: *Italic* page numbers refer to figures.